GOVERNORS STATE UNIVERSITY LIBRARY

W9-AAG-993

3 1611 00209 9510

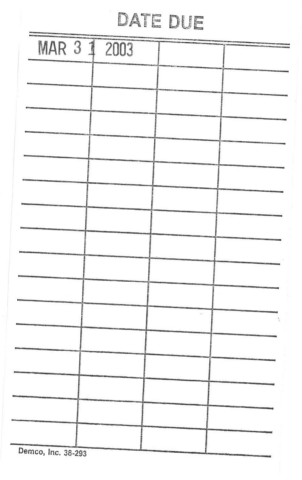

DATE DUE

MAR 3 1 2003			

Demco, Inc. 38-293

AGRICULTURAL RESTRUCTURING AND SUSTAINABILITY
A Geographical Perspective

Sustainable Rural Development Series

Series Editors:
David Pitt, Commission on Environmental Strategy and
Planning, International Union for Conservation of Nature and
Natural Resources (IUCN), Switzerland.
Jose I. dos R. Furtado, Economic Development Institute, The
World Bank, Washington DC, USA and Centre for Integrated
Development, London, UK.
Guy M. Robinson, Professor of Geography, Kingston
University, UK.

Agricultural Restructuring and Sustainability

A Geographical Perspective

Edited by

Brian Ilbery

Department of Geography
Coventry University
UK

Quentin Chiotti

Institute for Environmental Studies
University of Toronto
Canada

and

Timothy Rickard

Department of Geography
Central Connecticut State University
USA

GOVERNORS STATE UNIVERSITY
UNIVERSITY PARK
IL 60466

CAB INTERNATIONAL

S 494.5 .S86 A374 1997

Agricultural restructuring
and sustainability

CAB INTERNATIONAL
Wallingford
Oxon OX10 8DE
UK

Tel: +44 (0)1491 832111
Fax: +44 (0)1491 833508
E-mail: cabi@cabi.org

CAB INTERNATIONAL
198 Madison Avenue
New York, NY 10016-4341
USA

Tel: +1 212 726 6490
Fax: +1 212 686 7993
E-mail: cabi-nao@cabi.org

©CAB INTERNATIONAL 1997. All rights reserved. No part of this publication
may be reproduced in any form or by any means, electronically, mechanically, by
photocopying, recording or otherwise, without the prior permission of the
copyright owners.

A catalogue record for this book is available from the British Library, London, UK.

ISBN 0 85199 165 3

Library of Congress Cataloging-in-Publication Data
Agricultural restructuring and sustainability : a geographical
    perspective / edtied by Brian Ilbery, Quentin Chiotti, and Timothy
    Rickard.
          p.   cm.
        Selected and revised papers from a conference held in North
    Carolina.
        Includes index.
        ISBN 0-85199-165-3 (alk. paper)
        1. Sustainable agriculture--Congresses. 2. Agricultural
    geography--Congresses. 3. Sustainable development--Congresses.
    I. Ilbery, Brian W. II. Chiotti, Quentin. III. Rickard, Timothy J.
    S494.5.S86A374 1997
    338. 1--dc21   97-13059
                                                       CIP

Typeset in 10/12 Baskerville by Solidus (Bristol) Limited
Printed and bound in the UK by Biddles Ltd, Guildford

Contents

SECTION V: Sustainable Agriculture and Environmental Policy

SECTION VI: Sustainability and Restructuring the Agricultural System

Contributors

C. Andrews,
Department of Geography,
University College Chester,
Cheyney Road,
Chester CH1 4BJ,
UK

M.R.J. Battershill,
Department of Geography,
University of Exeter,
Exeter EX4 4RJ,
UK

I. Bowler,
Department of Geography,
University of Leicester,
Leicester LE1 7RH,
UK

G. Blunden,
Department of Geography,
The University of Auckland,
Private Bag 92019,
Auckland,
New Zealand

M. Brklacich,
Department of Geography,
Carleton University,
1125 Colonel By Drive,
Ottawa,
Ontario,
Canada K1S 5B6

C. Bryant,
Département de Géographie,
Université de Montréal,
C.P. 6128,
Succursale Centre-ville,
Montréal, Québec,
Canada H3C 3J7

Q. Chiotti,
Environmental Adaptation
Research Group,
Institute for Environmental Studies,
University of Toronto,
33 Willcocks Street,
Toronto,
Ontario,
Canada M5S 3E8

G. Clark,
Department of Geography,
Lancaster University,
Lancaster LA1 4YB, UK

C. Cocklin,
Department of Geography and
Environmental Science,
Monash University,
Wellington Road,
Clayton,
Victoria 3168,
Australia

A. Crockett,
Department of English Local
History,
University of Leicester,
Leicester LE1 7QR, UK

J.M. Curry-Roper,
Department of Geology,
Geography and Environmental
Studies,
Calvin College,
Grand Rapids,
Michigan 49546,
USA

J. Dumanski,
Research Branch,
Agriculture and Agri-Food Canada,
Central Experimental Farm,
Building #74,
Ottawa,
Ontario,
Canada K1A 0C6

L.A. Duram,
Department of Geography,
Southern Illinois University,
Carbondale,
Illinois 62901-4514,
USA

B. Ebel,
Department of Economics,
The University of Lethbridge,
4401 University Drive,
Lethbridge,
Alberta,
Canada T1K 3M4

N.J. Evans,
Department of Geography,
Worcester College of Higher
Education,
Henwick Grove,
Worcester WR2 6AJ, UK

O.J. Furuseth,
Department of Geography,
University of North Carolina at
Charlotte,
North Carolina 28223,
USA

A.W. Gilg,
Department of Geography,
University of Exeter,
Rennes Drive,
Exeter EX4 4RJ, UK

L.M.B. Harrington,
Department of Geography,
Kansas State University,
Manhattan,
Kansas 66506,
USA

M. Healey,
Department of Geography and
Geology,
Cheltenham and Gloucester
College of Higher Education,
Francis Close Hall,
Swindon Road,
Cheltenham,
GL50 4AZ, UK

J. Higginbottom,
Department of Geography,
University of Coventry,
Priory Street,
Coventry CV1 5FB,
UK

S.G. Hilts,
Department of Land Resource
Science,
University of Guelph,
Guelph,
Ontario,
Canada N1G 2W1

G. Hollander,
Department of Geography,
University of Iowa,
Iowa City,
Iowa 52242,
USA

B. Ilbery,
Department of Geography,
University of Coventry,
Priory Street,
Coventry CV1 5FB,
UK

T. Johnston,
Department of Geography,
The University of Lethbridge,
4401 University Drive,
Lethbridge,
Alberta,
Canada T1K 3M4

D. McNabb,
Department of Geography,
Carleton University,
1125 Colonel By Drive,
Ottawa,
Ontario,
Canada K1S 5B6

W. Moran,
Department of Geography,
The University of Auckland,
Private Bag 92019,
Auckland,
New Zealand

C. Morris,
Department of Geography,
University College Chester,
Cheyney Road,
Chester CH1 4BJ,
UK

R. Munton,
Department of Geography,
University College London,
26 Bedford Way,
London WC1H 0AP,
UK

D. Napton,
Department of Geography,
South Dakota State University,
Brookings,
South Dakota 57007-0648,
USA

M.D. Nellis,
Department of Geography,
Kansas State University,
Manhattan,
Kansas 66506,
USA

C. Potter,
Environmental Section,
Wye College,
University of London,
Wye,
Ashford,
Kent TN25 5AH,
UK

T. Rickard,
Department of Geography,
Central Connecticut State
University,
Connecticut 06050,
USA

R. Roberts,
Department of Geography,
University of Iowa,
Iowa City,
Iowa 52242,
USA

G.M. Robinson,
School of Geography,
Kingston University,
Penrhyn Road,
Kingston upon Thames,
Surrey KT1 2EE,
UK

A. Shaw,
Department of Geography,
University of Leicester,
Leicester LE1 7RH,
UK

J. Sheeley,
Department of Geography,
Kansas State University,
Manhattan,
Kansas 66506,
USA

B. Smit,
Department of Geography,
University of Guelph,
Guelph,
Ontario,
Canada N1G 2W1

J. Smithers,
Department of Geography,
University of Guelph,
Guelph,
Ontario,
Canada N1G 2W1

M. Troughton,
Department of Geography,
University of Western Ontario,
London,
Ontario,
Canada N6A 5C2

G. Walker,
Department of Geography,
York University,
Toronto,
Ontario,
Canada M3J 1P3

Preface

In August 1995, an international conference was convened in North Carolina by Owen Furuseth, with assistance from Timothy Rickard, Duane Nellis and Lisa Harrington. The conference brought together 44 rural geographers from Canada, the United Kingdom, the United States of America and one invited representative from New Zealand. This conference represented the second meeting of rural scholars from the three principal countries, the first being held in the United Kingdom in 1991. The general theme of the conference was rural systems and geographical scale, and the papers reviewed a wide range of issues pertaining to rapid changes in contemporary rural areas in each of the four respective countries. Two more specific themes emerged from the conference, with one series of papers focusing on agricultural restructuring and sustainability and another addressing the economic and social changes occurring in the broader rural system.

This volume is based on the papers presented at the conference which focused on agricultural restructuring and sustainability. The chapters are organized according to six sub-themes: (i) conceptualizing agricultural restructuring and sustainability; (ii) family farming and farming culture; (iii) diversification and alternative agriculture; (iv) agricultural sustainability and climate change; (v) sustainable agriculture and environmental policy; and (vi) sustainability and restructuring the agricultural system. Chapters in the book examine, at various spatial scales, the broad processes and structural changes that are common to all rural systems in developed countries. Different geographical contexts are used to illustrate the uneven development of these processes and the implications for sustainable agricultural and rural systems. Drawing upon varying research experiences in their respective countries, contributors review the relevant literature for their national contexts and support their arguments with original research. The resulting book covers a wide range of issues pertaining to agricultural restructuring and sustainability and provides a geographical perspective from different developed market economies.

Introduction

Brian Ilbery[1], Quentin Chiotti[2] and Timothy Rickard[3]
[1]Department of Geography, University of Coventry, Priory Street, Coventry CV1 5FB, UK; [2]Environmental Adaptation Research Group, Institute for Environmental Studies, University of Toronto, 33 Willcocks Street, Toronto, Ontario, Canada M5S 3E8; [3]Department of Geography, Central Connecticut State University, Connecticut 06050, USA

Rural areas in developed market economies have been undergoing considerable change in the postwar period in response to changes in the broader social, political and economic system. This includes changes in agriculture, whereby its restructuring has been caused by powerful forces that transcend all scales of analysis. In this book, emphasis is placed on changing agricultural structures and relationships, and the sustainability of these changes.

The emergence of sustainability issues as a significant topic of inquiry, or more specifically concern over the sustainability of responses to the globalization of agriculture, is a relatively recent phenomenon. In *Contemporary Rural Systems in Transition, Volume 1: Agriculture and Environment* (Bowler *et al.*, 1992), for example, the central themes were the consequences and responses to the uneven development produced by agricultural restructuring. Issues pertaining directly to sustainability were then only an emerging, if not peripheral, theme, as illustrated by the single section devoted to sustainable agriculture as a policy option. Five years later, however, interest in sustainability issues has risen dramatically whereby they now form the primary focus of this book. The patterns and processes resulting from agricultural restructuring, and the sustainability of these changes, are central themes of this book; they are examined at various spatial scales and in four national (plus numerous regional) contexts. A number of critical questions are raised, including, among others: How are agricultural systems changing? How sustainable is modern commercial agriculture? What evidence exists for sustainable agriculture? Is there an uneven pattern of sustainable agriculture? How sustainable are current farm-level and policy responses? What factors affect the adoption of a sustainable pattern of agricultural development? Will present agricultural responses to restructuring be sustainable under a changing global environment?

No single volume can address such a wide range of issues, although the chapters in this book manage to provide numerous insights into many of them. This introductory chapter develops a framework with which to view agricultural restructuring and sustainability from a geographical perspective, thereby serving as a backdrop to the contributions in the book. The dynamics of agricultural restructuring are considered first, emphasizing the shift from a productivist to a postproductivist phase during the past decade. The concept of sustainable agriculture is then outlined, in terms of its origins, meanings and ideological underpinnings. This is followed by a discussion on uneven development that provides the unifying thread linking agricultural restructuring and sustainable agriculture. Against this background, consideration is given to various manifestations of agricultural restructuring and sustainability, drawing upon the contributions to the volume.

The Dynamics of Agricultural Restructuring

Agriculture in developed market economies has undergone substantial restructuring in the postwar period, from which two major phases of change can be identified (Ilbery and Bowler, 1997). The first, from the early 1950s to the mid-1980s, was a productivist phase, characterized by a continuous modernization and industrialization of agriculture, with emphasis on raising farm output. Since the mid-1980s, a second postproductivist phase has emerged, characterized by the integration of agriculture within broader rural economic and environmental objectives, with emphasis on reducing farm output. In the 1990s, both productivist and postproductivist farming systems coexist, with intensive, high input–high output farming and its emphasis on food *quantity* and low costs now being complemented by low input–low output farming, with an emphasis on sustainable farming systems and food *quality*.

Although the nature of postproductivist agriculture has yet to be fully defined by governments and society in developed market economies, a number of attributes can be identified for what has been referred to as the postproductivist transition (PPT) (Lowe *et al.*, 1993; Shucksmith, 1993). This transition is characterized by a reduced output of food, the progressive withdrawal of state subsidies, the production of food within an increasingly competitive global market and the growing environmental regulation of agriculture. Thus the PPT represents the progressive reversal of the trends that dominated the preceding productivist era, challenging the sustainability of a system that was geared towards increasing outputs of food, a prominent degree of state intervention, highly protected and subsidized markets and a proclivity for environmental degradation.

The PPT can be conceptualized according to three general shifts in agricultural production, from: (i) intensification to extensification; (ii) concentration to dispersion; and (iii) specialization to diversification (Bowler and Ilbery, 1996). At the farm level, these shifts are reflected in various development strategies adopted by farm businesses, such as:

1. the continuation of a full-time, profitable food production farm business;
2. the diversification of the income base of the business through pluriactivity; and
3. the marginalization of the farm as a profitable business (Bowler, 1992).

As farm businesses seek to develop new sources of income through different types of agricultural production and non-agricultural diversification, there are important environmental implications, such as the reduction in levels of environmental pollution and in some cases the restoration of natural habitats.

The PPT is also reflected in, and being driven by, significant changes in state intervention. Most notable among these have been the Canada – United States Free Trade Agreement in 1988 (which was extended into the North American Free Trade Agreement in 1992), the reforms of the Common Agricultural Policy (CAP) in 1992 and the GATT agreement on world agricultural trade in 1993. One of the main forces behind the liberalization of trade and re-regulation is global pressure to reduce public expenditure on agriculture. Although these policies have been predominantly geared towards trade, the environment has also received attention, particularly with the package of 'accompanying measures' associated with reform of the CAP in the European Union (EU). The Agenda 21 declaration from the 1992 'Earth Summit' at Rio de Janeiro also emphasized the necessity to promote sustainable agriculture and rural development (Grubb *et al.*, 1993).

The Concept of Sustainable Agriculture

The concept of sustainable agriculture is closely linked to the concept of sustainable development as promoted by the World Commission on the Environment and Development (1987), although concern over the adverse environmental impacts associated with productivist agriculture has much earlier origins (e.g. Carson, 1962). Sustainable agriculture has also received considerable political support on an international (e.g. Agenda 21, Chapter 14) and national (e.g. Agricultural Canada) basis.

While the concept of sustainable agriculture has been subject to various interpretations, most definitions subscribe to the view that 'sustainability' must mesh economic with environmental concerns, recognizing that the continued neglect of the physical and biological resources is undermining the long-term health of the agricultural system. Pierce (1992), for example, has identified three basic propositions:

1. Rates of renewable resources should not exceed their rates of regeneration.
2. Rates of use of non-renewable resources should not exceed the rate at which sustainable substitutes are developed.
3. Rates of pollution emission should not exceed the assimilative capacity of the environment.

Brklacich *et al.* (1990) attempt to translate such propositions into more tangible agricultural terms, whereby sustainable agriculture provides three long-term, simultaneous goals:

1. *Environmental sustainability*: the capacity of an agricultural system to be reproduced into the future without unacceptable pollution, depletion or physical destruction of its natural resources such as soil, water, air and natural and semi-natural habitats.
2. *Socioeconomic sustainability*: the capacity of an agricultural system to provide an acceptable economic return to those employed in the productive system.
3. *Productive sustainability*: the capacity of an agricultural system to supply sufficient food to support the non-farm population.

To this list, Bowler (1992) has added two further criteria:
4. *Budgetary criteria*: the capacity of an agricultural system to produce food while absorbing an acceptable proportion of state (public) expenditure.
5. *Political criteria*: the capacity of an agricultural system to maintain the political support of society.

Whereas these propositions and criteria provide a useful framework from which to define 'sustainable agriculture', they offer little guidance on how to achieve such goals. Translating these broad concepts on 'sustainability' into tangible agricultural systems can be achieved in different ways. From an 'idealist' (ecocentric/alternative agriculture) viewpoint, the only viable, long-term option for human society is to radically reorganize socio-economic systems into *no* or *low* growth modes. On a societal level, this would require radical changes in consumption patterns and individual lifestyles, whereas at the farm level it would require the adoption of alternative farming practices, such as diversified land use, nutrient recycling and low inputs of agrochemical and biological disease control. An 'instrumentalist' (technocentric/conventional agriculture) viewpoint challenges the above goals as being practically and politically unrealistic, and instead defines 'sustainable' as a contextual process involving incremental changes towards a more sustainable, yet unattainable, ideal. Integrated crop management, diversified agriculture and conservation agriculture are examples of this type of agricultural model.

During the PPT, one will undoubtedly come across farmers who continue to subscribe to unsustainable land management practices typical of productivist agriculture. However, most farmers will hopefully be adopting some form of sustainable agriculture. Of these, a majority will follow 'instrumentalist' models of sustainable agricultural development, whereas a minority will pursue 'idealist' models of sustainable agriculture. In the latter context, although growing in importance, organic farming still only contributes a very small percentage of food to diets in developed market economies.

The Uneven Development of Agricultural Restructuring

Capitalism is known to impart an uneven spatial pattern of development or socioeconomic health upon the agricultural landscape. Uneven development arises from flows of capital and labour, and the geographical concentration of capital accumulation. In the case of agriculture, uneven development is reflected in differential degrees of capital penetration into the food production process, created through the interaction of dialectical processes, such as external capitals and internal farm production processes (Marsden *et al.*, 1989) or technology and social processes (Roberts, 1992).

Patterns and processes of uneven development have been an integral part of the agricultural landscape throughout the productivist phase of agriculture, resulting in marked differentiation between countries, regions and individual farms. With the relatively recent emergence of the PPT, uneven development is likely to become even more pronounced. It seems likely that under trade liberalization market forces will encourage the continued concentration of food production in areas of comparative advantage, thereby deepening the uneven development of the productivist regime. Similarly, with new forms of regulation emerging and farmers adopting a variety of different pathways of business development, the PPT can be expected to lead to increasing differentiation within the agricultural sector and rural areas. This may be particularly so among farmers and farming systems adopting alternative models of sustainable agriculture. Overall, therefore, the agricultural landscape will become even more highly differentiated, with some regions (and within regions, between individual farms) following productivist methods of farming, others following the 'instrumentalist' model of agricultural development and yet others following 'idealist' food production methods.

The Structure of the Book

The 21 contributions to this book examine various dimensions of agricultural restructuring and sustainability. They cover a range of spatial scales, from international to local, and draw upon different concepts from geography and other social sciences. Chapters 1–3 are concerned with different aspects of *conceptualizing agricultural restructuring and sustainability*. Richard Munton conducts a critical review of rural land use policy and sustainable development in the United Kingdom and argues that a new political process has been initiated that will be difficult to ignore. The concept of spatial scale is at the heart of the second chapter by Chris Cocklin, Greg Blunden and Warren Moran. They suggest that an examination of sustainable agricultural systems requires an analysis of relations at the levels of the field, farm, locality, region, nation and internationally. Geographical theory relating to property rights, regulation and industry structures is seen as being particularly appropriate for conceptualizing the main dimensions of agricultural restructuring among the spatial scales. Chapter 3 is firmly located at the macro-scale and Guy Robinson outlines how agriculture is increasingly being influenced by global-scale processes. Particular attention is paid to the role of transnational corporations and the growth of agrienvironmental policies and it is suggested that globalization and sustainability are key aspects of an emerging third food regime.

Chapters 4–7 discuss *family farming and farming culture* and Rebecca Roberts and Gail Hollander start by arguing that decisions to adopt sustainable agricultural practices should be examined at the farm household level, where social systems of farm succession, household income strategies and within-family relations intersect with political–economic systems. In particular they call for more ethnographic analyses of farming households in order to provide critical insights into agricultural restructuring and sustainability. Clive Potter in Chapter 5 continues this theme by exploring the role of the farm family life cycle in agricultural restructuring in Britain. Typical successional pathways followed by farm families are identified and the author suggests that these influence both the timing and trajectory of environmental change on family farms. In a thought-provoking Chapter 6, Carol Morris and Charlotte Andrews call for more research on the relationship between how farmers construct their own particular 'versions' of the environment and their 'farming culture'. These versions are likely to be at variance with those of policy-makers and conservationists and so have important implications for making sense of the responses made by farmers to environmental demands placed upon them. Ethnic and religious culture forms the focus of attention in Chapter 7, Janel Curry-Roper's study of agriculture in the American Midwest. She explores attitudes towards conventional and alternative (sustainable) agricultural systems by different religious and ethnic groups and finds a growing interest in an alternative agricultural paradigm that emphasizes local control, cooperation, community life and imitation of natural ecosystems.

Chapters 8–10 are concerned with *diversification and alternative agriculture* and discuss alternative farming systems (AFS) in England and Colorado. Using the concept of the mode of social regulation, Gordon Clark, Ian Bowler and Brian Ilbery, in Chapter 8, examine the interrelationships between the development of AFS and the institutional regulatory framework. The diversity of local responses to new policy changes is illustrated, but institutions are shown to be 'reactive' rather than 'proactive' towards AFS. One particular aspect of AFS – on and off-farm business diversification – is explored by Brian Ilbery, Michael Healey and Julie Higginbottom in Chapter 9. An 'extensive' survey of farm businesses highlights differences in the nature and extent of business diversification between different areas and shows that this relates to such characteristics as farm size, farm type, tenure and household composition. Leslie Duram, in Chapter 10, concludes by demonstrating the differences between 'conventional' and 'alternative' farmers in terms of operational, decision-making and attitudinal factors. A spectrum is shown to exist between these two groups of farmers, with the key differential components including farmers' external or internal locus of control, their perceptions of agriculture and ecology and the level of operational diversification.

Chapters 11–13 cover *agricultural sustainability under climate change* and explore the nature of agricultural adjustment to environment stress. In Chapter 11, John Smithers and Barry Smit develop a general model of agricultural adjustment to environmental variability that is applicable at different geographical scales and then apply it in a study of agricultural response to climate change in Ontario. The strong Canadian emphasis in this section is continued in Chapter 12 by Michael Brklacich, David McNabb, Chris Bryant and Julian Dumanski. They call for more work that explores the relationships between climate change, agricultural decision-making and agricultural adjustment; their farm level study in Renfrew County reveals that perceived changes to the prevailing climatic regime do not always result in farm level adjustments. In Chapter 13, Quentin Chiotti, Tom Johnston, Barry Smit and Bernd Ebel draw upon concepts from political economy and humanism to develop a framework for understanding land management responses to climate change. Farm-level responses to climatic variation in southern Alberta are presented and the concept of locality is advocated for more informed appreciation of climate change impacts.

Chapters 14–17 examine *sustainable agriculture and environmental policy*. Starting in the United States, in Chapter 14, Duane Nellis, Lisa Harrington and Jason Sheeley discuss the positive and negative aspects of the Conservation Reserve Programme. Analysis at different spatial scales supports the success of the Reserve Programme in developing appropriate and sustainable agricultural systems. Chapter 15 turns to England, and Nick Evans examines the uptake of particular agrienvironmental schemes among farmers. His analysis reveals that variations in scheme design, situations facing farm businesses, the nature of agricultural restructuring and the existence of other conservation measures influence adoption rates. This theme is continued in Chapter 16 by Andrew Gilg and Martin Battershill, who show that farmers participating in various environmentally friendly farming schemes in Devon are distinguished by behavioural and attitudinal traits rooted in their personal histories and circumstances. Chapter 17 moves to Canada, and Stewart Hilts argues that the rapid development of private stewardship programmes necessitates a greater understanding of land-use decisions at the landowner level. This has important implications for the (changing) roles of both landowners and government staff in relation to the sustainability of rural land management.

Chapters 18–21 cover various issues, at different spatial scales, concerning the *sustainability and restructuring of the agricultural system*. Chapters 18 and 19, by Michael Troughton and Owen Furuseth respectively, examine agricultural trends at both national and regional scales. Troughton argues that farming remains the key to rural sustainability; however, the increasingly heterogeneous and polarized systems of farming developed in Canada since 1950 have worked against sustainability. Similar polarization, this time in the US hog farming industry, geographical concentration and increases in the scale of production are the themes developed by Furuseth, who suggests that such changes have serious implications for rural and agricultural sustainability. Gerald Walker, in Chapter 20, argues that the social and cultural context of agriculture is intimately linked to the processes of capitalist transformation. In a study of the 'urban fringes' of Ontario, the modernization of agriculture is shown to alter fundamentally the settlement pattern, through processes of ruralization and urbanization of the countryside. The final chapter, by Darrell Napton, is concerned with restructuring in the marginal agricultural region of south Dakota. Over 70 years of federal programmes have failed to produce a sustainable economy and Napton examines the conflicts between individuals and communities as they strive towards sustainability.

References

Bowler, I.R. (1992) Sustainable agriculture as an alternative path of farm business development. In: Bowler, I.R., Bryant, C.R. and Nellis, M.D. (eds) *Contemporary Rural Systems in Transition, Volume 1: Agriculture and Environment.* CAB International, Wallingford, pp. 237–253.

Bowler, I.R., Bryant, C.R. and Nellis, M.D. (eds) (1992) *Contemporary Rural Systems in Transition, Volume 1: Agriculture and Environment.* CAB International, Wallingford.

Bowler, I.R. and Ilbery, B.W. (1996) The regional consequences for agriculture of changes to the Common Agricultural Policy. In: Laurent, C. and Bowler, I.R. (eds) *CAP and the Regions: Building a Multidisciplinary Framework for the Analysis of the EU Agricultural Space.* INRA, Versailles, pp. 103–118.

Brklacich, M., Bryant, C.R. and Smit, B. (1990) Review and appraisal of concepts of sustainable food production systems. *Environmental Management* 15, 1–14.

Carson, R. (1962) *Silent Spring.* Houghton Miflin, Boston.

Environment Canada (1990) *The Green Plan Supply and Services.* Ottawa.

Grubb, M., Koch, M., Thomson, K., Munson, A. and Sullivan, F. (1993) *The 'Earth Summit' Agreements: a Guide and Assessment.* Earthscan, London.

Ilbery, B.W. and Bowler, I.R. (1997) From agricultural productivism to post-productivism. In Ilbery, B.W. (ed.) *The Geography of Rural Change.* Longman, London (forthcoming).

Lowe, P., Murdoch, J., Marsden, T., Munton, R. and Flynn, A. (1993) Regulating the new rural spaces: the uneven development of land. *Journal of Rural Studies* 9, 205–222.

Marsden, T., Munton, R., Whatmore, S. and Little, J. (1989) Strategies for coping in capitalist agriculture: an examination of the response of farm families in British agriculture. *Geoforum* 20, 1–14.

Pierce, J.T. (1992) Progress and the biosphere: the dialectics of sustainable development. *The Canadian Geographer* 36, 306–320.

Roberts, R. (1992) Nature, uneven development and the agricultural landscape. In: Bowler, I.R., Bryant, C.R. and Nellis, M.D. (eds) *Contemporary Rural Systems in Transition, Volume 1: Agriculture and Environment.* CAB International, Wallingford, pp. 119–130.

Shucksmith, M. (1993) Farm household behaviour and the transition to post-productivism. *Journal of Agricultural Economics* 44, 466–478.

1 Sustainable Development: A Critical Review of Rural Land-use Policy in the UK

RICHARD MUNTON
Department of Geography, University College London,
26 Bedford Way, London WC1H 0AP, UK

Introduction

The purpose of this chapter is to provide a brief and critical assessment of the political salience of sustainable development as an idea and as a programme of action, and to do so in the specific context of rural land-use policies put forward by the British Government following its participation in the Earth Summit at Rio in 1992. The debate surrounding the objectives of sustainable development is substantial and strongly contested, and full justice cannot be done to it here. Equally, the adoption by government of the language of sustainable development is extensive but shallow, and its recency must make assessments of its impacts on policy preliminary. None the less, some idea of the tensions within government created by a sustainable development perspective, and between government and other interests, is beginning to emerge. Without doubt, obtaining public support for such a major shift in priorities is being made more difficult by the current rate of social and economic change and its attendant uncertainties, a problem exacerbated by the falling level of trust within society for those wielding political and economic power and responsibility. Mobilizing support for environmental concerns where these are one step removed from daily experience is increasingly difficult, especially now these concerns appear to have slipped down the British public's order of priority for government action since the late 1980s (Worcester, 1993; Witherspoon, 1994; Harrison *et al.*, 1996).

The extensive use of a sustainable development discourse by the UK Government makes this issue ripe for analysis, and this chapter describes some recent attempts to develop a broad-ranging strategy. Attention is restricted to aspects of rural land-use policy and to target setting as a means of measuring 'progress' and of moving matters forward. There is much more to implementing strategy than target setting but this step is especially important as it informs the range and nature of policy itself, and can promote public communication and participation (Local Government Management Board, 1993, 1994; Macnaghten *et al.*, 1995). To illustrate these points, this chapter draws upon a series of recent government statements, including the 1990 White Paper, *This Common Inheritance* (Department of Environment (DoE) *et al.*, 1990), *Sustainable Development: The UK Strategy* (DoE *et al.*, 1994a), the White Paper *Rural England* (DoE, 1995), and the recently published reports by the House of Lords Select Committee on *Sustainable Development* (House of Lords, 1995a,b,c) and the House of Commons Environment Committee *Rural England: The Rural White Paper* (House of Commons, 1996).

General Background

No attempt is made here to review the enormous literature on what is meant by sustainable development. There is no agreed position on what it should mean or what relative weight should be placed on its social, economic or environmental objectives (see, for example, Redclift, 1987; Norgaard, 1988; Cocklin, 1995). The dominant discourse argues that the future social and economic condition of humankind represents its central goal, a goal to be achieved in ways that do not lead to the aggregate, long-term deterioration in either the quality or quantity of the global natural resource stock. The stock, and the environmental processes that sustain and describe its condition, frequently form the objects of scientific study, but they constitute no more than integral components, possibly limiting ones, to the realization of economic and social goals. This position is made abundantly clear by the 27 Principles that underlie the Rio declaration (UNCED, 1992; see also Brundtland Report, 1987) and represents a significant shift in position from the debates of the 1960s and 1970s that made environmental limits the focus of attention.

These earlier debates challenged the competence of industrial progress, technological advance and administrative control to underpin the future. Those advocating limits not only insisted upon the immediacy of environmental catastrophe but also, more generally, promoted uncertainty and conflict over global futures (see Norgaard, 1994; Torgerson, 1995). These uncomfortable prognostications were strongly contested and gradually replaced by the 'win-win' arguments of the Brundtland Committee which suggest that economic growth and environmental protection can be achieved through technological advance and managerialist solutions enacted within existing institutional structures. This 'solution' appealed to business, to a managerial elite of environmental experts, and to most mainline political interests. It has not gone unchallenged, however, both at the practical level of how, in the current state of scientific ignorance, one can possibly manage such uncertain environmental futures, and at a deeper level where environmentalists insist upon the need for 'limits' rather than 'trade-offs', the use of moral rather than instrumental reasoning, and the acceptance of judgemental rather than calculative decision-making.

Today's arguments over nature–society relations are only 'new' in the sense of a heightened understanding of their truly global character, and the realization that a programme of action cast in terms of social gain rather than environmental threat is more likely to engage with the aspirations of national governments, and frequently their citizens. Translating global concerns into meaningful *local* action is the central issue. As a social and political construction, sustainable development is *necessarily* going to be subject to differential interpretation according to issue, interest group and local environmental circumstances. At the level of local action, therefore, there can be few givens. It is neither sufficient, nor even helpful, to argue as O'Riordan does that the notion of sustainable development is vague and internally inconsistent. The very indeterminacy of the term, he suggests:

> enables it to transcend the tensions inherent in its meaning. It has staying power, but no-one can properly put it into operation, let alone define what a sustainable society would look like in terms of political democracy, social structure, norms, economic activity, settlement geography, transport, agriculture, energy use and international relations.
>
> (O'Riordan, 1995, p. 21)

Few would question this judgement, especially at present levels of knowledge and agreement, but it is the 'staying power' that is crucial. It is preferable to examine the justification, constitution and implementation of programmes of action, and to acknowledge that sustainable development 'is not so much an idea but a convoy of ideas' (British Government Panel on Sustainable Development, 1995, p. 1) that has to be fed into individual life styles and the decisions of interest groups, business and governments, than to reject the claims of sustainable development because of the definitional problems they pose for 'rational' analysis. One can agree with Healey and Shaw when they say that 'the making of judgements is an issue for politics not techniques, and planning decisions cannot be left to an administrative/technical nexus, nor to lawyers, economists or natural scientists' (1994, p. 434), all the while examining the use of such principles as precaution, environmental limits and local assessment. Indeed, in practical terms, can one do better than decide whether the use of particular resources (and the distribution of the associated costs and benefits) is being moved in a more environmentally and socially sustainable direction than at present? As Scottish Natural Heritage argues in its evidence to the House of Lords Select Committee,

> Under present conditions of knowledge of both human and natural systems, we will rarely know whether or not sustainability has been achieved. It may therefore be more helpful to think in terms of whether any possible change in policy and practice would help to make the activity *more* or *less* sustainable than at present.
>
> (House of Lords, 1995c, p. 146)

Scottish Natural Heritage should have gone on to observe that all knowledge is contested and that its disputed nature is as much part of one's 'lack' of knowledge as the objective availability of data. Science only informs; it does not solve social questions. None the less, its pragmatism is commendable if only because it focuses attention upon objectives, leading logically to the identification of indicators of 'success' and targets of 'achievement'. These can always be adjusted in the light of experience and changed social values. It is much easier politically, as the Brundtland Committee acknowledged, to move matters forward on the basis of a rolling programme towards a generalized goal than to seek agreement at the outset on the detailed character of the end result.

There is, however, at least one major weakness with this approach. It runs the risk of unduly diminishing the importance of 'irreplaceable' resources. Even allowing for the fact that one may adjust the definition of what is irreplaceable on the basis of advancing knowledge and technology, such an approach can lead to minimalist definitions of the nature and extent of 'critical natural capital', or those natural (and heritage) assets regarded as irreplaceable (e.g. unique ecosystems; the ozone layer; historic landscapes). Being irreplaceable, by definition, makes such capital part of the end result. The importance of this point is revealed by a careful analysis of the evidence submitted to the House of Lords Select Committee (House of Lords, 1995b,c). This reveals that it is the *combination* of views taken over 'critical natural capital' and how the 'precautionary principle' should be applied that discriminates most powerfully between interest groups. Those who seek minimal definitions of what is irreplaceable also believe in the power of technology to find solutions. They are the greatest risk-takers with the environment, place least credence on non-materialist, non-instrumental environmental values, and are most convinced that objective measures of risk or economic value can (and should) form the basis of rational judgements over resource use. This position is subject to considerable critique and is central to the crucial debate over what discourse should frame an understanding of sustainable development (e.g. Owens, 1994; Adams, 1995; Irwin, 1995; Lash *et al.*, 1995; Harrison *et al.*, 1996).

Rio and the UK Response

The UK Government can claim to have taken an international lead in responding to the commitments it entered into at Rio. Following the Earth Summit, the Prime Minister, John Major, pressed the G7 group of industrialized countries and, at the Lisbon Council, the UK's EU partners, to draw up national strategies in response to Agenda 21 by the end of 1993. *Sustainable Development: the UK Strategy* (DoE *et al.*, 1994a) appeared in January 1994 and it remains one of only a small number of comprehensive national strategies to have been published. A major reason for the prompt and detailed response was the publication in 1990 of the White Paper *This Common Inheritance* (DoE *et al.*, 1990). This policy document contains a number of principles that get repeated almost word for word in the 1994 *UK Strategy* (hereafter *Strategy*). The White Paper was the first comprehensive review of environmental policy and it cut across the remits of ten government departments. It notes 'The Government needs to ensure that its policies fit together in every sector; that we are not undoing in one area what we are trying to do in another; and that policies are based on a harmonious set of principles rather than a clutter of expedients' (para. 1.6, pp. 8–9). For the first time, environmental issues were not to be seen as the sole province of the Department of the Environment.

Four years later, the *Strategy* was presented to Parliament by no less than 16 Departments of State. After allowing for government restructuring since 1990, it is interesting to note that the additions included the Foreign Office, the Treasury, Defence and Overseas Development. With the exception of a greater recognition of global considerations, it is difficult to see what effect these additional ministries had on reshaping policy. Both policy statements articulate a position based upon linkages between the economy and environment even if the term sustainable development is rarely used in the White Paper. More significantly, both are strongly influenced by the Government's neo-liberal ideology. Each indicates a preference for market measures over regulation, qualifications to the application of the precautionary principle on the basis of cost, and a firm determination to see that environmental policies neither inhibit wealth creation nor weaken the international competitiveness of the UK economy. Both statements challenge the argument that economic growth and environmental protection are necessarily in conflict and they present a policy position that many (e.g., Pearce *et al.*, 1989; Collis *et al.*, 1992) would recognize as 'very weakly sustainable'. For example, while the *Strategy* says 'What matters is that decisions throughout society are taken with proper regard to their environmental impact' (1994a, para. 3.2, p. 32), emphasis is placed upon the trade-offs between environment and economic development and only rarely is primacy given to environmental considerations. Continued economic growth is seen as the means of realizing the greater quality of life that environmental protection can bring.

Finally, both documents argue for a partnership between central government, local government, industry, environmental organizations and individual citizens in promoting and achieving a more sustainable development path. The Government engaged in extensive consultation before publishing both documents and accepts the need to persuade both producers and consumers to change their production methods and life styles. It prefers persuasion to regulation, and led it to establish a number of new advisory mechanisms. For example, the Government identified 'Green Ministers' in every government department, and set up an Advisory Panel, a consultative Round Table, a citizen-based campaign ('Going for Green'), and a range of other advisory bodies (see House of Lords, 1995a, pp. 55–57). But leadership, in the sense of ensuring the empowerment of others to *act*, as implied by the Rio declaration under Agenda 21, as opposed to consulting and encouraging expressions of local opinion, remains largely missing.

This omission raises interesting issues in the light of Macnaghten *et al.*'s (1995) research findings (see also Irwin, 1995; Harrison *et al.*, 1996). Among the more important of these are latent public support for the aims of sustainable development but a profound scepticism about the willingness of government and other corporate interests to achieve them; a much greater commitment among citizens to very localized issues that affect their daily lives than broader global concerns; a disinterest in issues where they feel powerless to effect change; and a greater concern for future employment and crime-related issues than for the environment. These conclusions may reflect the general unpopularity of the UK Government in 1994 and the declining attention given to environmental causes in the media since the late 1980s (see Burgess, 1993), but they also reveal the inability of government to communicate and stimulate local *community* action. This contrasts, perhaps, with local *government* activity. Local authorities are responding readily to Local Agenda 21, not only at the behest of central government to produce local plans by the end of 1996 but also because they view policies for sustainable development as providing a new rationale for local government now that the Conservative Government has watered-down their traditional powers and responsibilities (see House of Lords, 1995b, pp. 595–6).

The Strategy: Its Approach to Targets

The *Strategy* is regularly criticized by witnesses giving evidence to the House of Lords Committee for not taking matters forward since the 1990 White Paper. This criticism is rebutted by Government and seeks credit instead from the identification in the White Paper of '350 detailed commitments to action' (1994, para. 1.17), consolidated by two subsequent reports of progress prior to 1994 (DoE *et al.*, 1991, 1992). These reports list, sector by sector, the commitments entered into in the White Paper, action to date, and new initiatives. In practice, critics have not been inclined to let the Government get credit twice for virtually the same set of initiatives. Thus, whereas the original White Paper and the follow-up annual reports were well received, the *Strategy* was viewed as disappointing because it largely reiterated existing commitments. This lack of momentum can be interpreted as a lowering of environmental priorities in Government during the 1990s, stimulated by its inability to generate a 'feel good' factor and the perceived need to encourage economic growth for electoral reasons.

Throughout, the Government has remained reluctant to specify quantitative targets by which their attempts to achieve greater sustainability can be monitored and assessed. Many observers regard this as the logical next step to the detailing of policy commitments. On appearing before the House of Lords Select Committee, Ministers were pressed over this issue. The Environment Secretary argued that targets cannot be set profitably until objectives are clear and mechanisms are in place to deliver them – a frank if unintended admission of where the Government has (or has not) reached in the policy process (House of Lords, 1995b, pp. 656–657). The Minister of State for Agriculture suggested that targets encourage achievement of the lowest common denominator, are readily overturned by technological advance and can reduce the managerial flexibility of producers (p. 643). Neither line of reasoning is accepted by the Select Committee (1995a, pp. 16–17). The Committee holds to the view that targets give policy a clear sense of direction and the political consequences of 'underperformance' are worth risking. The Committee also argues that 'targets can make explicit those aspects of policy that otherwise might remain opaque. The greater the transparency of the process, the more acceptable targets are likely to be' (1995a, p. 16). The Committee's report also reflects the range of opinion it received on targets and is especially sensitive to those who argue in favour of market forces rather than regulation. For this reason it accepts that targets can have different functions and discriminates between 'indicative' and 'imperative' targets. Indicative targets, based upon rolling assessments of what can reasonably be achieved, should be used extensively, whereas imperative targets, akin to limits, should be absolute, used comparatively rarely and be backed-up by statute (House of Lords, 1995a, ch.3).

But in support of the Environment Secretary, it should be observed that an increasing number of new targets have been set by the Department of the Environment since 1994, including ones for air and water quality, recycling and waste disposal, all of which demand improvements on present circumstances. Whatever the limitations of the targets and the timescales set for their achievement, they will now be reported on annually and are available as pressure points for action by environmental interests (see DoE *et al.*, 1995a). Furthermore, in its response to the Select Committee Report, the Government expresses interest in developing and employing the distinction made between indicative and imperative targets, but once again exhibits extreme caution about specifying targets, constantly retreating to the suggestions that it is prepared to 'consider' setting targets and to set them 'where appropriate' (DoE *et al.*, 1995b). It is much happier discussing potential *indicators* of sustainable development. In 1996, it produced a set of 120 indicators in '21 families' that range from the economy to fish resources to landscape to radioactivity (DoE, 1996). Their individual and aggregate value must be open to question as their inclusion is more dependent upon data availability – data often collected for other purposes – than any review of what is required. None the less, they represent another collated source of information available for public debate and review[1].

Perhaps of greater interest is the wide variability in target setting between different sectors of activity. In the case of rural land-use, for example, firm targets are being set for biodiversity alongside a much greater reluctance to do so for agriculture and forestry. Why might this be so? It can be argued that targets are difficult to set because most rural land in the UK is in multiple use providing a range of benefits, some public and some private, at the same time (e.g., food, amenity, nature protection). Establishing targets for the differing relative importance of each benefit according to location is complicated, other than through a process of statutory designation that not only targets places but also the nature of the benefits expected. This difficulty may help to explain the greater progress made towards effective targeting in the area of biodiversity than in food and fibre production, the location and nature of agricultural and timber production being expected to respond to market forces (including state subsidy). In contrast, ever since the 1947 National Parks and Access to the Countryside Act, there has been a policy of statutory designation for environmental resources (amenity, nature and historic landscapes)[2]. This long-established practice not only provides a 'minimum' base from which environmental interests can seek to confirm and extend their control over land management, i.e. there is a statutory expression of their interests in the landscape which provides a tangible basis for lobbying activity, but also a professional civil service, incorporating statutory bodies such as English Nature and the Countryside Commission, with resources and powers. It also reflects the power of several major environmental groups with large memberships, including the National Trust, the Royal Society for the Protection of Birds and the Council for the Protection of Rural England, and their longstanding commitment to amenity and nature conservation[3].

This political and administrative base, combined with international obligations (e.g. RAMSAR sites; the EU Natural Habitats Directive) and detailed evidence of biodiversity loss in recent decades, created the impetus for a Biodiversity Action Plan (DoE *et al.*, 1994c) containing 59 'broad targets'. Indeed, some of these targets are quite precise (e.g. to 'ensure that summary management plans are prepared and, where possible, implemented for each biological SSSI by the year 2004'), and contributed to the setting up of the Biodiversity Action Plan Targets Sub-Group. This Group was charged with preparing before the end of 1995 fully costed draft action plans for various habitats and species up to the years 2000 and 2010. This it has achieved (Biodiversity Steering Group, 1995). The group has produced individual action plans for 116 of the UK's most threatened and endangered species (plans for a further 286 species will be produced in the next three years) and for 14 key habitats that have been suffering decline or for which the UK has international obligations. Annual estimated public costs of data collection, land purchase and site management to meet the needs of these action plans is expected to rise from £17.8 million in 1997 to £40.9 million in 2010 (Biodiversity Steering Group, 1995, p. 7). These figures compare with approximately £3000 million spent annually by the UK Government on agricultural support (£100 million of which goes to support agrienvironment schemes) and £100 million allocated to nature conservation agencies.

Witnesses appearing before the House of Lords Committee on behalf of agriculture and forestry frequently opposed targets. As the House of Lords report notes, in the case of agriculture 'targets were seen as "difficult" ... partly because the formulation of agricultural policy lay largely in the EU[4], partly because of a commitment to a voluntary system of environmental incentives, and partly because of a wish not to undermine the international competitiveness of United Kingdom agriculture by imposing tough targets' (p. 22). Instead, a sustainable agriculture is expected to emerge once an appropriate financial and regulatory framework is set up. In the case of forestry it is argued, the uncertainties inherent in its long production cycle work against simplistic targets. Annual planting targets, for example, are seen as too sensitive to short-term changes in fiscal policy, the profitability of competing land-uses, technological substitution and world market prices. In consequence, the ability of producers to respond to market forces should be constrained as little as possible by land-use targets. Targets, such as they are, are to be financial targets laid down by the Treasury for bodies such as the Forest Authority (see, for example, Scottish Office, 1994).

The plausibility of these arguments is recognized by the Committee but not accepted as valid reasons for failing to have at least indicative targets. Indeed, these are coming anyway through, for example, Indicative Forestry Plans, targeted woodland subsidies and the targeted nature of the Common

Agricultural Policy's agrienvironment measures. In effect, the right of individual freedom to manage for profit and to collect public subsidy is increasingly to be balanced against the collective obligations implicit in the managerialist version of sustainable development. Despite the Government's neo-liberal philosophy, including its attempts to privatize environmental goods, producers of food and fibre are slowly losing ground to collective environmental interests. The steady arrival of designations, quotas and other targets, whatever the private financial gains each may bring, are seeing to that. It is one of the paradoxes of contemporary rural change in the UK that just when commodity *markets* are being extended and liberated, the *processes* of production and consumption are being increasingly regulated (see Lowe *et al.*, 1993; Munton, 1995).

Assessment

The *Strategy* was only published in 1994 and specific responses to it are not yet evident in land-use terms. The consequences of the *Strategy* cannot be easily separated from those of the comprehensive approach to the environment set in motion by the 1990 White Paper and the introduction of other environmental legislation, 80% of which is now decided in the EU (House of Lords, 1995a, p. 61). It is possible to note, however, a series of shifts in policy that will affect the land-use pattern and may be seen as reducing the impact of farming and forestry on the environment, such as the drive towards multipurpose forestry and the extensive Environmentally Sensitive Areas Scheme, but whether these measures can be said to mark a definable shift in present practice towards more sustainable development is uncertain. They may aid the rural environment, and even increase constant natural assets in the countryside, but their social and economic gains and losses to the whole economy are yet to be comprehensively assessed. They do not suggest a greater commitment to definable limits, merely a greater recognition of the need for trade-offs between development and environment.

Important and disappointing though this may be, for the moment, this is not the main question. The primary concern must be with the political conditions that will allow the goals of sustainable development, as outlined at Rio, to have 'staying power', or even assume a dominating position over the UK's development path. Only then will an appropriate legislative framework be forthcoming. It is easy to be very critical of the Government's present position; but this may prove too gloomy a view. There are new institutional mechanisms in place; there is a firm commitment to report annually to Parliament on progress; and targets are, under pressure, being increasingly if unevenly established. Monitoring change and publishing the results may be an unexciting academic pursuit but if information is power then all these reporting mechanisms provide new opportunities for environmental interests to bring pressure to bear on government. It is just possible to share Owens' view that 'sustainability, eagerly endorsed by governments and at least partially encoded in legislation, may yet prove to be a Trojan horse admitting radical environmental values' (1994, p. 451).

New kinds of partnership between agents of the state, private bodies, local communities and individual citizens will be essential to the next steps. But how is responsibility and power to be devolved to local levels without a loss of national direction or a sense of the global imperatives? This question raises another paradox identified by recent research (Macnaghten *et al.*, 1995; Harrison *et al.*, 1996). Despite its mistrust of government motives, the public appears to want government to provide a firm lead, assuming this brings genuine involvement in decision-making, compliance among all groups to what is agreed and continuity of purpose. A deep sense of collective fairness and responsibility are involved that stands in opposition to the individualistic, market-driven ideology of the current UK administration. This means that tapping the latent support for sustainable development will not be easy while the Government appears remote and detached from these aspirations. It does, however, suggest that concern for the environment remains more solid than the immediate replies given to recent opinion surveys might indicate, and this should be interpreted optimistically by those committed to a more radical approach to sustainable development than the Government has, so far, been prepared to endorse.

Notes

1. The general reluctance to set targets is revealed again in the White Paper *Rural England* (DoE and MAFF, 1995). The White Paper is almost totally devoid of quantitative targets despite embracing an integrated approach to rural policy and a sustainable development discourse from the outset. Only two quantitative targets are identified in its 146 pages, one being the proposed doubling of the proportion of rural land under forestry from 7.5% to 15% by 2045, the other the wish to build 50% of new homes on re-used urban sites by 2005 (47% in 1992). This weakness is immediately identified and criticized by the House of Commons Environment Committee in its report on the White Paper (House of Commons, 1996).

2. In a recent review of how to identify critical natural capital (CNC) in the terrestrial environment, consultants to English Nature argue that there should not be a one-to-one relationship between CNC and current designations. Although this argument is based on the supposition that current levels of biodiversity are minimal and must be expanded, the paper left scope for a debate as to whether all that is currently designated constitutes CNC (and is therefore irreplaceable) or form just part of the stock of constant natural assets (Gillespie and Shepherd, 1995).

3. These bodies contributed to a report in 1993, *Biodiversity Challenge*, which insisted on measurable targets for species and habitats (see Adams, 1996, pp. 47–50).
4. The EU under its Fifth Environment Action Programme is quite happy to lay down targets.

References

Adams, J. (1995) *Risk*. UCL Press, London, 228 pp.

Adams, W.M. (1996) *Future Nature: A Vision of Conservation*. Earthscan, London.

Biodiversity Steering Group (1995) *Biodiversity: The UK Steering Group Report, Volume I: Meeting the Rio Challenge; Volume II: Action Plans*. HMSO, London.

British Government Panel on Sustainable Development (1995) *First Report*. DoE, London.

Brundtland Commission (World Commission on Environment and Development) (1987) *Our Common Future*. Oxford University Press, Oxford.

Burgess, J. (1993) Representing nature: Conservation and the mass media. In: Goldsmith, F.B. and Warren, A. (eds) *Conservation in Progress*. John Wiley, Chichester, pp. 51–64.

Cocklin, C. (1995) Agriculture, society and the environment: discourses on sustainability. *International Journal of Sustainable Development and World Ecology* 2, 240–56.

Collis, I., Heap, J. and Jacobs, M. (1992) *Strategy Planning and Sustainable Development*. Paper prepared for English Nature, Peterborough.

Department of the Environment (DoE) *et al.* (1990) *This Common Inheritance: Britain's Environmental Strategy*, Cm 1200. HMSO, London, 291 pp.

Department of the Environment and Ministry of Agriculture, Fisheries and Food (1995) *Rural England: A Nation Committed to a Living Countryside*, Cm 3016. HMSO, London, 146 pp.

Department of the Environment *et al.* (1991) *This Common Inheritance: The First Year Report*, Cm 1655. HMSO, London.

Department of the Environment *et al.* (1992) *This Common Inheritance: The Second Year Report*, Cm 2068. HMSO, London, 192 pp.

Department of the Environment *et al.* (1994a) *Sustainable Development: The UK Strategy*, Cm 2426. HMSO, London, 267 pp.

Department of the Environment *et al.* (1994b) *Climate Change: The UK Programme*, Cm 2427. HMSO, London, 80 pp.

Department of the Environment *et al.* (1994c) *Biodiversity: The UK Action Plan*, Cm 2428. HMSO, London, 188 pp.

Department of the Environment *et al.* (1994d) *Sustainable Forestry: The UK Programme*, Cm 2429. HMSO, London, 32 pp.

Department of the Environment *et al.* (1995a) *This Common Inheritance: UK Annual Report 1995*, Cm 2822. HMSO, London, 190 pp.

Department of the Environment (1995b) *Government Response to the House of Lords Select Committee on Sustainable Development*, Cm 3018. HMSO, London.

Department of the Environment (1996) *Indicators of Sustainable Development for the United Kingdom*. HMSO, London.

Gillespie, J. and Shepherd, P. (1995) Establishing criteria for identifying critical natural capital in the terrestrial environment: a discussion paper, *English Nature Research Reports*, 141. English Nature, Peterborough.

Harrison, C.M., Burgess, J.A. and Filius, P. (1996) Rationalising environmental responsibilities: a comparison of lay politics in the UK and in the Netherlands. *Global Environmental Change* (in press).

Healey, P. and Shaw, T. (1994) Changing meanings of 'environment' in the British planning system. *Transactions of the Institute of British Geographers* 19, 425–438.

House of Commons (1996) *Rural England: The Rural White Paper.* Environment Committee, 163–I, Volume 1. HMSO, London.

House of Lords (1995a) *Report from the Select Committee on Sustainable Development,* Volume 1 – Report, HL Paper 72. HMSO, London, 82 pp.

House of Lords (1995b) *Report from the Select Committee on Sustainable Development,* Volume 2 – Oral Evidence, HL Paper 72–I. HMSO, London, 684 pp.

House of Lords (1995c) *Report from the Select Committee on Sustainable Development,* Volume 3 – Written Evidence, HL Paper 72–II. HMSO, London, 170 pp.

Irwin, A. (1995) *Citizen Science: A Study of People, Expertise and Sustainable Development.* Routledge, London.

Lash, S., Szersynski, B. and B. Wynne (1995) *Risk, Environment and Modernity: Towards a New Ecology.* Sage, London.

Local Government Management Board (1993) *A Framework for Local Sustainability.* Luton, LGMB.

Local Government Management Board (1994) *Sustainability Indicators Research Project: Report of Phase One.* Luton, LGMB.

Lowe, P., Murdoch, J., Marsden, T., Munton, R. and Flynn, A. (1993) Regulating rural spaces: issues arising from the uneven development of rural land. *Journal of Rural Studies* 9, 205–222.

Macnaghten, P., Grove-White, R., Jacobs, M. and Wynne, B. (1995) *Public Perceptions and Sustainability in Lancashire: Indicators, Institutions and Participation.* Centre for the Study of Environmental Change, University of Lancaster, Lancaster.

Munton, R.J.C. (1995) Regulating rural change: property rights, economy and environment – a case study from Cumbria, UK. *Journal of Rural Studies* 11, 269–284.

Norgaard, R.B. (1988) Sustainable development: a coevolutionary view. *Futures* 20, 606–620.

Norgaard, R.B. (1994) *Development Betrayed: The End of Progress and a Coevolutionary Revisioning of the Future.* Routledge, London.

Owens, S. (1994) Land, limits and sustainability: a conceptual framework and some dilemmas for the planning system. *Transactions of the Institute of British Geographers* 19, 439–456.

Pearce, D.W., Markandya, A. and Barbier, E. (1989) *Blueprint for a Green Economy.* Earthscan, London.

Redclift, M. (1987) *Sustainable Development: Exploring the Contradictions.* Methuen, London.

Scottish Office (1994) *Our Forests – the Way Ahead: Enterprise, Environment and Access.* Conclusions from the Forestry Review, Cm 2644. HMSO, Edinburgh.

Torgerson, D. (1995) The uncertain quest for sustainability: public discourse and the politics of environmentalism. In: Fischer, F. and Black, M. (eds) *Greening*

Environmental Policy: The Politics of a Sustainable Future. Paul Chapman, London, pp. 3–20.

United Nations Conference on Environment and Development (UNCED) (1992) *Earth Summit '92.* Regency Press, London, 240 pp.

Witherspoon, S. (1994) The greening of Britain: romance and rationality. In: Jowell, R., Curtice, J., Brook, L., Prior, G. and Ahrendt, D. (eds) *British Social Attitudes: The 11th Report.* SCPR, Dartmouth, pp. 107–139.

Worcester, R.M. (1993) Public and elite attitudes to environmental issues. *International Journal of Public Opinion Research* 5, 315–334.

2 Sustainability, Spatial Hierarchies and Land-based Production

CHRIS COCKLIN[1], GREG BLUNDEN[2] AND WARREN MORAN[2]
[1]Department of Geography and Environmental Science, Monash University, Wellington Road, Clayton, Victoria 3168, Australia. [2]Department of Geography, The University of Auckland, Private Bag 92019, Auckland, New Zealand

Introduction

Geographers have participated widely in the discourses on sustainability/ sustainable development. Yet, a significant contribution that is distinctly geographical is not clearly evident in the literature. There are several avenues through which the disciplinary contribution might be confirmed, however. One is by emphasizing the potential strength that resides within the discipline to forge meaningful linkages between human activities and the environment in which we live. The sustainability/sustainable development issue presents a vital opportunity to rekindle the essential unity of geography, by reaffirming the importance of human–environment relations. Few other disciplines are as well placed to achieve this.

Geographers have the opportunity to make an important contribution also, by asserting the critical importance of space and geographical scale in understanding sustainability. That spatial relations and the issue of scale are important has been acknowledged quite widely in the literature on sustainability (for example, Lowrance *et al.*, 1986; Conway and Barbier, 1988; LeFroy *et al.*, 1991; Pierce, 1993; Smit and Smithers, 1993). The acknowledgment of importance has not been accompanied by a sound understanding of spatial relations, however. Conceptual and theoretical frameworks that have been developed from within geography present the potential to understand these spatial relations better, even though much of this theory has no explicit recourse to sustainability itself. One of the challenges, and thence the contribution, will lie in affirming the relevance of geographical thought and understanding to the improved conceptualization of space and scale *vis-à-vis* sustainability/sustainable development. The sustainability/sustainable development theme thus presents geographers with cause to refocus on human–environment relations and to make a distinctive disciplinary contribution through confirmation of the critical importance of geographic space in understanding sustainability.

Tentative steps towards this are taken here, primarily by considering the potential relevance of some aspects of geographical theory to the understanding of spatial relations. In particular, the focus is on aspects of tenure and ownership, industry structures, regulation, and enterprises and agro-commodity chains. Whereas the links to sustainability/sustainable development are not always explicit, an attempt is made to show the potential of aspects of geographical theory to understand better spatial relations in the context of sustainability. This chapter closes with some reflections on the problematic nature of defining spatial scale.

Sustainability and Geographical Scale

The theme of spatial scale is well developed and recurrent in geography, even though it may have suffered from neglect in the development of some geographies. In recent discourses on sustainability within both the geographical literature and in other literatures there has been frequent reference to the importance of considering geographic scale. This is evident particularly in the literature on the sustainability of rural systems and land-based production (e.g., Lowrance *et al.*, 1986; Conway and Barbier, 1988; LeFroy *et al.*, 1991; Pierce, 1993; Smit and Smithers, 1993). A common element in these essays is the proposition that it is useful to consider sustainability in the context of a geographical hierarchy that ranges from the individual field, to the enterprise or farm, to the local community, to the region, to the nation, and through to the international level.

The conceptualization of agriculture with reference to spatial hierarchies is certainly not novel. Clarence Olmstead (1970) developed this theme in his explication of a systems approach to the study of agriculture. Olmstead simplified the spatial hierarchy through reference to the two levels of the 'circumjacent environment' and the 'distance centred environment' of the farm, but clearly acknowledged a more intricate 'layering' of geographic space. Moreover, he suggested that the influences operating among different levels were 'the most significant key to understanding not only the complex areal differentiation of agriculture over the world but also the continuous change in that pattern' (p. 40).

The importance of scale has been reasserted strongly in the recent literature on sustainable agriculture and rural systems. According to some authors, scale is important because the factors that define and influence sustainability are emphasized differentially at each spatial resolution.

LeFroy *et al.* (1991) suggested: 'Looking at agriculture in this way helps remove some of the confusion caused by the use of the term sustainability, as it becomes apparent that different values and constraints (agronomic, economic, ecological, and social) are seen, at present, to dominate at different levels' (p. 215).

That issues of sustainability take on alternative expressions at different geographic scales was considered also by Smit and Smithers (1993). Their contribution is summarized by matching interpretations of agricultural sustainability to different geographic scales (Table 2.1). They comment that '... most empirical research on sustainable agriculture adopts a particular scale and interpretation of agriculture. It is no wonder that there is little agreement on the meaning of "sustainable agriculture"' (Smit and Smithers, 1993, p. 503).

In another expression of the importance of scale in the analysis of sustainable agriculture, Conway and Barbier (1988) acknowledged the existence of a spatial hierarchy which ranges from the environment of the individual plant or animal through to the global economy. The detail of their discussion is focused at three levels – the international constraints to sustainable and equitable agriculture, the influence of national-level policies, and the needs of rural households in terms of livelihoods and farming systems.

At the international level, sustainable and equitable agricultural development is affected by international relationships and patterns of world trade (Conway and Barbier, 1988). Soaring population numbers and

Table 2.1. Meanings of agriculture by dimensions and scale.

	Scale		
Dimension	Field/farm	Locality/region	Country/global
Natural resource base	Field level soil fertility, moisture ...	Agroecosystems, regional land capability ...	Continental water and land resources, global climate ...
Crop production	Field yield, management ...	Regional production, land-use patterns ...	Global food supplies ...
Economic return	Farm level production costs, viability ...	Regional economy, value of production ...	Trade marketing, policies ...
Rural community	Farm level tenure, family involvement ...	Rural community size and function, access to food ...	Global poverty, hunger, equity ...

Source: After Smit and Smithers (1993).

the consequent expansion in global food needs is a stress on the production system. Arguably, what is needed is an effort to bring the scale of human enterprise back into line with the capacity of the Earth's resource base (Erlich, 1993). Without the necessary reduction in population and energy use, sustainability will remain an elusive goal.

Conway and Barbier (1988) suggest that in the face of the continued population increase, which is not expected to level-off until the middle of the next century, developing countries will have to rely increasingly on food imports. Food security will be compromised even further, due to a shortfall in purchasing power. Despite possible continuing global food surpluses, ever more people in the developing world will suffer from under-nourishment. Conway and Barbier (1988) point also to various external stresses and shocks that will impinge severely upon developing countries. These include falling international commodity prices, protectionism, international terms of trade and the cost and availability of credit and capital. Some of these are internal to agriculture systems, whereas others are inherent characteristics of the world economy.

Conway and Barbier then note that how international forces play out in different countries is very much a function of domestic policies. Four main types of national policy are discussed in turn; pricing and macro-economic policy, embracing trade, fiscal and monetary strategy, agricultural input and output pricing, and agricultural stabilization policies; exchange rate policies; food pricing policies; and input subsidization, particularly from the perspective of how this may impose ecological costs.

The third level in the hierarchy is the individual farm. As Conway and Barbier noted: '. . . in the final analysis, sustainable agriculture depends on the individual, day-to-day actions of millions of farmers and their families, pursuing a variety of strategies aimed at securing their livelihoods' (Conway and Barbier, 1988, p. 663). Conway and Barbier hinge their discussion on the concept of 'sustainable livelihoods', where this is defined as: '. . . the totality of resources, activities and products that go to securing a living. It relies on ownership of, or access to, resources and access to produce or income generating activities, and is measurable in terms of both the stocks, that is the reserves and assets, as well as the flows of food and cash' (p. 665).

These views on the influence of geographic scale on sustainability are captured partly in Fig. 2.1, which is adapted from a similar diagram presented by Lowrance *et al.* (1986). In Fig. 2.1, the different levels of the geographical hierarchy are represented, along with some of those factors that define and influence sustainability at each level, and examples of the interactions among spatial scales. National-level economic policy, for example, has a direct influence on the decisions of the individual landholder.

Although it has been acknowledged that issues to do with sustainability are variably relevant at different spatial scales, much less has been said

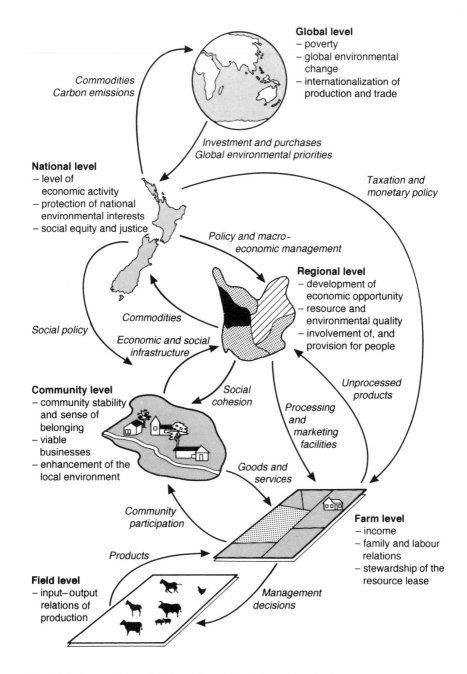

Fig. 2.1. A nested spatial hierarchy and land-based production.
(Source: Adapted from Lowrance *et al.*, 1988.)

about how the actors at each of the different levels relate to, and influence, sustainability at each of the other geographic levels. It is undoubtedly the case, though, that decisions and actions taken at one scale influence sustainability at some or all other scales (Fig. 2.1). As suggested above, macroeconomic policy is part of the structure in which the individual landholder takes decisions, for example. In some cases, processes or decisions taken at one scale may help to promote or reinforce sustainability at other levels; conversely, they may undermine the pursuit of sustainability at one or more of the other scales. In many cases, the effect will be mixed in terms of the different dimensions of sustainability (i.e. economic, social, ecological) and in terms of the different geographic levels.

That there are interconnections among the levels in the geographic hierarchy is captured in the term 'nested spatial hierarchy'. The notion of a nested hierarchy helps to convey the idea that decisions and actions taken at each level are influenced by, and have an influence upon other levels in the hierarchy. Pierce (1993), summarizing the work of a group of ecologists (Lowrance *et al.*, 1986), referred to 'a nested hierarchy originating at the field/farm level and progressing successively to larger ecological and economic units. It follows that systems which emphasize one of these components to the detriment of others are to be considered less sustainable' (p. 383). Similarly, Munton observed: 'Thus, given that environmental change arises from the actions of individual farmers, it remains necessary to analyze changing production relations at the level of the individual business in a manner that allows for particular family circumstances, but to do so with a clear perspective on the structural conditions within which farmers take their decisions' (1987, p. 55).

In spite of this recognition, there has been little progress in understanding the nature of the interconnections among geographic levels, neither has there been much applied research that explicitly works between and among levels in the spatial hierarchy. It is of course a complex undertaking, not the least because scale itself is a construct susceptible to variable interpretations – levels in the arrangement of space are not fixed entities, but are human constructs shaped by context and agency.

An improved understanding relies in part upon the conceptualization of processes at a defined spatial scale and involving the integration of social, economic and environmental considerations (the three main dimensions of 'sustainability'). It relies also on an understanding of the processes working among levels in the spatial hierarchy, again with a view to the social, economic and environmental aspects. There is a temptation to embark on a search for some kind of theoretical universal that might achieve this integration. Success in such an endeavour is most unlikely, however. None the less, one can look to existing theory for some useful insights in regard to the spatial interconnections. In the next section reference is made to selected examples to illustrate the point.

Interdependencies Between Geographic Scales

It has been suggested that a nested spatial hierarchy is a useful way of conceptualizing the interdependencies among geographic scales when considering the question of sustainable land-based production. In the remainder of this chapter, attention is focused on three arenas of geographic theory in an effort to show how these can inform an improved understanding of spatial relations *vis-à-vis* sustainability. The work used relates to the nature of ownership and tenure, regulation, and enterprises and agrocommodity chains. Examples are taken from New Zealand, where unitary governments of various political persuasions have pursued a neo-liberal restructuring ideology while at the same time implementing one of the most wide-ranging and firmly integrated environmental management regimes in the world.

Land ownership and tenure

Any consideration of sustainability and land-based production must give recognition to the groups of people who have the rights to use and have access to parcels of land. To geographers this may seem self-evident but in much of the sustainability literature originating from the physical sciences the sustainability of the natural environment is treated as if natural resources are not socially contestable. Although the maintenance of the physical environment is one central objective and attraction of sustainability it can be achieved only through the agency of people, families, and corporations – the organizations that are using the land – and the regulatory frameworks within which these organizations operate.

The geographic literature has long argued the interdependence of people and the physical environment (Hartshorne, 1959; Harvey, 1969; White, 1994). The literature on land-based production systems also contains excellent examples that demonstrate the specific need to conceptualize systematically the relationship of land tenure and ownership to other resources that are part of rural production systems (Olmstead, 1970; Barlowe, 1986; Berkes, 1989). Once the maps of tenure and ownership at the scale of individual properties (the cadastre) are overlaid on maps of physical resources the lack of coincidence between natural systems and the boundaries of the units that have rights to use the land immediately becomes apparent – natural systems have little respect for cadastral boundaries.

In considering tenure and access to land at a variety of scales the distinction needs to be drawn between land in fee simple ownership (or its equivalent), land in public (state) ownership and land that is held by groups of people (often indigenous) in various forms of social organization.

Alongside these tenure categories must be considered the bundles of rights of those other people who have access to the land or products from the land. At the smallest scale, different nation states have quite different configurations of these categories of land and access. These differences in the tenure categories have direct implications for sustainability because the rules and regulations about the use of land and access to it are quite different for each tenure category. So too are the groups of people and organizations using the land.

    The pressure to limit the power of private landowners is a manifestation of both the international accords relating to the protection of the environment (including the increased emphasis given to this by research funding agencies) and the lobbies (both local and international) that have supported these moves. Recent debate within Federated Farmers of New Zealand (the main farmers lobby group) over property rights is an example of the local negotiation that will be necessary in any move towards sustainable land-based production. After lengthy discussion their national meeting rejected an internal document that was seen by delegates to impose unacceptable limitations on their freedom to use their land as they wished (Moran *et al.*, 1994).

Regulatory influences

Regulatory analysis has emerged strongly as a fertile basis for under-standing rural systems and their evolution in contemporary western societies (e.g. Lowe *et al.*, 1993; Flynn and Marsden, 1995; Goodwin *et al.*, 1995; Munton, 1995). Of particular significance to this discussion is a recognition that regulatory analysis provides a foundation for analysing the social construction and contestation of sustainability (e.g. Marden, 1990, Flynn and Marsden, 1995).

    Land-based production takes place within the context of regulation. The most extensive level of regulation in New Zealand, as in other unitary countries, is legislation enacted by the nation's parliament or house of representatives. In New Zealand, legislation at this level has currency across the country, although a notable recent exception is the Resource Manage-ment Act (1991) wherein the implementation of planning and environ-mental legislation is devolved to the regions for interpretation within broad national guidelines (Furuseth and Cocklin, 1995). Land-based producers now operate within the context of regional policy statements and soil and water plans administered by regional councils. International accords such as Agenda 21 and global trade agreements such as the GATT also influence land-based production through global, national and regional regulation. In addition, the range of spatial jurisdictions is complicated by the diversity of regulation that affects land-based producers. In New Zealand, this includes legislation controlling producer marketing boards, the statutory role of government agencies in the regions, and legislation such as the Commerce Act, which, while appearing to contribute little directly to issues concerning sustainability, has enormous influence on economic activity and overrides other legislation specifically concerned with land and land-based production.

The concept of real regulation (Clark 1992; Marden 1992) provides a framework for analysing the rules governing economy, society and environment, and the processes by which these rules evolve (Moran *et al.*, 1996). Following Munton (1993), Moran *et al.* (1996) emphasize that real regulation is a social practice. Here, we emphasize that regulation operates at a range of scales, and that real regulation can be used to deconstruct the various geographic scales and the interactions between these levels. Regulation is a contested terrain because it is the outcome of interaction between sections of society that each seek to have their position adopted within particular regulatory frameworks, cultural norms and resource bases. One of the tensions between different viewpoints is demonstrated by the inclusion of the somewhat conflicting aims of economic development and sustainability in the Resource Management Act in New Zealand, the 1991 Act that brought together over 70 planning and land management statutes and that takes as its purpose the 'sustainable management' of resources (Furuseth and Cocklin, 1995).

The effects of regulation on the sustainability of land-based production take many forms and operate across several different scales, some predictable, some less so. One example from the Resource Management Act is the variation that is emerging in the regional and district interpretations of the act. This variation, in theory, reflects different regional resources, problems and priorities but also can be explained by the different political make-up of local governments. In some districts, such as Western Bay of Plenty in the North Island, subdivision of farmland is relatively uncontrolled and is no longer defined in terms of economic use, whereas in other districts like Franklin (which is similarly adjacent to a large metropolitan area) subdivision of farmland below 40 hectares is now prohibited in order to retain economic farm size. At another level, land use in many small, individual parts of Northland is likely to require consents because the Regional Policy Statement proposes the use of land capability as a criterion to manage areas highly susceptible to erosion.

A second example is the differential power with which particular groups are able to influence resource management. A range of societal views emerge during both the pre-statutory phase of legislation and in the contestation stage when, for example, appeals are lodged against decisions made under legislation. Although all people and groups are subject to the same resource and planning regulation, it is likely that some groups may influence proceedings more than others, principally through having more resources (usually money), and coincidentally, because of the larger scale at which these organizations operate (for example, large forestry corporations compared with family farms). Not surprisingly, power is correlated with the size of enterprises and the geographic scales at which they operate. Thus, the implications for sustainability even under innovative legislation such as the Resource Management Act are far from certain. The interpretation of sustainability is still contested and the outcomes of this contestation are contingent in part upon prevailing power relations.

Enterprises and agrocommodity chains

The sustainability of any farm, locality or region cannot be ascertained without consideration of how these levels are linked into global trading systems. Here one is concerned primarily with the nature of the enterprises involved in land-based production. The essential question for land-based producers is how they earn enough money to sustain their enterprises. This requirement varies by the type of enterprise. Corporations, which often operate at more than one scale, must return suitable profits to their shareholders or withdraw from the industry whereas producers based on the family have a different, longer-term set of imperatives that centre on the reproduction of both household and enterprise, and attachment to the land and farming as a way of life rather than a necessary return on capital as the essential structural requirement.

In earning a return for their enterprises, land-based producers are, in general, producers of undifferentiated commodities and are consequently price-takers. Again, this is variable by type of enterprise. Family producers are more limited by the scale of their operations compared with corporate producers, which may have sufficient scale to overcome this disadvantage through further processing and control of marketing channels. This weakness of small producers has been reduced in industries where cooperative processing and marketing exist as in the dairy, apple and pear and kiwifruit industries in New Zealand (Moran *et al.*, 1996). In the dairy industry, the New Zealand Dairy Board maintains ownership of dairy products (for farmers) through to the wholesale distribution networks in major customer countries. In this way, the Dairy Board is similar to other large multinational companies such as Nestlé, Dôle and Kraft as it has a high degree of vertical integration. Yet the implications are quite different – the Dairy Board is owned by small producers who retain the benefits of scale while the more clearly capitalist enterprises own the entire production and distribution network apart from the actual production facility, and return their profits to shareholders. A third type of enterprise (other than family-based and corporate) is becoming more visible in the settler countries as indigenous people retrieve customary rights and reclaim their lands. In New Zealand, Maori tribal incorporations and trusts are significant land owners and participants in land-based production. Yet the scale of these organizations, while larger than family-based producers, is still much less than the corporations involved in land-based production, and their level of influence upon agrocommodity chains is consequently limited.

The types of enterprises and the structure of agrocommodity chains have several implications for the sustainability of land-based production. For the economic aspects of sustainability, these centre on price-determination, production decisions, regional multipliers and the expatriation of profits. Where land-based production is principally for export, prices are determined largely outside the locality, region or country. This in itself implies a certain level of dependency for regions producing primary goods under these conditions. Where producers have more control, as in cooperative systems such as the dairy industry and possibly tribal incorporations, benefits are retained by the regions through the producers. However, for those regions that have industries owned and operated by enterprises external to the region the benefits are less clear, because, whereas these enterprises deliver economic activity, they also expatriate profits and control over land-based production.

Conclusions

That geographic space and scale are important in respect of understanding the dimensions of sustainability has been acknowledged quite widely. The notion of a nested spatial hierarchy, for example, has been used by several people to express the view that at different geographic scales sustainability takes on alternative emphases and interpretations. The theme is an important one. What is lacking, though, is a sufficient understanding of the relations at the respective geographic levels and, as importantly, what relations exist among the different levels in the geographic hierarchy.

Geographers can make a distinctive disciplinary contribution to the discourse on sustainability by providing a more refined and elaborate conceptualization of space and scale. There is a great deal of insight in geographical thought that provides the foundation for an improved theoretical understanding that explicitly acknowledges the spatial dimension. Established theory such as central place, for example, provides a framework for exploring the relations among different geographic levels of activity. As argued above, other theoretical developments, such as those relating to ownership and tenure, trade and comparative advantage, forms of production, regulation and industrial structure offer other insights into spatial relations. One of the challenges lies in establishing more clearly the conjunction between these ideas and those relating to sustainability itself.

Understanding the social, economic and environmental relations at and across geographic scales is complicated because, as geographers have long argued, space and scale are elusive concepts. There are at least two reasons for this. First, relations within the geographic hierarchy are diverse and complex. If one fixes the perspective at the farm level, for example, it is evident that there are many and varied relations between the individual enterprise and other levels in the geographic hierarchy. The farmer who is involved in growing trees for production, for example, engages with an agrocommodity chain that extends in its spatial characteristics from local companies (e.g. silviculture contractors, transportation firms) through to international commodity markets for wood products. The economic viability of the woodlot is defined not only by the qualities of the farmer's resource base (e.g. the land) and the farmer's management of the woodlot but also by the international terms of trade in wood products. Moreover, the landholder may have entered into joint ownership and management agreements with other investors, therein adding the complexity of property rights, with their own attendant geographic configurations. In the very next field, the same farmer may be involved in dairy production, and therefore with a commodity chain with a spatial character that is quite different from that of wood products. As well as this, the two commodities may well be governed by different sets of legislation. The institutional and political structure with which the landholder engages will also be spatially differentiated by commodity. It is not sufficient, then, to suggest simply that sustainability is embedded within a hierarchy that is arranged geographically. The character of this spatial arrangement is highly differentiated.

The second observation is that levels in the arrangement of space are not fixed entities, but are produced and reproduced by human context and agency. The debate that persists in geography over the interpretation of regions demonstrates this. Two examples from the New Zealand context illustrate the point. First, investment patterns of corporate capital can reshape regions and place. It has been proposed, for example, that a forestry company will soon invest in wood processing facilities in Northland, New Zealand. The character, spatial arrangement, and sustainability of land-based production in this region will be affected fundamentally by such a decision. Indeed, they have been already. The changed spatial character of the commodity chain will further emphasize the viability of timber production in the Northland region, with attendant social, economic and environmental consequences that will be wide-ranging. As a result of the regionalization of timber processing, local and regional government will have to negotiate outcomes (e.g. in relation to environmental quality) that are satisfactory in local terms.

The second example comes from the reorganization of local government. As part of the widespread restructuring of the state and the economy in New Zealand over the last decade, the territorial jurisdictions of local and regional governments have been extensively reorganized. Whereas several criteria were prescribed, the redrawing of the boundaries was dictated primarily by concerns for economic efficiency (e.g. in terms of service delivery), and with a view to the shift in responsibility for resource management to local and regional government (Furuseth and Cocklin, 1995). The first of these two considerations was reflected in the delineation of relatively large territories (i.e. spatial amalgamation) while the second consideration eventuated in the use of water catchments as a basis for the definition of jurisdictions. One important outcome was the greater administrative integration of urban centres and their rural hinterlands, which under previous government organization had been functionally separate (Moran, 1991).

The new administrative map has been strongly contested in many parts of the country. Local government amalgamation has been equated with the loss of identity and of influence by people over their local body representatives. The boundary controversy persists in several places and might be described as a 'crisis of rationality' (Britton *et al.*, 1992), wherein the logic or rationale of change promoted by government at the centre is rejected at the locality. It goes to issues of community identity and hence to issues of social sustainability.

Sustainability is a natural domain of interest for geographers because it is about people in their environment. The fact that social, political, economic and environmental interests necessarily converge under the rubric of sustainability makes it a logical and indeed traditional focus of interest for geographers. Moreover, as argued here, there is a spatial dimension that is an essential characteristic of our interpretations of sustainability. It is in this dimension that geographers offer the promise of improved conceptualization and theoretical understanding. Both traditional and more recent theoretical contributions from within the discipline offer considerable promise in this respect.

References

Barlowe, R. (1986) *Land Resource Economics: The Economics of Real Estate.* 4th edn. Englewood Cliffs, NJ and London, Prentice-Hall.

Berkes, F. (ed.) (1989) *Common Property Resources: Ecology and Community-based Sustainable Development.* London, Belhaven Press.

Britton, S., Le Heron, R. and Pawson, E. (eds) (1992) *Changing Places in New Zealand: A Geography of Restructuring.* New Zealand Geographical Society, Christchurch, NZ.

Clark, G.L. (1992) 'Real' regulation: the administrative state. *Environment and Planning A* 24, 615–627.

Conway, G. and Barbier, E. (1988) After the green revolution: sustainable and equitable agricultural development. *Futures* 20, 651–678.

Ehrlich, P. (1993) The scale of the human enterprise. In: Saunders, D. Hobbs, R. and Ehrlich, P. (eds) *Nature Conservation 3: Reconstruction of Fragmented Ecosystems.* Beaty and Sons, Surrey, pp. 3–8.

Flynn, A. and Marsden, T. (1995) Guest editorial. *Environment and Planning A* 27, 1180–1192.

Furuseth, O. and Cocklin, C. (1995) An institutional framework for sustainable resource management: the New Zealand model. *Natural Resources Journal* 35, 243–273.

Goodwin, M., Cloke, P. and Milbourne, P. (1995) Regulation theory and rural research: theorising contemporary rural change. *Environment and Planning A* 27, 1245–1260.

Harvey, D. (1969) *Explanation in Geography.* Edward Arnold, London.

Hartshorne, R. (1959) *Perspective on the Nature of Geography.* Chicago, London.

Lefroy E., Salerian, J. and Hobbs, R. (1991) Integrating economic and ecological considerations: a theoretical framework. In Hobbs, R. and Saunders, D. (eds) *Reintegrating Fragmented Landscapes: Towards Sustainable Production and Nature Conservation.* New York: Springer Verlag, pp. 209–244.

Lowe, P., Murdoch, J., Marsden, T., Munton, R. and Flynn, A. (1993) Regulating the new rural spaces: the uneven development of land. *Journal of Rural Studies* 9(3), 205–222.

Lowrance, R., Hendrix, P. and Odum, E. (1986) A hierarchical approach to sustainable agriculture. *American Journal of Alternative Agriculture 1*, 169–173.

Marden, P. (1990) *Modern Capitalist Society and a Theory of Regulations: Searching for the Elusive Archimedean Point,* Working Paper No. 31. Department of Geography and Environmental Science, Monash University, Melbourne, 80 pp.

Marden, P. (1992) 'Real' regulation reconsidered. *Environment and Planning* A, 24, 751–767.

Moran, W. (1991) Local government reform. In: Britton, S., Le Heron, R. and Pawson, E. (eds) *Changing Places in New Zealand: A Geography of Restructuring.* Christchurch, NZ: New Zealand Geographical Society.

Moran, W., Blunden, G., Bradly, A. and Workman, W. (1994) Sustainable agriculture and family farming. In: Whittaker, W. (ed.) *Proceedings of the Seventeenth Conference of the New Zealand Geographical Society at Victoria University of Wellington, Aug. 30–Sept 2.* Christchurch, New Zealand Geographical Society, pp. 300–306.

Moran, W., Blunden, G. and Bradly, A. (1996) Empowering family farms through cooperatives and producer marketing boards. *Economic Geography* 72, 161–177.

Munton, R. (1987) The conflict between conservation and food production in Great Britain. In: Cocklin, C., Smit, B. and Johnston, T. (eds) *Demands on Rural Lands: Planning for Resource Use.* Boulder, Colorado, Westview Press, pp. 47–60.

Munton, R. (1993) Rural accumulation and property rights: sustaining the means. Paper presented at the Association of American Geographers Annual Meeting, Atlanta, Georgia, 1993.

Munton, R. (1995) Regulating rural change: property rights, economy and environment – a case study from Cumbria, UK. *Journal of Rural Studies* 11, 269–284.

Olmstead, C. (1970) The phenomena, functioning units and systems of agriculture. *Geographia Polonica* 19, 31–34.

Pierce, J. (1993) Agriculture, sustainability and the imperatives of policy reform. *Geoforum* 24(4), 381–96.

Smit, B. and Smithers, J. (1993) Sustainable agriculture: interpretations, analyses and prospects. *Canadian Journal of Regional Science* 16(3), 499–524.

White, S.E. (1994) Ogallala oases: water use, population redistribution, and policy implications in the High Plains of Western Kansas, 1980–1990. *Annals of the Association of American Geographers* 84(1), 29–45.

3

Greening and Globalizing: Agriculture in 'the New Times'

GUY M. ROBINSON
School of Geography, Kingston University, Penrhyn Road, Kingston upon Thames, Surrey KT1 2EE, UK

Introduction: Food Regimes

The concept of food regimes links international relations of food production and consumption to forms of accumulation and regulation under capitalist systems from the 1870s onwards (Friedmann and McMichael, 1989). It is a concept that considers macro-forces of demand and supply acting as a 'growth ensemble in the international food sector', varying from country to country and 'differentiated from slower growing or declining food industries' (Le Heron and Roche, 1995, p. 24). The principal proponents of the concept recognize three regimes, the third of which is in the early stages of development (Table 3.1).

'Friedmann and McMichael explain national economic development in terms of sectoral changes within and between nations via the emergence of a collection of liberal nation states and through the internationalization of agriculture, food production and consumption' (Roche, 1994, p. 12). This stresses the role of white settler agriculture in the late nineteenth century as part of a new international order alongside tropical colonial exports.

The first food regime represents production for the world's metropolitan core in North America and Western Europe, with the settler colonies supplying unprocessed and semi-processed foods and materials to the core. The introduction of refrigerated ships in the 1880s increased both the range of produce that could be supplied by distant colonies and the distance over which perishables such as butter and meat could be transported (Peet, 1969; Schedvin, 1990). This regime was subsequently undermined after World War I as agriculture in the core competed with imported produce, often under the protection of trade barriers. Its

Table 3.1. Food regimes.

Character	First regime	Second regime	Third regime
Products	Grain, meat	Grain, meat, durable food	Fresh, organic, reconstituted
Period	1870s–World War I	1920s–1980s	1990s–
Capital	Extensive	Intensive	Flexible
Food systems	Exports from family farms in settler colonies	Transnational restructuring of agriculture to supply mass market	Global restructuring, with financial circuits linking production and consumption
Characteristics	Culmination of colonial organization of precapitalist regimes; Rise of nation states	Decolonization; consumerism; growth of forward and backward linkages from agriculture	Globalization of production and consumption; disintegration of national agrofood capital and state regulation; 'Green' consumers

Source: Based on Roche (1994) as derived from Friedmann and McMichael (1989).

replacement reached its apogee in the 1950s and 1960s. This second food regime was based on the development of agroindustrial complexes focusing on production of grain-fed livestock, fats and durable foods. The regime incorporated production in both the developing and developed worlds and was associated with increased specialization and geographical segmentation of production systems. According to Friedmann (1987, p. 253), this regime reached a crisis point in 1973 as USA–USSR grain deals eliminated the American wheat surplus, as export competition between the United States and the European Community grew, and as traditional patterns of trade with Developing Countries were disrupted (see also Friedmann, 1982).

Some analysts contend that this second regime is now giving way to a successor as the established agroindustrial complexes lose their growth potential (Goodman, 1991). For example, McMichael (1992a) argues that in addition to restructuring performed by those transnational corporations (TNCs) involved in food processing and marketing, the changing role of the International Monetary Fund (IMF) and the General Agreement on Tariffs and Trade (GATT) is providing a new regulatory structure in which the new regime will emerge and operate. One aspect of this third regime recognized in some quarters is production of 'fresh' fruit and vegetables for a market often described as 'global' (e.g. Roche, 1994). This is part of a complex mix of processes relating to globalization and sustainability that could lead to the dominance of a clearly recognizable third food regime in the first decades of the twenty-first century.

McMichael's (1992a, p. 359) interpretation of the third food regime is of agrocommodity production 'characterized by the reconstitution of food through industrial and bio-industrial processes via flexible global sourcing of generic crops and increasing importance of affluent foods (animal protein, processed foods, fruit and vegetables) (Roche, 1994, p. 13). In the New Zealand context, Le Heron and Roche (1995) suggest there is evidence that the third regime involves a conflation of globalization and sustainability tendencies. This may involve a reorientation from 'basic' foods and export crops to a more important role for the supply of inputs for 'elite consumption in the north' (McMichael and Myhre, 1991, p. 100). For example, McMichael (1992b, p. 113) refers to the internationalization of Pacific Rim food systems leading to 'the reconstruction of supply zones such as the United States, Australia and Thailand, especially in the development of integrated animal protein-complexes sponsored by Japanese capital'.

In arguing that geographers should give greater attention to the concept of food regimes, Le Heron and Roche (1995) contend that tensions and contradictions between the second and third regimes are represented in widespread differences in consumption trends and in the impacts of both globalization and sustainability. They note the uneven development of both the first two food regimes while suggesting that more use should be made of the idea of the regime as an organizing concept or framework. In more practical terms they consider the impacts of key aspects of the third food regime as it emerges in New Zealand, concluding that the concept offers an organizing scheme that can transcend simplistic international comparisons and country case studies. This scheme can explicitly include aspects of the spatial reorganization of production and consumption inherent within the notion of regimes. A suggested focus is upon patterns of investment within each regime and how these have developed differentially within the controlling context of prevailing regulatory controls.

If food regimes are to occupy an important epistemological role within studies of rural development, more work will be required to clarify the nature of the concept and both its theoretical and practical dimensions. As part of this work, this chapter considers two of the central characteristics of a possible incipient third regime, namely globalization and sustainability.

Globalization

Is the world moving towards a globalized rather than an international economy? Can it be argued that one is moving to a situation in which there will be no national products or technologies, no national corporations or national industries as every factor of production can move effortlessly across borders (Reich, 1991)? The development of globalization is qualitatively different from the internationalization of the first food regime, which involved the extension of a firm's activities across national boundaries. Globalization implies a degree of purposive functional integration among geographically dispersed activities. As conceptualized by Bartlett and Ghoshal (1989), TNCs with a globalizing strategy centralize strategic assets, resources, responsibilities and decisions, but operations, which are based in several countries, are aimed at tapping a unified global market. However, the home base or country or origin of a TNC may remain important to the nature of its operations and affects the character of its globalization.

Bonnano (1993) refers to a crucial aspect of globalization as 'trans-nationalization' or 'the recomposition of production processes across national boundaries in such a way that they transcend the locus determined by the physical limits of the nation state' (p. 342; see also Friedland, 1991). This goes beyond the multinational capitalism of the second food regime as it affects the international organization of productive sectors at both spatial and sectoral levels. It has intensified agricultural specialization at regional and production-unit levels, whereas, at sectoral level, it has brought increased transformation of agricultural products from items destined for immediate consumption into inputs for the greater food manufacturing system (e.g. Constance and Heffernan, 1991). This has also involved increased differentiation of linkages between production and consumption, offering new opportunities for study of the spatial and social variability that are major characteristics of this differentiation.

In recognizing the growth of globalizing tendencies in the claimed recent emergence of a new world food order, Friedmann and McMichael (1989) refer to the declining significance of national regulation in shaping the character of agricultural production, and that has even extended to previously heavily protected products (e.g. Robinson, 1995). This view stresses the global implications of what could now be termed the 'post-GATT era', referring to the conclusion of the Uruguay Round of the negotiations of the GATT, which focused on trade in agricultural produce (McMichael, 1992a). The Uruguay Round represents an appropriate starting point for a consideration of the globalization of agriculture and of the production of agrocommodities. This is because it can be argued that underlying various multinational conflicts in the negotiations was the internationalization of productive and financial capital, symbolized in the growth of transnational food companies and their global food production and processing operations (Grant, 1993; McMichael, 1993). These global trends have provided a stimulus towards trade liberalization as part of a shift from the state to capital as the major force reshaping regional agricultures. Whereas this is challenging existing agricultural and food policy regimes, it is also threatening the alliance between the state and farmers that has been in place since World War II (Ufkes, 1993a).

The strong farm lobby has an influence on many national governments, with a strong tradition in several European countries: it has had a direct influence on EU officials, and has had a wider effect by enlisting the support of affiliated EU lobbying organizations. There are close relationships between most national Farm Ministers and domestic farm organizations. There is a long history of farm organizations delivering political legitimacy and political support for governments, and, in return, farm organizations have been allowed to participate in policy-making. There has been public support for this in some cases via the agrarian myth in which rural life is regarded as a vital ingredient in European culture, and the continuation of inefficient peasant producers is seen as a national priority. However, recent reforms to the Common Agricultural Policy (CAP) reflect some decline of the farm lobbies' political power and fragmentation of agricultural lobbies both in individual states and supranationally. Moreover, with the farm lobby being unable to block the conclusion of the Uruguay Round, the transformation of European agriculture is likely to continue apace, driven by a new international structure, with TNCs growing in importance and a smaller number of farm operators surviving. Yet, there is still a continued high level of support for small farmers in some quarters and a greater prominence given to certain environmental considerations.

In the Uruguay Round there was a tremendous contrast between the wishes of most TNCs and European farmers who are dependent on EU farm policy for their viability. The farmers' protest at the prospect of deregulation reflects the greater competitive conditions that deregulation may bring through greater exposure of farmers to world prices. Industrial countries of the EU, North America and Japan could be freed from a traditional but expensive alliance with national farm sectors while also freeing controls on the investment and sourcing activities of transnational agrofood capital.

Globalization processes are not solely a phenomenon of the 1990s. It can be argued that globalization tendencies were boosted when the previous postwar norms started to disappear with the demise of the Bretton Woods system in the early 1970s and the decoupling of the US dollar from gold in 1971, ending national currency regulation via the gold and dollar standard. Subsequently, there has been a growth of flexible accumulation assisted by new institutional mechanisms of control and cooperation in the global movement of capital. For example, the International Monetary Fund

(IMF) has assumed increased importance as some aspects of policy-making have been displaced from the state to supranational institutions.

A GATT-based free trade regime could further assist institutionalization of the mechanisms and norms of a system of global regulation. For example, unifying the market for efficient producers should help global accumulation strategies by TNCs, enhancing capital mobility and eliminating control of market and production sectors by national companies. Trade liberalization will probably also enhance the farm commodity exports of the USA and efficient producers of particular commodities (Ufkes, 1993b).

Within the Uruguay Round, efforts to reduce national subsidies to farm production and trade helped to foster an environment that greatly enhances the geographical scope of agrofood capital. Friedmann and McMichael (1989) argue that TNCs were restricted in the previously unliberalized agrocommodity markets, so therefore they have been strong supporters of trade reform in agrocommodities whereby economic trade relations would be regulated by global institutions favouring global accumulation.

The dismantling of national regulatory structures also represents a shift from national to international policies, possibly removing agricultural policy further from public scrutiny. On the other hand, the attention focused on the Uruguay Round and the proposed reforms to the CAP have actually appeared to bring agricultural policy under closer public scrutiny, partly because of the influence of other processes largely antithetical to globalization, for example what Buttel (1992) refers to as the processes of greening and environmentalism.

Greening

Greening is the process by which modern environmentally related symbols have become increasingly prominent in social discourse. Environmentalism is the greening of institutions and institutional practices, that is the trend towards environmental considerations being increasingly brought to bear in political and economic decisions in educational and scientific research institutions, and geopolitics. These considerations include the growing emphasis upon sustainability within economic and rural development planning, and in the market-place. The notion of sustainability as an important component of agricultural development is one that has moved into mainstream political thought only recently. Impetus has come from the United Nations 'Earth Summit' at Rio de Janeiro in 1992 and from ideas put forward by the World Commission on Environment and Development (the Brundtland Report) in 1987. Thus, sustainable growth that pays due regard to environmental considerations is incorporated as a principal policy goal of the EU in Article 2 of the Treaty of the European Community as amended by the Treaty of the European Union. Article 130R requires environmental protection to be integrated into the definition and implementation of EU policies. Moreover, promotion of sustainability is a central element in the EU's Fifth Environmental Action Programme, 'Towards Sustainability', adopted in March 1992 and intended to operate from 1993 to 2000. This Programme bases its actions on the concept of sustainable development as put forward in the Brundtland Report, emphasizing the need to address root causes of environmental degradation before they affect economic growth and efficiency.

Various academics have noted the major paradox inherent in the term 'sustainable development', namely the combination of the contradictory ideas of limits to growth and active promotion of growth (e.g. Redclift, 1990, 1994; Korten, 1994; Willers, 1994). It can be argued that reconciliation of this contradiction may be sought via the twin goals of economic self-sufficiency and production for limited use, but the tensions between 'sustainable' and 'development' are generally inadequately recognized in policies promoting sustainability (Pearce *et al.*, 1989). Nevertheless, sustainable development is now enshrined as an attainable goal in many articles of legislation in various countries, including the EU and the USA (Clinton and Gore, 1992; EPA, 1992).

One dimension of 'sustainability', which may be an adjunct to farm-based production in the third food regime, is the modification of agricultural policy in favour of a broader role for farmers (Robinson and Ilbery, 1993). This role embraces the production of countryside as well as traditional assemblages of crops and livestock. In the EU this has been promoted by changes in the basis of farm support and by specific policies encouraging both environmentally friendly farming, for example the establishment of Environmentally Sensitive Areas (ESAs), and a broader aim of sustainable development (Robinson, 1994b). Yet there are several contradictions within these policy reforms, the vital ones being whether the championing of the role of the farmer as steward of the land is not really just a policy for farm survival in marginal regions rather than a concerted attempt at landscape enhancement (Webster and Felton, 1993), and, second, just what EU policy-makers regard as 'sustainable'.

To date, policies promoting sustainability have tended to pursue one of two broad themes:

1. simply giving environmental considerations greater weight, but balancing these against the benefits of economic development;
2. using environmental capacities to plan a constraint on economic activity so that environmental concerns are uppermost.

Increasingly, the idea of natural capital has been incorporated within the concept, recognizing that some aspects of the environment incorporate a

critical natural capital that is essential to human survival, and is irreplaceable. But just what is 'critical' and how should it be preserved and handed-
on to future generations? And what of non-critical natural capital and its
relationship to economic imperatives? These questions are at the heart of
current academic debates on sustainability (see Owens, 1994), but they are
also already embedded in policy implementation that is underway. Thus
there is an urgent need for investigation into how sustainable development
policies are being implemented. This research needs to investigate ways in
which sustainability is operationalized in environmental management
through more concrete objectives such as combating pollution, maintaining agriculture and other economic activities, especially in marginal areas.
This work needs to rectify the relatively limited attention given to the
contribution (or lack of it) of agrienvironmental policy measures to their
stated goals of environmental and economic sustainability.

In a European context the recent enlargement of the EU, through the
accession of Austria, Finland and Sweden, offers increased scope for
investigation of these problems. The two new Scandinavian members in
particular present new opportunities for comparative work on agrienvironmental policies because of their combination of agriculture and timber
production on farms, the practice of afforestation of set aside land, a
history of use of environmental subsidies to restrict damaging practices in
areas of agricultural intensification, and a strong element of local community planning, especially in northern peripheral areas.

For example, a prime source of subsidized support for Finnish agriculture is a system of environmental payments. The largest subsidies are given
to farmers in areas where they are not eligible for other financial support,
primarily the areas of most intensive arable production in the south and west
of the country. Here, on a voluntary basis, farmers can participate in a scheme
promoting the extension of environmentally friendly farming practices,
including cessation of cultivation on 3 m wide strips of land between fields
and rivers/lakes; 30% of a farm's field area to be unprepared and allowed to
regenerate during winter; restrictions on the use of fertilizers; other
measures to reduce pollution of watercourses; and maintenance of landscape, cultural heritage and biodiversity. The support for this programme is
nearly equivalent to the price support Finland will receive under the CAP and
so contrasts sharply with the total level of payments under the ESAs scheme
available in designated areas in the United Kingdom (UK) and Denmark.
Take-up rates in the upland ESAs in the UK have been high (e.g. Robinson,
1994a), but the extent to which this type of scheme or the similar ones in
Scandinavia and Germany contribute to a broader goal of sustainability
remains to be seen. In terms of the development of a new food regime, the
role of sustainability needs much greater investigation in countries like
Finland where the 'freshness' dimension recognized for New Zealand may be
replaced by other characteristics.

Conclusion: Whither Food Regimes?

Le Heron and Roche (1995, p. 25) argue that aspects of globalization and sustainability have combined to deepen and intensify the potential exposure of agriculture to large-scale capitalist forces, with the adoption of 'international norms, standards and practices as well as production to meet consumer preferences'. Their interpretation of the third food regime extends the concept of sustainability to include a growing emphasis upon freshness of food, high quality and consumer preference for healthy eating. This may include an extension of technology associated with the second regime, such as the use of chemicals to simulate naturalness and control ripening of packaged fruit (Arce and Marsden, 1993). It can also include the production of 'organic' fruit and vegetables to tap the rising wave of 'green' consumerism (Clunies-Ross, 1990) and, as suggested above, the 'greening' of agriculture via the development of environmental dimensions within agricultural policy.

For New Zealand, several examples can be cited of how both globalization and sustainability dimensions of the third food regime are apparent in evolving agroproduction systems (e.g. Robinson, 1993). More generally, this chapter suggests that combinations of globalizing tendencies and concerns for sustainable development have been central to international dialogue in the GATT and within the reforming agenda pursued with respect to the CAP. Aspects of a new food regime are apparent in the ways in which European farmers are being asked to fulfil new and often contradictory roles. The result is a patchwork of productivist and environmentalist scenarios that are often present to different degrees on neighbouring farms. It is apparent that corporate and economic restructuring has affected most countries' political economies, often producing similar problems for localities and localized plans for effecting change (Dicken, 1994). And the food regimes concept implies that, despite the variation in national contexts, there is a common framework that can be utilized to help elucidate trajectories of local change. However, it is not very clear from Le Heron and Roche's work just how or whether the contours of the new regime can be clearly delimited from the previous one nor of how these general concepts advance understanding of the tensions between unique occurrences in particular locations.

Nevertheless, the concept of food regimes does offer opportunities for further development of research on both globalization and sustainability, explicitly in terms of the linkages between land-based production and its suppliers and markets. In other words, food regimes can be viewed as part of an evolving geography of agriculture that, having embraced political economy (Bowler and Ilbery, 1987), and having extended beyond a concern for production on farms to one dealing with 'the geography of food' (Le Heron, 1993), now includes explicit consideration of a broad spectrum of agrocommodities within an historico-political context that links geopolitics, business economics, consumption patterns, retailing, processing, farming and supply industries. Indeed, it has been asserted that the political economy perspective has reached its empirical and conceptual limits and that new approaches to rural development problems are already in place that emphasize commodification – 'relating farmers' variable actions to an understanding of simple commodity production' (Arce and Marsden, 1993, p. 296; Friedmann, 1980). Other work, developing similar themes has focused on the relationship between the exchange process, social life and its politics (e.g. Appadurai, 1986) and, increasingly, on environmental considerations including sustainable development, valuation of environmental assets, the importance of environmental accounting and the use of market indicators. To this could be added the growing volume of work on the impact of new forms of retailing and business organization upon food production and marketing (e.g. Wrigley, 1992).

In summarizing the opportunities provided by a focus on food regimes, Arce and Marsden (1993) recognize three prime research areas:

1. the relationship between international food regimes and agricultural structures, in particular to highlight the formation of 'commodity complexes' (as referred to by Le Heron and Roche with respect to fruit and vegetable production in New Zealand in the third food regime) and the consequences of the new international division of labour;
2. analysis that extends beyond a concern for agricultural production to examine consumer behaviour and changes in marketing, distribution chains and food processing;
3. the 'macro-explanation' of food systems, linking accumulation strategies of international and national capital to employment and consumption norms, types of state regulation and legislation, and household livelihood and consumption practices (e.g. Goodman and Redclift, 1991).

Thus there is potential for the general framework of the food regime to offer a basis for comparison between countries, regions and localities experiencing different aspects of the transition from the second to the third regime. This may be a valuable organizing concept as it is clear that a single and fixed spatio-temporal divide between different regimes is not a reality. Instead there are mixes of all three regimes present in different localities. Society may be in a transition period away from the second regime to the third, but it is a contested transition over which businesses and governments are vying for control. In developing a more complete understanding of this process one will need to ensure that the theoretical armoury is well honed and that concepts like food regimes are subjected to more thorough scrutiny to yield maximum benefit.

References

Arce, A. and Marsden, T.K. (1993) The social construction of international food: a new research decade. *Economic Geography* 69, 293–311.

Appadurai, A. (1986) Introduction: commodification and the politics of value. In: Appadurai, A. (ed.) *The Social Life of Things*. Cambridge University Press, Cambridge, pp. 1–15.

Bartlett, C.A. and Ghoshal, S. (1989) *Managing Across Borders: The Transnational Solution*. Harvard Business School Press, Boston.

Bonnano, A. (1993) The agro-food sector and the transnational state: the case of the EC. *Political Geography* 12, 341–360.

Bowler, I.R. and Ilbery, B.W. (1987) Redefining agricultural geography. *Area* 19, 327–332.

Buttel, F. (1992) Environmentalization and greening: origins, processes and implications. In: Harper, S. (ed.) *The Greening of Rural Policy: International Perspectives*. Plenum Press, London and New York, pp. 12–26.

Clinton, B. and Gore, A. (1992) *Putting People First: How We Can All Change America*. Time Books, New York.

Clunies-Ross, T. (1990) Organic food: swimming against the tide. In: Marsden, T.K. and Little, J. (eds) *Political, Social and Economic Perspectives on the International Food System*. Avebury, Aldershot, pp. 200–214.

Constance, D. and Heffernan, W.D. (1991) El complejo agroalimentario global de las aves de corral. *Agricultura y Sociedad* 60(3), 63–92.

Dicken, P. (1994) The Roepke Lecture in Economic Geography. Global–local tensions: Firms and states in the global space-economy. *Economic Geography* 70, 101–128.

Environmental Protection Agency (EPA) (1992) *Sustainable Development and the Environmental Protection Agency: Report to Congress*. Report 230-R-93-005, Washington DC.

Friedland, W. (1991) The transnationalisation of production and consumption of food and fibre: challenges for rural research. In: Aimas, R. and Withi, N. (eds) *Final Futures in an International World*. University of Trondheim Press, Trondheim, Norway, pp. 115–135.

Friedmann, H. (1980) Household production and the national economy: concepts for the analysis of agrarian formations. *Journal of Peasant Studies* 7, 158–184.

Friedmann, H. (1982) The political economy of food: the rise and fall of the international food order of the post-war era. *American Journal of Sociology* 88S, 248–266.

Friedmann, H. (1987) The family farm and the international food regimes. In: Shanin, T. (ed.) *Peasants and Peasant Societies: Selected Readings*, 2nd edn. Basil Blackwell, Oxford, pp. 247–258.

Friedmann, H. and McMichael, P. (1989) Agriculture and the state system: the rise and fall of national agricultures, 1870 to the present. *Sociologia Ruralis* 29, 93–117.

Goodman, D. (1991) Some recent tendencies in the industrial reorganisation of the agri-food system. In: Friedland, W.H., Busch, L., Buttel, F.A. and Rudy, A.P. (eds) *Towards a New Political Economy of Agriculture*. Westview Press, Boulder, Colorado, pp. 37–63.

Goodman, D. and Redclift, M. (1991) *Refashioning Nature*. Blackwell, Oxford.

Grant, R. (1993) Against the grain: agricultural trade policies of the United States, the EC and Japan at the GATT. *Political Geography* 12, 247–262.

Korten, D.C. (1994) Sustainable development case studies. In: Meffe, G.K. and Carroll, C.R. (eds) *Principles of Conservation Biology*. Sinauer Associates Inc., Sunderland, Massachusetts.

Le Heron, R.B. (1993) *Globalized Agriculture: Political Choice*. Pergamon Press, Oxford.

Le Heron, R.B. and Roche, M.M. (1995) A 'fresh' place in food's space. *Area* 27, 23–33.

McMichael, P. (1992a) Tensions between national and international control of the world food order: contours of a new food regime. *Sociological Perspectives* 35, 343–365.

McMichael, P. (1992b) Agro-food restructuring in the Pacific Rim: a comparative-international perspective on Japan, South Korea, the United States, Australia and Thailand. In: Palat, R. (ed.) *Pacific Asia and the Future of the World-system*. Greenwood, Westport, pp. 103–116.

McMichael, P. (1993) World food system restructuring under a GATT regime. *Political Geography* 12, 198–214.

McMichael, P. and Myhre, I. (1991) Global regulation versus the nation state: agricultural food systems and the new politics of capital. *Capital and Class* 43, 83–105.

Owens, S. (1994) Land, limits and sustainability: a conceptual framework and some dilemmas for the planning system. *Transactions of the Institute of British Geographers, new series* 19, 439–456.

Pearce, D.W., Markandya, A. and Barbier, E. (1989) *Blueprint for a Green Economy*. Earthscan, London.

Peet, R.J. (1969) The spatial expansion of commercial agriculture in the 19th century: a von Thunen interpretation. *Economic Geography* 45, 283–301.

Redclift, M.R. (1990) *Sustainable Development: Exploring the Contradictions*. Methuen, London.

Redclift, M.R. (1994) Reflections on the 'sustainable development' debate. *International Journal of Sustainable Development and World Ecology* 1, 3–21.

Reich, R. (1991) *The Work of Nations: Preparing Ourselves for 21st Century Capitalism*. Vintage Books, New York.

Robinson, G.M. (1993) Trading strategies for New Zealand: The GATT, CER and trade liberalisation. *New Zealand Geographer* 49, 13–22.

Robinson, G.M. (1994a) The greening of agricultural policy: Scotland's Environmentally Sensitive Areas (ESAs). *Journal of Environmental Planning and Management* 37, 215–225.

Robinson, G.M. (1994b) Dimensions medioambientales de la politica agricola comun en el Reino Unido. *Agricultura y Sociedad* 71, 127–151.

Robinson, G.M. (1995) The deregulation and restructuring of Australia's cane sugar industry. *Australian Geographical Studies* 33, 212–227.

Robinson, G.M. and Ilbery, B.W. (1993) Beyond MacSharry: reforming the CAP. *Progress in Rural Policy and Planning* 3, 95–107.

Roche, M.M. (1994) 'Britain's farm to global producer? Food regimes and New Zealand's changing links within the Commonwealth'. Unpublished paper presented at the Commonwealth Geographical Bureau Silver Jubilee Symposium on Global Change and the British Commonwealth, The Chinese University of Hong Kong, 5–7 December, 1994.

Schedvin, C.B. (1990) Staples and regions of Pax Britannica. *Economic History Review, 2nd series* 43, 533–559.

Ufkes, F. (1993a) The globalisation of agriculture. *Political Geography* 12, 194–197.

Ufkes, F. (1993b) Trade liberalisation, agro-food politics and the globalisation of agriculture. *Political Geography* 12, 215–231.

Webster, S. and Felton, M. (1993) Targeting for nature conservation in agricultural policy. *Land Use Policy* 10, 67–82.

Willers, B. (1994) Sustainable development: a new world deception. *Conservation Biology* 8, 1146–1148.

Wrigley, N. (1992) Antitrust regulation and the restructuring of grocery retailing in Britain and the USA. *Environment and Planning A* 24, 727–749.

4 Sustainable Technologies, Sustainable Farms: Farms, Households and Structural Change

REBECCA ROBERTS AND GAIL HOLLANDER
Department of Geography, University of Iowa, Iowa City, Iowa 52242, USA

Introduction

'Sustainable agriculture' is an ambiguous term for a countersystemic set of technological production practices and social forms. Much of the literature on the possibilities for a transition to sustainable agriculture has reduced the integration of social and technological forms to the purely technological and has applied the conceptual models developed to study systemic change. Sustainable agriculture has been defined as the adoption of particular production practices, and adoption decisions have been investigated with the tools used to explain the diffusion of mechanization and the hybrid seeds/agrochemical complex under the aegis of the United States land grant research and extension system. The resulting behavioural models have emphasized two distinct sets of variables: the information delivery system and the characteristics of the farm and operator. However, the application of these models to sustainable technologies has produced little capability to explain adoption decisions and has generated few insights or excitement. The most salient insight is that many sustainable agriculture practices are not 'profitable' in the short run and therefore cannot be expected to follow the trajectory of the mechanization, hybrid seed and agrochemical technologies.

The inadequacies of the diffusion model have yielded critiques that suggest three alternative routes of conceptual development. The most fully

explored and well articulated route – that of political economy – emphasizes the development of systemic forces underlying the movement away
from more sustainable technologies and social forms. Earlier work in this
vein described the coalition of state, capital, and science that resulted in the
government-sponsored subsidy programmes and agricultural research and
extension system that have transformed the agricultural sector in the
United States and other developed countries. More recent work details the
shift to privately controlled research and product development that is
increasingly marginalizing the role of natural forces, the integrity of the
food product and the family farm that integrates them. The power and
validity of such analyses is undeniable, yet leaves little room for optimism
or for an understanding of how countersystemic development might occur,
in large part because they leave little room for agency. Two alternative
routes, not as well mapped conceptually, centre on agency at the level of the
farm and the social movement. Approaches based on agency at the
individual farm level run the risk of reducing social change to behaviourism
but also have the capability to explore important sources of change through
individual experimentation with alternative practices and identities. Such
alternatives represent an essential raw material for social movements at
larger scales. Major pitfalls to productive analysis lie in ignoring political
economic forces or in granting them exclusive sway; the former error is
evident where the social context is reduced either to that of the farm firm
or ignored by focusing on attitudes as behavioural determinants. Farms
represent a complex nexus of environmental, technological and social
relations; a key requisite for understanding the range of social possibilities
lies in better understanding the social forces influencing this nexus.

This chapter focuses on the possibilities for sustainability inherent in
the diversified Corn Belt family farm. Analyses informed by the earlier
'diffusion' research have concentrated too exclusively on the single
dimension of the 'innovativeness' of the (male) farmer concerned with
farm–firm modernization and short-term profitability. Analyses in the
political-economy mould have overemphasized the constraints to experimentation derived from systemic trends. This chapter contributes to a more
complex and nuanced understanding of social forces by resituating
decisions to explore, adopt or preserve more sustainable practices in the
farm household where a number of social systems intersect. In addition to
political–economic forces, these include systems of farm succession and
reproduction, the integration of farm development and household income
strategies with household needs over the life cycle, and social relations
within the farm family. Finally, this analysis raises the tantalizing possibility
that sustainable practices provide an 'edge' to diversified family farms that
are able to use biological processes to their advantage, while family farms
in turn provide a social basis for achieving sustainability. This interrelation-
ship between the social and technical also means that the loss of diversified

family farms will inhibit the potential for achieving agricultural sustainability in the US Midwest.

A Starting Point: Diffusion and Adoption Research

The literature on the adoption of agricultural technologies is at a crossroads, with one phase concluded and the future route not clearly defined. Perhaps as a result of the need to take stock, there are several excellent reviews and assessments of this literature (Fliegel and van Es, 1983; Buttel *et al.*, 1990; Duff *et al.*, 1992; Fliegel, 1993). The starting point is the paradigmatic 'adoption/diffusion' approach that emerged out of publicly funded programmes to encourage the adoption of hybrid corn. In the 1950s and 1960s, a sociopsychological framework coalesced that dominated diffusion/adoption research for years. Though social structure was not entirely ignored, emphasis was placed on farm operators' characteristics and the linkages between the individual and various sources of influence and information. Most versions of the 'classic' diffusion model assumed that the overall course of adoption could be described by an S-shaped curve that categorized farmers on a continuum from 'early adopters' to 'laggards'. Major attention was devoted to assessing the ability of various modes of communication – e.g. mass media, agricultural experts, neighbours – to deliver effective information to the individual farmer's decision process.

Interest in diffusion research has rebounded as of late, primarily driven by concern for the environmental impacts of current agricultural practices. The primary focus has shifted to conservation practices and more sustainable technologies. In this sense, this research is of an entirely different order than the 'first wave', which assumed that economic efficiency was the goal of technological innovation. The classic model still frames the research questions, although significant effort has been devoted to rethinking its limitations. The central issue is the extent to which important differences between production-enhancing and conservation practices will change the nature of the adoption decision as envisioned by the classic diffusion model (Pampel and van Es, 1977; Lasley and Bultena, 1986; Nowak, 1987; Napier and Sommers, 1994; Saltiel *et al.*, 1994). Particular aspects of this general question that have generated the most interest include:

1. the implications for effective communication and information delivery (Korsching and Nowak, 1983; Contant, 1990);
2. the potentially retarding influence of older farmers through kin-based farming relationships (Carlson and Dillman, 1983; van Es and Tsoukalas, 1987; Warriner and Moul, 1992; Lighthall, 1993; Ward and Lowe, 1994; Lighthall and Roberts, 1995);

3. the influence of farm scale (Abd-Ella *et al.*, 1981; Heffernan and Green, 1986; Lighthall, 1993; Lighthall and Roberts, 1995);
4. the importance of farmer attitudes towards environmental issues (Buttel *et al.*, 1981; Lasley and Bultena, 1986; Gillespie and Buttel, 1989; Beus and Dunlap, 1991; Napier and Brown, 1993).

Despite its continued importance, the sociopsychological approach underlying this research agenda is seriously limited by the extent to which the origin of innovation is ascribed to the attitudes and situation of the individual farmer (blaming the farmer) to the exclusion of a more structural assessment of relations among state, farm economy and individual operator. Agency is emphasized, but an impoverished form of agency without a realistic assessment of the structural environment in which agency operates. The failure to adequately incorporate both structure and agency becomes particularly critical when the practices considered run counter to the dominant structural forces, as for many conservation practices and sustainable technologies. The ability of the classic model to explain successfully adoption while ignoring structure derived paradoxically from its original focus on innovations that were carrying systemic structural transformations into the countryside. The inappropriateness of the model for counter-systemic practices is demonstrated by its lack of success in explaining their adoption. One of the striking features of quantitative analyses of conservation practice adoption using modifications of the classic model is the very low level of explanation achieved; R^2 values uniformly fall below 0.25. Despite a decade of effort, the diffusion model has been unable to generate significant insights into decisions to adopt sustainable technologies.

Putting Sustainable Practices in a Structural Context

The chief limitation of the adoption-diffusion approach is its failure to set the choice of production technology within a structural context. Without an adequate structural framework, the role of farmer agency was reduced to that of mechanical, passive behaviouralism. The sense of farmers creatively and actively attempting to develop successful farm enterprises within a context of rapidly changing structural forces over which they had no immediate control was lost. Particularly in its applications to conservation technologies, the adoption-diffusion approach conceptualized the problem as one of adoption of new technologies, with little attention to how such technologies interacted with the strategic efforts of farmers to simultaneously manage structural forces and development of a family enterprise. Forgotten in the emphasis on *adoption* was the massive *loss* of more sustainable practices as farmers responded to structural forces that led to less diversified, more industrialized operations. Both socially and technologically, the agricultural revolution brought changes in the organization of farm enterprises that reduced the capacity of farmers to adopt new sustainable practices or to maintain old ones. This section seeks to recover the ways in which structural change and sustainable practices are intimately entwined.

Agricultural exceptionalism and structural change

The power of capital to transform production and capital accumulation is no less profound for agriculture than for other economic sectors. Central means to capitalist control in all sectors have been commodification of production and exchange and the rationalization of the labour process necessary to the commodification of labour. Yet the transformation process in agriculture has proceeded along different pathways than for most production sectors, pathways that leave open possibilities for the future of agriculture that have been foreclosed elsewhere. These possibilities reside in the persistence of petty commodity production – the family farm – despite exogenous change in the internal organization of family production and its articulation with forces beyond the farm gate.

Its biological and land-based character sets agriculture apart from industrial production processes in ways that have maintained the role of petty commodity production in the food system (Goodman *et al.*, 1987; Mann, 1990). The divergence between labour time and production time, the craft required in managing ecological processes, the extensive character of production and the problem of supervising labour over extensive space have prevented industrialization through an assembly-line rationalization of the labour process. The long capital turnover times imposed by the annual production cycle and the inefficiencies of securing and managing an intermittent labour force further reduce the attractiveness of agricultural production to capitalist enterprises.

Faced with these barriers to full commodification of labour, industrial capitals have taken three different routes to subsuming agriculture to capital accumulation. The first is to reduce the natural constraints to commodification by substituting industrially produced commodities for natural processes and farmer-produced inputs – e.g. fertilizer for naturally managed fertility or marketed hybrid seeds for farmer-produced seed (Goodman *et al.*, 1987; Kloppenburg, 1988). As discrete elements of agricultural production are 'appropriated' by industrial capital and reincorporated as purchased inputs the locus of accumulation is shifted to the industrial sector away from the family farm. Second, linkages between farm production and human dietary requirements that had accorded the farmer power in the food sector are being broken by industrial technology that is increasingly able to substitute agricultural raw materials or even eliminate the agricultural basis for food. Highly processed proteins, fats, and sugars from a variety of agricultural and non-agricultural sources are now substitutable in industrially produced foods. Third, new animal production technologies are breaking down the barriers to assembly-line rationalization of the labour process. Continuous production confinement technologies, supported by advances in disease prevention, have virtually eliminated the petty commodity producer from poultry production and are making inroads in cattle and hog production. These technologies even out the demand for labour over time, eliminate much of the craft basis of production, confine labour to a supervisable site, increase input commodification, and provide a more rapid turnover of capital. These changes have increased the ability of capital to subsume directly the petty commodity producer, either by turning the farmer into a propertied labourer through contract production or making the large, wage–labour factory farm possible.

These three tendencies have transformed the family farm. The technology treadmill has encouraged farmers to mechanize and adopt production-enhancing technologies that, when generalized, have led to an unrelenting cost–price squeeze. As a result, agrarian differentiation has proceeded rapidly as advantageously placed farmers exploit returns to scale. Scale and the increasing capital intensity of production have encouraged specialization in crops or livestock. Rationalization of the labour process in livestock production threatens the unity of capital and labour in petty commodity production. In sum, the mid-sized, family-labour farm and the diversified crop-livestock farm have declined in importance and are everywhere threatened. The significance of these changes to the use of more sustainable technologies is explored next.

The diversified family farm

The relationship between structural change and sustainable technologies is well illustrated by the compatibilities between family production and sustainable practices for the diversified Corn Belt family farm. These compatibilities imply that the question of sustainability cannot be addressed separately from the consequences of structural change in the family farm sector, where the loss of diversified farms is associated with the loss of both existing sustainable practices and the flexibility to adopt sustainable practices.

The diversified Corn Belt farm is characterized by the production of crops fed to hogs or cattle, where the possibilities for on-farm production of inputs and recycling of nutrients provide a basis for the protection of existing and development of new sustainable practices. On diversified farms, manure is a resource rather than an environmentally dangerous waste product, thereby reducing fertilizer and soil improvement expenses. Corn–corn and corn–soybean crop rotations give way to more complex rotational systems, often incorporating a small grain (typically oats) and/or alfalfa because of their value in livestock production. Oats and alfalfa provide feed; oats are particularly beneficial in preventing hog 'scours'. Small grain straw serves as animal bedding, and the spring-harvested small grain land provides summer ground for manure-spreading. Other advantages of more complex rotations incorporating small grains and/or alfalfa are that they provide nitrogen for crop production, reduce pesticide applications by breaking up disease and insect cycles, and provide erosion control. In addition, some farmers report reduced weed pressure as a result of feeding their own crops, thereby avoiding external seed sources.

The defining characteristic of the sustainable farm is greater reliance on agroecological forces rather than industrialized inputs. Family-based production offers important compatibilities here. An owner-operator is more likely to have the motivation, knowledge, and interest to undertake the more management intensive ecologically based production technologies. Kloppenburg (1991) and Ehrenfeld (1987) have focused on local knowledge as a key to a revitalization of the family farm sector and the transition to a more sustainable agriculture. Local knowledge is by no means a static tradition, but rather an actively constructed knowledge that is gained through experimentation, collaboration with neighbours and family members, and intimate experience with the land. Family-based production can also better supply the disciplined, flexible labour necessary to reliance on ecological forces of production, which may require intense applications of labour in short production windows defined by biological cycles. Families may be better able to deploy such a highly variable labour force by relying on all family members, including the extended family, during peak periods. Family producers, with their flexible labour supply, also have a greater ability to substitute labour for capital in livestock production through the use of lower-cost livestock production facilities. Such facilities often require more labour for feeding, manure handling, and animal handling. But they can also reduce the use of antibiotics and vaccines and achieve efficiencies by relying on natural ventilation and the biological capabilities of the animals in mothering, skeletal strength and immunity.

Diversification and reliance on agroecological production forces offers advantages to family producers that can form a basis for competitiveness relative to industrialized producers. The extensive on-farm production of inputs, recycling of nutrients, and use of biological forces reduces production costs, often with little impact on yields. Diversification provides means to use labour profitably throughout the year, reducing underemployment and providing cash flow. Finally, the ability to expand or contract livestock production allows production opportunities to be adjusted to labour

availability over the course of the life cycle as children and parents engage
or disengage from the enterprise; expanding and contracting the land base
imposes far greater risks.

These compatibilities between diversified family farms and more
sustainable technological choices do not imply a necessary connection
between sustainability and structural change. Flora (1992), for example,
argues that the local knowledge privileged by Kloppenburg is not necessarily sustainable. It is clear, however, that the diversified family farm
emphasizing on-farm cycling of inputs and waste products and management of biophysical processes presents opportunities for sustainable
production that are threatened by further commodification and rationalization.

Sources of resistance

There is no denying the power of systemic structural forces or the threat
that they pose for the complementarities between family farms and
sustainable practices described in the previous section. However, an
exclusive focus on systemic structural change obscures both the diversity of
responses to these forces and the possibilities for counter-systemic resistance. These possibilities will depend ultimately on the abilities of farmers
and their allies to organize social movements that will change the structural
framework in which farmers make decisions. However, the origins of such
movements will lie in the competitive strengths of family-based production
and the ability of family producers to maintain and develop the alternatives
that will serve as the sources of innovative change. Understanding the
possibilities for a transition to a more sustainable agriculture is therefore
closely related to understanding the sources of resistance to commodification and labour rationalization inherent in family forms of production.
Experimentation by family producers, both in production practices and
forms of social organization, constitutes a powerful source of emergent
change. Structural analyses that attribute the staying power of family farms
to processes beyond the farm gate ignore their abilities to influence their
own fate by mediating and negotiating their relationships with industrial
capitals.

An analysis of the competitive strengths of family producers suggests
three avenues of resistance to full labour commodification and subsump-
tion to agroindustrial capitals (Roberts and Mutersbaugh, 1996). Following
Chayanov (1925), these strengths derive from the different logic underlying decision-making resulting from the inability of family producers to
distinguish wages from profit and from the overriding influence of the
demographic life cycle on labour supply, investment strategies and farm
succession. Although these characteristics permit farm families to destruc-

tively 'self-exploit' by accepting lower standards of living and neglecting investment, they also provide capabilities allowing family producers to successfully compete with more fully commodified producers. The critical questions involve the circumstances and actions that shift outcomes towards one alternative or the other. The similarities between the following competitive advantages of family farms and the compatibilities between family farming and sustainable technologies described in the last section demonstrate the importance of linking these two issues.

First, farm families possess intrinsic advantages over capitalist producers in the deployment, surveillance, and monitoring of high quality agricultural labour. The point here is not that agricultural production is 'unattractive' to capital because of the variable labour demands, or that family members will work for lower wages (although they may), but that family producers are *better* at deploying and monitoring labour matched to production needs. These greater efficiencies in labour management then allow family farmers to take advantage of biological and natural forces to reduce production costs.

Second, an important corollary of high-quality labour deployment is the ability of family producers to generate efficiences through the development and application of 'local knowledge'. Agricultural science is a constructed knowledge that has emphasized increased yields on the basis of generalized conditions of production, accelerating the processes of commoditization through increasing input costs and labour rationalization (Kloppenberg, 1991). As such, it implicitly appropriates and replaces farmers' knowledge of their own capabilities and the ecological possibilities of their own farms. Not only is the farmer's craft and agency reduced but also short and long-run productive efficiencies are lost. Family farmers, through both individual and cooperative effort, are capable of reappropriating their own agency and knowledge, thus securing the accompanying economic efficiencies and shifting the locus of power in agricultural change away from the industrialized partnership of science and commodity.

Third, the strong commitment of many farm families to intergenerational succession confers competitive advantages in addition to problems. In many cases dedication to farm stewardship and the production of local knowledge is transferred from one generation to the next (Ehrenfeld, 1987). This ability to reproduce sustainable conditions of production through the union of local knowledge and land-productive capability may give family farms an edge over agrocapitals, especially over the course of generational cycles. Family succession may also play an instrumental role in structuring land rents in favour of family producers. Intergenerational succession results in periodic labour surpluses for family farms, which can then use their ability to flexibly allocate surplus (combined profits and wages) to pay more for land. In so doing, they both outcompete more capitalist producers and maximize total, long-term returns to family labour. The demographic cycle thus both recreates and valorizes conditions of family production (Roberts and Mutersbaugh, 1996).

Directions for research

As argued above, the focus on technology adoption in the sustainability literature has been conceptually inadequate, whereas overly structural explanations of technological change in agriculture leave little hope for sustainable systems. One way around this *impasse* is to broaden the question of agricultural sustainability to include social relations as well as technology adoption, bringing into focus the relationship between the persistence and viability of sustainable practices and the preservation of the diversified family farm. In this reconceptualization, farm family agency is recognized as a vital component of sustainable systems, albeit operating within the constraints of the larger political–economic agricultural system. As the analysis of sources of farmer resistance to full commodification suggests, social and environmental sustainability will be related to: (i) the logic of farms as family enterprises; (ii) labour deployment strategies; and (iii) role of life-cycle and intergenerational succession. The next section reviews what is known about these processes as a first step to integrating them into analyses of sustainable agricultural production.

Family Farming: The Farm as a Family Enterprise

The structures of the family, the farm firm and the larger agricultural political economy intersect to provide a ground where individual family members cooperate and struggle to develop farm strategies and goals. Of particular importance in understanding change over time are the *interactions* among the dynamics of enterprise development, the life cycle and the dynamics of the macro farm economy. Farmers' strategies integrate these different dynamic components in ways that are often both difficult to resolve successfully and so inherent to the mindset of any individual farmer as to frequently operate below the level of conscious decision-making.

It is, therefore, critical to examine the ways in which family and enterprise interrelate, creating opportunities and constraints that shape investment and production choices that affect farm scale, diversity and sustainability. Although most of the agricultural literature notes that family farms are unusual in advanced capitalist societies because the household and production unit remain integrated, most post World War II farm research emphasized production economics to the exclusion of household dynamics (Gasson and Errington, 1993). It is only in the past two decades that a series of ethnographic studies has reunited the enterprise and the family in the analysis of contemporary farming (Bennett, 1982; Colman and Elbert, 1984; Friedberger, 1988; Whatmore, 1991a; Salamon, 1992; Barlett, 1993). These insights into the ways family needs and capabilities shape farm strategies have clear implications for technology choice that have yet to be fully explored. From this literature, the particulars of household dynamics emerge as central to farm decision-making about technologies; 'opening' the household leads one to considerations of labour, intergenerational reproduction and gender.

Labour

The reliance on family labour has long been given as one of the reasons that family farms have been able to survive within industrial capitalism. Explanations emphasize either the ability of family enterprises to self-exploit, which leads ultimately to marginalization, or the efficiency of family labour, which provides a measure of competitiveness to family farms (Friedmann, 1978; Barlett, 1993; Gasson and Errington, 1993). Therefore, understanding potential economic and ecological efficiencies and their implications for choice of technology requires delving into the ways family labour is managed and deployed. Ethnographic analyses provide the most insight here, emphasizing that family labour efficiencies are not given but are constructed out of both cooperation and conflict. Researchers focusing on gender relations tend to highlight the patriarchal nature of labour management on family farms, whereas those concerned with the farm as a firm emphasize the way in which family labour is managed for the benefit of the enterprise. An important contribution that links these two approaches is that of Colman and Elbert (1984). Their analysis, based on 15-year intensive case studies of more than 30 farm families, points to the centrality of labour management and control for farm decision-making and for conflict among family members. Family labour provides considerable flexible, disciplined labour, but it also generates constraints of availability and quality associated with the life cycle; many practice and enterprise adoption decisions are strongly influenced by compatibility at a particular life-cycle stage between labour demands and labour supply. There is also an inherent and protracted conflict between the desires of the owner-operator (usually husband and father) to maximize profit and streamline labour coordination and the desires of the workers (wife and children) to maintain their own varying degrees of control over participation and share of the profits. Many farms evolve from a strictly family labour unit to one in which the male operator is involved in more sophisticated management and may 'lay off' family members. The possible trajectories of family and farm are delineated through a complex interweaving of both family cycles and farm cycles, as household and farm needs and goals compete and interact for family resources, i.e. capital and labour. Terrains of contest often emerge in which the goals and aspirations of individual family members conflict and, in turn, help to structure labour resources and technological choices.

Intergenerational reproduction

Long-term ethnographic studies of family farms provide essential insights into the mutually conditioning stages of family life cycle and enterprise development (Bennett, 1982; Colman and Elbert, 1984). As part of a ten-year study investigating the dynamics of household, enterprise and macro-structure interaction, Bennett (1982) and his team discovered that the problem of succession creates both opportunities and constraints that significantly structure decision-making according to stage in the life cycle. First, fundamental changes in management approach, enterprise mix, and technology choice tend to coincide with succession of the young operator or with expansion in order to accommodate incorporation of a child into the enterprise. Second, failure to expand to permit succession or to support rising living standards sets the enterprise on a course of conservative decision-making and disinvestment. A key insight of Bennett's extensive anthropological fieldwork is that farm firm decision-making does not respond primarily to short- or long-term trends in the market but rather to the cycles of growth and development of the enterprise/household combination within the context of the market; the conjoint development of enterprise and household present problems that are more difficult to manage than those of the uncertainties of the market. From their 15-year studies, Colman and Elbert 'became convinced that a distinctive feature of farming families is the *desire of women, as well as men, to effect intergenerational transfer of the family farm*' (Colman and Elbert, 1984, p. 65). However, they note that this is a contradictory process for women, on the one hand, fulfilling their goals for the farm, and on the other, possibly displacing them as a child begins to take over the partnership relationship they once held. Conflicting evidence has emerged from the various studies that have examined whether the desire to pass on the farm provides an incentive for better stewardship (Bennett, 1982; Carlson and Dillman, 1983; Potter and Lobley, 1992; Ward and Lowe, 1994).

The way that succession is achieved has been identified as critical to the ongoing economic viability of the family farm, as well as a method of characterizing two approaches to farm strategy, i.e. yeoman and entrepreneur. According to Salamon (1992), yeomen, whose agricultural influences came from their German heritage, have strong commitments to the land and to establishing their children in farming, resulting in conservative traditions. Entrepreneurs, of Yankee Anglo-Irish background, have greater expectations that their children will make their way independently, thereby encouraging them towards a more entrepreneurial, risk-taking management style. Friedberger (1989) and Barlett (1993) identify similar differences in farm cultures without making the strong ethnic connections that Salamon does. Both found considerable differences between operations that relied heavily on borrowed capital to expand for the purpose of succession compared with those in which equity came primarily from familial holdings accumulated over generations. Friedberger identifies the differing intergenerational transfer strategies of yeoman and entrepreneurs as the critical factor in whether a farm family survived the economic downturn of the 1980s in Iowa, concluding that 'the yeoman style of farm management and family organization – continuity, risk-averse financial practices, the limitation of expansion to fit family needs, intergenerational succession, and cooperation – which had guided families through a century of economic uncertainty in corn-belt agriculture, still applied' (Friedberger, 1989, p. 143).

To the extent that the 'yeoman' approach enables certain family farm enterprises to persist when others fail, both Gasson and Errington (1993) and Friedberger (1989) ask what that means for the overall structure of agriculture, expressing concern that only the risk-averse will survive, while the more innovative and progressive farmers may, in the long run, be pushed out of farming. The critical question here is *which* innovations; the adoption literature tends to lump all innovations together and to identify a general characteristic of farmers known as 'innovativeness'. Indeed, Kloppenburg (1991) suggests that the 'lost history' of the non-innovators, i.e. laggards and resisters, needs to be recovered as part of the project of 'reconstructing' agricultural science along more sustainable lines. Sustainable farming innovations may respond to a very different incentive structure than entrepreneurial, production-enhancing innovations. Salamon's (1992) work in the two cultures suggest that risk-averse yeomen may be much more willing to adopt sustainable techniques. Lighthall (1993) and Lighthall and Roberts (1995) concur, finding that mid-size yeomen operations are more conducive to the management- and operation-intensive sustainable ridge-till techniques than larger scale entrepreneurial operations. Barlett (1993) and Salamon (1992) found women more intensively involved in farm operations and decision-making in 'yeoman' farms than in industrial operations, opening the possibility for their concerns and preferences to influence technology choices.

Gender

Many of the attempts to integrate household and enterprise treat the household unit as 'an individual by another name' with a logic and interests of its own (Folbre, 1986; Whatmore, 1991a,b). Feminist analyses have opened the household to examine not only gender relations but also the relationships of cooperation and conflict among all family members that constitute it. For example, Whatmore's (1991a,b) extensive fieldwork demonstrates the complexities and patterns of social relations, decision-making, and work processes that make up the household-production complex. One of the most significant findings of the Cornell decision-making studies (Colman and Elbert, 1984) is the extent to which the interests of farm women and farm men tend to converge or diverge depending on the stage of the life-cycle and enterprise development. Qualitative analyses have demonstrated clearly that the household, and the gender roles of men and women in households, must be conceived as actively constructed through the meanings and practices of everyday social life embedded within the larger institutional fabric of society (Whatmore *et al.*, 1994).

The theoretical emphasis in much of the literature that explicitly incorporates gender is the patriarchal nature of the division of labour on the family farm (Sachs, 1983; Fink, 1986; Haney and Knowles, 1988; Whatmore, 1991b), but ethnographic studies have also emphasized the ways in which the strategic position held by women in the family farm modifies patriarchal power relations. Kohl (1977) finds that farm women have wide alternatives and important power in the family enterprise, a consequence of the fusion of family and enterprise. Farm women tend to be better educated than their husbands (Kohl, 1977; Bultena *et al.*, 1983; Rosenfeld, 1985) and to be more involved in farm accounting and bookkeeping (Kohl, 1977; Rosenfeld, 1985). However, Fink (1991) and Barlett (1993) found that women's economic autonomy has decreased as farms have become increasingly industrialized. Whereas older women had knowledge and control of the production and marketing of particular commodities, younger women are incorporated into agribusiness in the more subservient helper roles or have assumed more stereotypic home-maker roles. Salamon (1992) connects these differences to the ethnic traditions that distinguish yeoman and entrepreneurial farms.

The linkage of gender relations to technological choice and sustainable practices has yet to be carefully explored in the literature. Yet the close connection of both gender relations and technological choice to labour deployment, intergenerational reproduction, and accumulation strategy would suggest such a linkage. For example, the choice to pursue the advantages of scale or the advantages of intensive family management involve major decisions concerning investment of time, knowledge, land and equipment that are not independent of family goals such as lifestyle, children's opportunities, income strategies and farm succession opportunities open to the next generation. Similarly, the choices of various family members to take off-farm jobs, with consequences for both labour and capital availability, involve decisions about farm enterprise goals, individual aspirations and family lifestyle. These choices, the parameters of which vary over the course of the life cycle, have divergent significance for different family members. Farm men and women may concur on some decisions, yet find others a terrain of continuing interpersonal struggle.

Deborah Fink's (1986) Iowa case study is one of the few empirical analyses to connect technological choices and gender relations. She documents the transition from a dual-income farm household, in which women's poultry flocks created a separate income stream under their management, to the post World War II 'rationalization' of the broiler and egg industry under the control of male farmers. This restructuring of intra-household economics was accomplished through the 'diffusion' of production practices promoted by the USDA and Extension Service, resulting in less diversified farming systems.

Feldman and Welsh (1995) make a strong case for a more comprehensive incorporation of gender relations in understanding agricultural sustainability. Responding to Kloppenburg's (1991) and Ehrenfield's (1987) arguments that agricultural sustainability should be derived from farmer-based 'local knowledge', they charge that this formulation has limited analytical power because it fails to analyse how the local is constructed, invoking instead the perspective of 'a nongendered neutral other, often assumed to be the ideal male' (p. 28). They argue that the focus should be shifted from place, i.e. 'the local', to social relations, in particular the gendered household division of labour because of its relationship to choices about whether to seek or preserve intensive family management of biological processes rather than rationalized, industrial production. Local, working knowledge is in fact constituted by diverse and contradictory household relations based on both gender and generation. Such relations cannot be read as merely a structural feature of the farm, but are negotiated and struggled over in ways that affect the production possibilities open to the farm. This view of active engagement and dynamic exchange among family members provides a connection between social inequalities and power differentials and the productive practices that constitute local knowledge.

Concluding Remarks

An overriding impression gained from this review relates to the quality and depth of the insights derived from ethnographic analyses of farming households and family farms, insights not available from other research approaches. From these analyses it is clear that family farms are highly differentiated both agroecologically and socially, and that both of these dimensions influence significantly the technological possibilities open to the farm enterprise within the larger political economic context. In particular, issues of labour availability and management, intergenerational reproduction or transfer, and gender relations are closely interconnected in ways that influence the goals and strategies of farming families and the production practices that make sense to them. Labour mobilization and organization are critical to the viability of many sustainable practices because these practices tend to substitute labour and management for commoditized inputs. The ways that farm households work out these interrelated problems, in turn, have significant implications for the viability of the family-farm system. Greater attention to these smaller-scale social forces, and the resistance that they offer to the regime of capital, remains an important complement to political–economic analyses. Sustainability, in its interrelated social and technological forms, cannot be achieved except by exploring the possibilities of countersystemic movement inherent in existing forms of resistance.

References

Abd-Ella, M.M., Hoiberg, E.O. and Warren, R.D. (1981) Adoption behavior in family farm systems: an Iowa study. *Rural Sociology* 46, 42–61.

Barlett, P.F. (1993) *American Dreams: Rural Realities: Family Farms in Crisis.* University of North Carolina Press, Chapel Hill.

Bennett, J.W. (1982) *Of Time and the Enterprise: North American Family Farm Management in a Context of Resource Marginality.* University of Minnesota Press, Minneapolis.

Beus, C.E. and Dunlap, R.E. (1991) Measuring adherence to alternative vs. conventional agricultural paradigms: a proposed scale. *Rural Sociology* 56, 432–460.

Bultena, G., Hoiberg, E., Jarnagin, S. and Exner, R. (1992) Transition to a more sustainable agriculture in Iowa: a comparison of the orientations and farming practices of conventional, transitional, and sustainable farm operations. Unpublished paper, Department of Sociology, Iowa State University.

Buttel, F.H., Gillespie, G.W., Jr, Larson, O.W., III and Harris, C.K. (1981) The social bases of agrarian environmentalism: a comparative analysis of New York and Michigan farm operators. *Rural Sociology* 46, 391–410.

Buttel, F.H., Larson, O.F. III and Gillespie, G.W., Jr (1990) *The Sociology of Agriculture.* Greenwood Press, New York.

Carlson, J.E. and Dillman, D.A. (1983) Influence of kinship arrangements on farmer innovativeness, *Rural Sociology* 48, 183–200.

Chayanov, A.V. (1925, 1986) *The Theory of Peasant Economy*, translated by D. Thorner, R.E.F. Smith, B. Kentlay. University of Wisconsin Press, Madison.

Colman, G. and Elbert, S. (1984) Farming families: the farm needs everyone. *Research in Rural Sociology and Development* 1, 61–78.

Contant, C.K. (1990) Providing information to farmers for groundwater quality protection. *Journal of Soil and Water Conservation* 45, 314–317.

Duff, S.N., Stonehouse, D.P., Blackburn, D.J. and Hilts, S.G. (1992) A framework for targeting soil conservation policy. *Journal of Rural Studies* 8, 399–410.

Ehrenfeld, D. (1987) Sustainable agriculture and the challenge of place. *American Journal of Alternative Agriculture* 2, 184–187.

Feldman, S. and Welsh, F. (1995) Feminist knowledge claims, local knowledge, and gender divisions of labor: constructing a successor science. *Rural Sociology* 60, 23–43.

Fink, D. (1986) *Open Country, Iowa: Rural Women, Tradition and Change.* State University of New York Press, Albany.

Fink, V.S. (1991) What work is real?: changing roles of farm and ranch wives in south-eastern Ohio. *Journal of Rural Studies* 7, 17–22.

Fliegel, F.C. (1993) *Diffusion Research in Rural Sociology: The Record and Prospects for the Future.* Greenwood Press, Westport CT.

Fliegel, F.C. and van Es, J.C. (1983) The diffusion-adoption process in agriculture: changes in technology and changing paradigms. In: Summers, G.F. (ed.) *Technology and Social Change in Rural Areas: A Festschrift for Eugene A. Wilkening.* Westview, Boulder, Colorado, pp. 13–28.

Flora, C.B. (1992) Reconstructing agriculture: the case for local knowledge. *Rural Sociology* 57, 92–97.

Folbre, N. (1986) Hearts and spades: paradigms of household economics. *World Development* 14, 245–255.

Friedberger, M. (1988) *Farm Families and Change in Twentieth-Century America.* The University Press of Kentucky, Lexington.

Friedberger, M. (1989) *Shake-Out: Iowa Farm Families in the 1980s.* The University Press of Kentucky, Lexington.

Friedmann, H. (1978) Simple commodity production and wage labor in the American plains. *The Journal of Peasant Studies* 6, 71–98.

Gasson, R. and Errington, A. (1993) *The Farm Family Business.* CAB International, Wallingford.

Gillespie, G.W., Jr and Buttel, F.W. (1989) Understanding farm operator opposition to government regulation of agricultural chemicals and pharmaceuticals: the role of social class, objective interests, and ideology. *American Journal of Alternative Agriculture* 4, 12–21.

Goodman, D., Sorj, B. and Wilkinson, J. (1987) *From Farming to Biotechnology.* Basil Blackwell, London.

Haney, W.G. and Knowles, J.B. (eds) (1988) *Women and Farming: Changing Roles, Changing Structures.* Westview Press, Boulder, Colorado.

Heffernan, W.D. and Green, G.P. (1986) Farm size and soil loss: Prospects for a sustainable agriculture. *Rural Sociology* 51, 31–42.

Kloppenburg, J., Jr (1988) *First the Seed: The Political Economy of Plant Biotechnology.* Cambridge University Press, Cambridge.

Kloppenburg, J., Jr (1991) Social theory and the de/reconstruction of agricultural science: local knowledge for an alternative agriculture. *Rural Sociology* 56, 519–548.

Kohl, S.B. (1977) Women's participation in the North American family farm. *Women's Studies International Quarterly* 1, 47–54.

Korsching, P.F. and Nowak, P.J. (1983) Flexibility in conservation policy. In: Brewster, D.E., Rasmussen, W.D. and Youngberg, G. (eds) *Farms in Transition.*

Iowa State University Press, Ames, pp. 149–159.

Lasley, P. and Bultena, G. (1986) Farmers' opinions about third-wave technologies. *American Journal of Alternative Agriculture* 1, 122–126.

Lighthall, D.R. (1993) Chemicals, class, and sustainable agriculture: a corn belt case study. PhD dissertation, The University of Iowa, Iowa City, Iowa, USA.

Lighthall, D.R. and Roberts, R.S. (1995) Towards an alternative logic of technological change: insights from corn belt agriculture. *Journal of Rural Studies* 11, 319–334.

Mann, S.A. (1990) *Agrarian Capitalism in Theory and Practice.* University of North Carolina Press, Chapel Hill.

Napier, T. and Brown, D.E. (1993) Factors affecting attitudes toward groundwater pollution among Ohio farmers. *Journal of Soil and Water Conservation* 48, 432–438.

Napier, T.L. and Sommers, D.G. (1994) Correlates of plant nutrient use among Ohio farmers: implications for water quality initiatives. *Journal of Rural Studies* 10, 159–171.

Nowak, P.J. (1987) The adoption of agricultural conservation technologies: economic and diffusion explanations. *Rural Sociology* 52, 208–220.

Pampel, F., Jr and van Es, J.C. (1977) Environmental quality and issues of adoption research. *Rural Sociology* 42, 57–71.

Potter, C. and Lobley, M. (1992) Aging and succession on family farms: the impact on decision making and land use. *Sociologia Ruralis* 32, 317–334.

Roberts, R. and Mutersbaugh, T. (1996) On rereading Chayanov: understanding agrarian transitions in the industrialized world. *Environment and Planning A* 28, 951–956.

Rosenfeld, R.A. (1985) *Farm Women: Work, Farm, and Family in the United States.* The University of North Carolina Press, Chapel Hill.

Sachs, C.E. (1983) *The Invisible Farmers: Women in Agricultural Production.* Rowman & Allanheld, Totowa, New Jersey.

Salamon, S. (1992) *Prairie Patrimony: Family, Farming, and Community in the Midwest.* University of North Carolina Press, Chapel Hill.

Saltiel, J., Bauder, J.W. and Palakovich, S. (1994) Adoption of sustainable agricultural practices: diffusion, farm structure, and profitability. *Rural Sociology* 59, 333–349.

van Es, J.C. and Tsoukalas, T. (1987) Kinship arrangements and innovativeness: a comparison of Palouse and Prairie findings. *Rural Sociology* 52, 389–397.

Ward, N. and Lowe, P. (1994) Shifting values in agriculture: the farm family and pollution regulation. *Journal of Rural Studies* 10, 173–184.

Warriner, G.K. and Moul, T.M. (1992) Kinship and personal communication network influences on the adoption of agricultural conservation technology. *Journal of Rural Studies* 8, 279–291.

Whatmore, S. (1991a) *Farming Women: Gender, Work and Family Enterprise.* Macmillan Academic and Professional, London.

Whatmore, S. (1991b) Life cycle or patriarchy?: gender divisions in family farming. *Journal of Rural Studies* 7, 71–76.

Whatmore, S., Marsden, T. and Lowe, P. (1994) *Gender and Rurality*, David Fulton, London.

5 Environmental Change and Farm Restructuring in Britain: The Impact of the Farm Family Life Cycle

CLIVE POTTER
Environment Section, Wye College, University of London, Wye, Ashford, Kent TN25 5AH, UK

Introduction

The profound impact of farm business restructuring on the landscapes, habitats and biodiversity of rural areas has been recognized for some time. As Bowers (1995) recognizes, agriculture is central to any strategy for rural sustainability, agricultural modernization invariably bringing the depletion of genetic diversity and landscape value in its wake. In the UK, where public concern about the environmental implications of modern agriculture dates back to the late 1960s, early debate focused very quickly on the way the actions of farmers were bringing about landscape change (see Westmacott and Worthington, 1974). Positions polarized as, on the one hand, farmers were castigated for their 'theft of the countryside' (Shoard, 1980) or, on the other, defended as good stewards and distanced from the few 'bad apples' thought to be responsible for any conflicts which arose. Neither position could be sustained for long. Mounting evidence of the sheer pervasiveness of environmental change discredited those who claimed the problem was localized and small scale. Meanwhile, *ad hominem* criticisms of farmers began to seem less convincing as commentators looked beyond the farm gate for the driving forces behind agricultural change (Potter, 1986). According to the 'policy thesis' that began to emerge, 'the problem is not one of ill will and ignorance but of a system (chiefly the Common Agricultural Policy) which systematically establishes financial inducements to erode the countryside, offers no rewards to prevent market failure and increases penalties imposed ... on farmers who may want to farm in a way which enhances the rural environment' (Cheshire, 1985, p. 17). With gathering support for the idea that agricultural policy should be reformed on environmental grounds, circumstantial evidence such as that reviewed by Rae (1993) was marshalled to establish a causal link between levels of farm support and the nature, speed and direction of recent environmental change (see Bowers and Cheshire, 1983; Cheshire, 1985). The concern was to identify the broad driving forces behind environmental change in order to make the case for equally broad policy reform.

More recently the policy debate has moved on to new ground. With growing concern to secure a more sustainable use of rural land, agrienvironmental policies have been invented that pay farmers to adopt environmentally sensitive practices. This means that the emphasis is switching from generalities to specifics as policy-makers seek ways to use the agrienvironmental policy measures now in place in the UK and other EU member states to arrest, deflect, slow down or speed up environmental change. New kinds of policy knowledge are needed as broad-brush explanations couched in terms of driving forces become less useful than those seeking to identify the farm level tendencies and influences – behavioural, social, economic, as well as policy-related – which determine *where* change is taking place. In this latest coevolution of the policy and research agendas, the need is for analyses of the decisions farmers make, but more particularly, the reasons they make them. A basic assumption is that environmental change can ultimately be traced to actions taken to maintain or increase farm household income and ensure family continuity in farming. Another is that farm households with shared characteristics and pursuing similar survival strategies are likely to have similar environmental histories. In an early investigation into the processes of landscape change conducted along these lines, Potter (1986) commented that whereas it is clear some farmers are much greater contributors than others, researchers have still not succeeded in understanding the causal relationships involved. Later work conducted under the aegis of the ESRC's Countryside Change research programme in the UK nevertheless managed to give some pointers to the relationship between a farm household's economic trajectory, occupancy change and environmental change (Marsden and Munton, 1991; Munton and Marsden, 1991). A continuing methodological problem affecting this and other studies, however, has been the lack of reliable base line data describing the rate and extent of environmental change at the farm level. Researchers must typically rely on farmers' own estimates of change or, in the case of the Countryside Change work, on incomplete aerial photographic coverage of study areas. Focusing on study areas also limits the extent to which change can be observed and studied in a wide enough range of locations to gain insights into pattern and process.

Following completion of the UK Government's Countryside Survey 1990 (CS1990) – a nationwide survey of environmental change conducted by the Institute of Terrestrial Ecology (ITE) based on field surveys and the use of satellite imagery (see Barr *et al.*, 1993) – an opportunity presented itself to carry out a social survey with the aim of integrating farm survey data with environmental data from ITE's field surveys in the 1 km sample squares that are distributed across Great Britain (see Fig. 5.1). Environmental change data were available for 504 occupiers with land in these squares based on surveys undertaken in 1978, 1984 and 1990. The objective of the research was to relate the land cover and botanical changes measured by ITE to the actions and decisions of those who own or manage the land

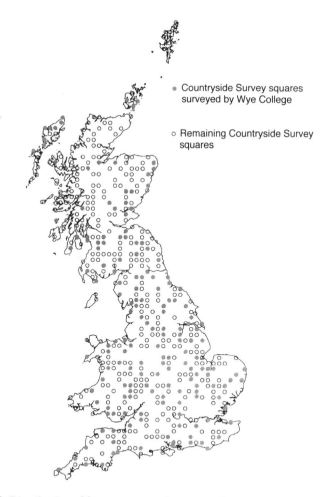

Fig. 5.1. Distribution of farm survey and countryside survey sample squares.

concerned. One of the main hypotheses is that environmental change, where it occurs, can be traced to groups of farmers with similar farm management and farm family histories. Implicit in this is the assumption that some farmers are much greater contributors to the total sum of environmental change than others, given their different environmental endowments and willingness and ability to undertake land use change. In this chapter, results from the research are drawn on to explore the idea that life cycle influences are particularly important in determining the timing and extent of environmental change on family farms. This chapter begins by describing the nature of recent environmental change on farms in Britain. It goes on to identify likely explanations for the different trajectories of change observed. By emphasizing the life cycle effect, this chapter underlines the need to understand family processes as well as policy and other structural influences in any explanation of environmental change on farms.

Unequal Endowments, Uneven Change

It is a safe assumption that some locations and farms are much richer in environmental terms than others, a reflection of uneven natural endowments and differing susceptibilities to change. From the survey, the area of semi-natural vegetation, deciduous woodland and extensively managed grassland in each ownership parcel in the sample squares was calculated in order to give an indication of the environmental stock present. Figure 5.2

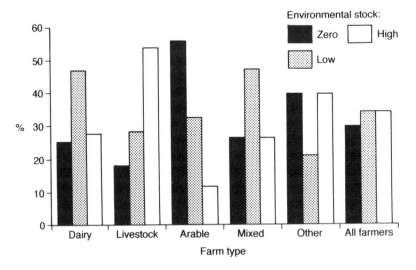

Fig. 5.2. Degree of environmental stock by farm type. (Source: Farm survey and countryside survey.)

shows the distribution of this stock by farm type, indicating an expected concentration on grassland farms and very little present on arable holdings. None of this is unexpected but the survey allows conventional knowledge to be quantified and produced the picture shown in Fig. 5.3 whereby 57% of the total areas of environmental stock present is found in upland landscapes, 28% in pastural landscapes and just 14% in arable landscapes.

Moreover, it is clear from Fig. 5.4 that environmental stock change has been far from evenly distributed by these landscape types. Movements of environmental stock have been greatest in pastoral landscapes, least in arable ones over the period studied. However, the significant finding here is the implied stability of upland landscapes, which make the smallest showing compared with the large areas of stock found there: less than 13% of the total area of deciduous woodland, semi-natural vegetation and extensive grass in upland parcels has been subject to a change compared with 57% of that found in pastoral and 53% in arable landscapes. Throughout the sample, it is the extensive grass component of environmental stock that is most subject to change, with movements into and out of semi-natural vegetation and deciduous woodland being fractionally small. As Fig. 5.5 reveals, 9% of the total area of extensive grass had become intensive grass by 1990; but in the

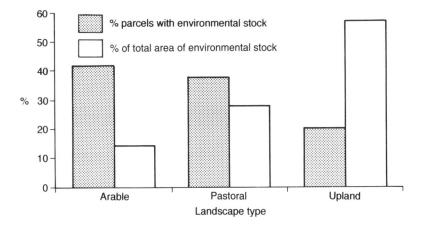

Fig. 5.3. Distribution of environmental stock by landscape type. The landscape type classification is derived by combining separate land classes into three distinct groups (see Barr *et al.* 1993). Briefly, arable landscapes are dominated by arable crops and intensive grass and are concentrated largely in the South, East Anglia and East Midlands. 'Pastoral' landscapes are characterized by large areas of pasture, small fields, hedgerows and small woods, and are typical of western Britain. In contrast, upland landscapes, found particularly in north and west Britain, are dominated by a mixture of low intensity dairy and forestry. This landscape contains extensive tracts of semi-natural vegetation.

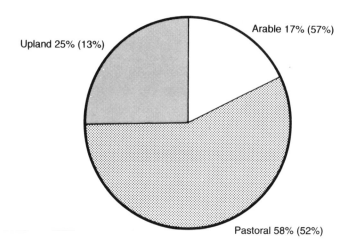

Fig. 5.4. Change in environmental stock by landscape.

other direction, 31% of the intensive grassland area had been reclassified as extensive grass (the only other shift of any magnitude being the expected one between tillage and intensive grass).

Clearly a fairly liberal definition of environmental stock is being employed here and it cannot always be assumed that a gain in the area of extensive grass is necessarily always beneficial from an environmental point of view. A reasonable conclusion, however, is that environmental change measured in these terms is coming about very largely because of a complex intensification and extensification of grassland management and that far from being in one direction, both gains and losses have occurred (see below). Figure 5.4 also underlines the extent to which change is concentrated in relatively few parcels, 55% of those in arable landscapes recording no land cover change of any sort. In fact, 90% of the area of environmental stock subject to a change here took place on just 17% of parcels. By farm type, 57% of the total area subject to a change was located on grassland farms with mixed farms accounting for a further 21%. Predictably, change is even more farm type specific elsewhere, with grassland farms accounting for the lion's share of changes taking place in upland and pastoral landscapes, contributing 78% and 93% respectively of the total area subject to a change.

The Processes of Change

Given evidence from the survey of much more widespread land-use change in the middle past, it would seem that the CS1990 has photographed the tail of a process that stretches back over several decades and was once much

1990: Shift to

1984: Shift from (% of area)	Tillage	Intensive grass	'Environmental stock': Extensive grass	'Environmental stock': Semi-natural vegetation	'Environmental stock': Deciduous woodland	Row total (%)
Tillage		20.3	2.0	0.5	0.3	23.1
Intensive grass	25.5		30.9	2.4	0.4	59.2
'Environmental stock': Extensive grass	1.6	8.9		2.3	0.1	12.9
'Environmental stock': Semi-natural vegetation	0.5	1.2	1.3		0.1	3.1
'Environmental stock': Deciduous woodland	0.1	0.2	0.3	0.2		0.8
Column total (%)	27.7	30.6	34.5	5.4	0.9	100.0

Fig. 5.5. Relative importance of different landcover shifts. (Source: farm survey and countryside survey.)

more continuous on some farms than others. Strong feedback effects appear to be operating, with the ability of farmers to carry out change in the recent past being partly a function of the changes made to their businesses in the more distant past. One explanation for those changes observed on previously quiescent farms since 1978 is that a 'catching up' process is at work here and that high change farmers are simply laggards in the diffusion of land improvement technologies, some of whom have only been active in the recent past because constraints to development have suddenly been removed by a change in personal or family circumstances.

An obvious influence is government support for agriculture. The expansionary agricultural policy of the 1960s and 1970s was a 'rising tide that lifts all boats', farm survey evidence reinforcing the impression of a high water mark in rates of land improvement at about this time. Over 60% of all field underdrainage, 68% of all hedgerows removed and 70% of all grass to arable conversions recorded on survey farms was telescoped into these two decades, for instance. Questions remain, however, about the specific circumstances of those who failed to float with the tide. It is also hard to explain the characteristic cycle of intensification and extensification exhibited on some survey farms wholly in these terms. A process common to all farm families is the family life cycle and a plausible hypothesis is that this has been important in determining the timing and duration of land improvement, intensification and hence environmental change on survey farms. More precisely, the processes of succession and inheritance that give rise to, and are the product of, certain life cycle events may define critical transitions when farm business restructuring, expansion and retrenchment is most likely to take place.

The Life Cycle Effect

Bennett (1982, p. 299) has pointed out that family farms, like most human enterprises, tend to have cyclical histories 'with a start, a middle and then perhaps a new start. The important question is how this historical sequence differs from one endeavour to another'. An overlapping view is that other factors, related to the life cycle but less easily assigned to particular phases, may be operating which cut across this life cycle effect. Recent work (Potter and Lobley, 1992) suggests that the succession status of the farm family household is particularly important in shaping the way farm businesses develop over time. At its simplest, the presence of a successor provides an incentive to expand the farm, invest in capital and increase output over longer periods than would be the case if succession is uncertain or has been ruled out. A 'succession effect' operates as some businesses are developed in anticipation of succession taking place, either to provide a living for the successor and his/her family on the same farm or (more rarely nowadays) to generate sufficient capital to set up the successor on a separate holding. With succession underway a 'successor effect' comes into play on successor farms as younger managers put through changes to the way the farm is managed and even embark on new rounds of expansion and restructuring. This new blood effect can be gradual or sudden, depending on the process of succession itself and the efficiency with which it is accomplished. Finally there is the retirement effect that obtains at the end of a farmer's career. Farmers lacking successors may at this point decide to run down their farms, or at least extensify production in some way, again with important environmental effects.

In order to assess how far the patterns of farm development observed in the sample can be attributed to a life cycle effect, respondents were allocated to one of five categories describing the trajectory of their businesses over the study period. Information about the management histories of farms collected from the farm survey was combined with that relating to patterns and trends in land and input use to distinguish between recent developers, consolidators, stabilizers, disengagers and withdrawers (see Box 5.1).

Box 5.1. The characteristics of farmer clusters.

Cluster 1: *Stabilizers*. This is a large cluster of apparently very stable farmers who have made few changes to parcel land over the study period. Found in all landscape types, this is a heterogeneous cluster representing most farm types although the majority are grass based (58%). Farmers in this cluster are the least likely to have erected new agricultural buildings (8%) and no members of this cluster increased fertilizer use or stocking density and none carried out field scale drainage works.

Cluster 2: *Livestock improvers and intensifiers*. A well-defined group of high change farms, concentrated in pastural and upland landscapes (46% and 36% found here respectively). The majority of farmers in this cluster are cattle and sheep farms (52%) while 23% operate dairy farms, and 25% are over 200 ha in size. These are the farms most likely to have carried out land improvement in the recent past: a high percentage of this group have removed internal field boundaries or carried out field underdrainage (86%), while 77% increased fertilizer applications on grass and/or increased stocking density.

Cluster 3: *Livestock intensifiers*. A group of farmers clearly intensifying production alongside little or no land improvement. 54% of this cluster are found in pastoral landscapes, 52% operate livestock farms while 30% are responsible for dairy businesses. This is reflected in the high proportion intensifying grassland use – 56% increased fertilizer use on grass and 85% increased stocking density.

Cluster 4: *Arable intensifiers*. A group of high change farms concentrated in arable landscapes. This cluster is made up mostly of arable and mixed farms (52% and 26% respectively) and 30% of the holdings in this cluster are in excess of 200 ha. Members have been active in erecting new agricultural buildings (23%), removing field boundaries (29%) and carrying out field underdrainage (33%), but are particularly distinguished by large increases in the intensity of land use on their farms in the recent past. 84% have increased fertilizer use on arable land and 69% increased use of pesticides and/or herbicides.

Cluster 5: *Livestock extensifiers*. A very distinct group of small, grassland farms found in pastoral and arable landscapes (57% are less than 50 ha and 64% are livestock or dairy farms). The distinguishing characteristics of this group are low levels of change and moves towards extensification. For example, less than 2% have removed field boundaries and only 9% have carried out field underdrainage but 67% have reduced stocking densities since 1978. Farmers in this cluster are the least likely to have increased fertilizer use on grass, indeed 51% have actually decreased fertilizer use.

The picture presented in Figs 5.6 and 5.7, showing that succession status is strongly associated with an expansionary and 'professionalizing' pattern of farm development, clarifies how the life cycle effect may actually operate. For instance, farmers with a successor present are significantly more likely than the sample average to have seen a net increase in the area farmed (54%) and an increase in the intensity with which the land is farmed (65%). Most currently manage medium or large holdings and over half operate a business with a turnover in excess of £100,000. These are evidently the consolidators in the middle stage of the family life cycle, two-generation farms subject to both a successor and a new blood effect. Figure 5.8 confirms this by showing the high proportion of consolidators who have a successor present on the farm. They are the continuous developers in the sample, the people who, having carried out significant land improvement in the 1960s and 1970s, are still energetically upgrading, intensifying and professionalizing their business, invariably in partnership with a successor. Note also the expansionist profile of farmers who say it is 'too early to tell' in Figs 5.6 and 5.7. Figure 5.8 shows that a very high percentage of farmers in this succession class are developers and further cross-tabulation (not shown) reveals that they are mostly young farmers who have come to or taken over management control of their farms in the last ten years (in other

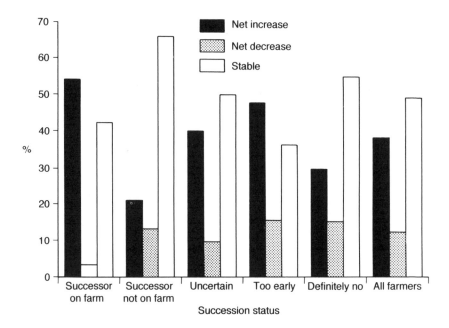

Fig. 5.6. Change in holding area (1978-1993) by succession status. The association between farm succession status and change in holding area was significant at 5% level using chi-square. (Source: farm survey.)

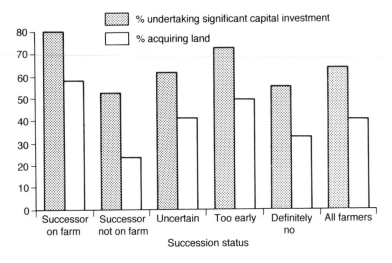

Fig. 5.7. Investment and land acquisition by succession status. The association between farm succession status and land acquisition was significant at 5% level using chi-square. (Source: farm survey.)

words, farms in the early stages of the life cycle on which a new blood effect is likely to be much in evidence). Case study evidence suggests that it is on such farms that the most dramatic land-use changes have taken place in recent years.

The profile of farmers with successors not on the farm is more stable than those who have definitely ruled out succession (either because they have no heirs or successors or none willing to take over the farm or have actively dissuaded them from doing so). Indeed, the latter group is most likely of any in the sample to be actively disengaging or withdrawing from full-time agriculture. Closer examination of these cases shows that non-successor farms tend either to be risk-taking entrepreneurs, who embark on new enterprises and follow the most expansionist trajectories of any farmers in the sample, or the least managerially dynamic people who, lacking any incentive to expand and increasingly convinced that the family farm will die with them, run down their farms in retirement and old age. Relatively few (35%) have farm debt and a significant minority (46%) operate farms of 50 ha or less. Interestingly, while 35% operate a business with a turnover of less than £25,000, 31% are responsible for businesses with a turnover of over £100,000.

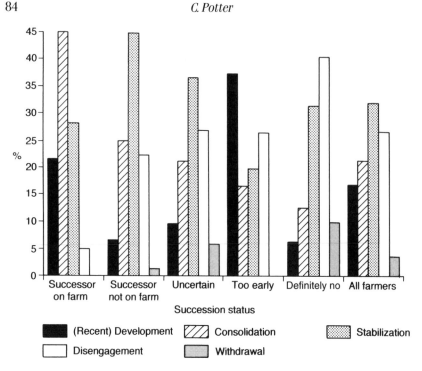

Fig. 5.8. Farm trajectory by succession status. The association between farm trajectory and succession status was significant at 5% level using chi-square. (Source: farm survey.)

Conclusions

The analysis of CS1990 results presented here shows a countryside in which environmental change is more farm and location-specific than ever before. Geographically, the pattern of change is uneven, with the most significant land cover shifts taking place in pastoral landscapes. In arable landscapes recent land cover changes have further fragmented an already residual conservation resource. The uplands are, by comparison, much more stable. Where change has occurred, it is confined to a relatively small number of high change farms, particularly in arable landscapes. Grassland farms account for most of the significant land cover shifts recorded, which are chiefly attributable to changes in grassland management.

CS1990 though offers only a snapshot of change between 1984 and 1990 and in seeking to investigate the causes of change, it is essential to take a long view. An analysis of the management profiles of survey farms distinguishes between high change farms with long histories of land improvement and intensification and those on which land-use change is confined to the recent past. Equally there are farms in the sample apparently unchanged in land use terms since the early 1960s and others where recent stability follows upgrading and intensification in the middle past. Finally, some farmers, having built up their businesses in the past, are now proceeding to run them down.

Behind these different trajectories of farm business (and, by implication, environmental change) stand a number of important influences. Government policy has obviously been important in bringing about the very widespread improvement of land and intensification of production that the management histories of respondents suggest took place throughout the 1960s and 1970s. However, a more farm-specific analysis is required to explain both the catching up process, which appears to have precipitated recent land cover change on many survey farms, and the trend to extensification after years of intensification evident on others. There is some indirect evidence to support the idea that the family life cycle plays a role in switching on and switching off phases of expansion, stability and retrenchment. To the extent that they operate much more strongly on some farms than others, these go some way to explaining both the uneven pattern and very different time profiles of environmental change observed in the British countryside in the recent past.

All of this has important implications for future public policies designed to promote sustainable rural land-use. If, as seems likely, traditional farming practices often survive because of accidents of inheritance and succession, their continuation beyond the present generation of farmers must now be increasingly in doubt. Moreover, policy change, by affecting the pattern of succession and making it harder to ensure the transfer of economically marginal family run businesses to the next generation, may have environmental consequences that are presently hard to discern. Sustaining landscapes and habitats may yet be seen to depend on sustaining the farm family structures within which so much of Europe's rural land is managed and deployed.

Acknowledgements

The research project on which this chapter is based was funded by the Economic and Social Research Council and the Department of the Environment's Wildlife and Countryside Directorate. I am grateful to Matt Lobley for his comments on an earlier draft of this chapter.

References

Barr, C., Bunce, R., Clarke, R., Fuller, R., Furse, M., Gillespie, M., Groom, G., Hallam, C., Hornung, M., Howard, D. and Ness, M. (1993) *Countryside Survey*

1990: Main Report. Department of the Environment, London.

Bennett, J. (1982) *Of Time and the Enterprise – North American Family Farm Management in a Context of Resource Marginality.* Minneapolis: University of Minnesota Press.

Bowers, J. (1995) Sustainability, agriculture and agricultural policy. *Environment and Planning A* 27, 1231–1243.

Bowers, J.K. and Cheshire, P. (1983) *Agriculture, the Countryside and Landuse: an Economic Critique.* Methuen, London.

Cheshire, P. (1985) The environmental implications of European agricultural support policies. In: Baldock, D. and Conder, D. (eds) *Can the CAP fit the Environment?* CPRE, IEEP, London, pp. 9–18.

Gasson, R. and Errington, A. (1993) *The Farm Family Business.* CAB International, Wallingford.

Marsden, T. and Munton, R. (1991) The farmed landscape and the occupancy change process. *Environment and Planning A* 23, 663–676.

Morris, C. and Potter, C. (1995) Recruiting the new conservationists: farmers adoption of agri-environmental schemes in the UK. *Journal of Rural Studies* 11, 51–63.

Munton, R. and Marsden, T. (1991). Occupancy change in the farmed landscape: an analysis of farm level trends, 1970–85. *Environment and Planning A* 23, 499–510.

Potter, C. (1986) Processes of countryside change in lowland England. *Journal of Rural Studies* 2(3), 187–195.

Potter, C. and Lobley, M. (1992). Ageing and succession on family farms: the impact on decision making and land use. *Sociologia Ruralis* 32, 317–334.

Potter, C. and Lobley, M. (1995) *The Causes and Processes of Countryside Change: An Analysis Relating to the Countryside Survey 1990.* Department of the Environment Countryside Survey Series, London.

Rae, J. (1993) Agriculture and the environment in the OECD. In: Williamson, C. (ed.) *Agriculture, the Environment and Trade.* International Policy Council on Agriculture and Trade, Washington, pp. 82–114.

Shoard, M. (1980) *The Theft of the Countryside.* Temple-Smith, London.

Westmacott, R. and Worthington, T. (1974) *New Agricultural Landscapes.* Countryside Commission, Cheltenham.

6

The Construction of Environmental Meanings Within 'Farming Culture' in the UK: The Implications for Agrienvironmental Research

CAROL MORRIS AND CHARLOTTE ANDREWS
Department of Geography, University College Chester,
Cheyney Road, Chester CH1 4BJ, UK

Introduction

The environmental impacts of agriculture in the UK continue to cause concern (e.g. Barr *et al.*, 1993). However, the farming community has not been insensitive to this concern. One measure of this is the increasing numbers of farmers participating in a range of recently implemented voluntary incentive schemes designed to facilitate the development of agricultural systems more in harmony with the environment (Morris and Potter, 1995). Whereas this represents a step in the right direction, many farmers remain unresponsive to these schemes in particular and environmental issues in general. Given their control of the majority of land in the UK, this has important implications for the development of sustainable land-use practices. Understanding the constraints to the integration of environmental practice on farms is likely, therefore, to remain an important part of the agricultural geographer's research agenda. Work to date (largely in relation to the adoption and non-adoption of agrienvironmental schemes) has started to tell us about these constraints, and, together with work concerned with agricultural restructuring, has highlighted the importance of retaining a sensitivity in research to investigating farm level processes of change. In spite of this it has been limited in its consideration of how farmers come to view conservation, nature and environment from the perspective of the farmers themselves. This suggests the need for a more in-depth investigation of how farmers come to construct environmental meanings within the context of farming 'culture' and how these influence environmental practice on farms, how these meanings are time and place specific, and how these constructions are being challenged through the influence of policy, rural social change, etc. As McEachern (1992, pp. 161–162) argues, 'the level at which various practices and logics are so culturally embedded that they seem natural to farmers is a very important one.... The knowledge that these taken for granted meanings can change with new circumstances is also important'. The central purpose of this chapter is to begin to explore the theoretical and methodological implications of utilizing a concept of culture within agrienvironmental research. This culturally sensitive framework is currently being developed in research within the locality of Cheshire, North West England.

Understanding Agrienvironmental Decision-Making: A Brief Review

Constraints placed upon farm businesses by the reduction in price support for traditional agricultural products through the reform of the Common Agricultural Policy (CAP), and by other parts of the food supply system, have created a context that is far from conducive to enabling farmers to prioritize the environment within their management strategies. On the other hand, the implementation of policy initiatives designed to facilitate the development of 'alternative and sustainable agricultures' (Bowler, 1992) (largely an outcome of the increase in the consumption demands placed upon rural areas) has presented new opportunities and new means of support to farm businesses, conditional upon the production of environmental goods. At the present time policy-makers favour the use of financial incentives and voluntary codes of practice over a range of other conservation policy tools such as regulation (although cross-compliance is currently receiving more attention (MAFF, 1995)). These general considerations provide the context for research that has set out to investigate environmental practice on farms.

Some of this work has described the differences in farmers' attitudes to the environment and conservation in general terms (e.g. Newby *et al.*, 1977; Westmacott and Worthington, 1984; Carr and Tait, 1991), whereas considerable attention has been directed at analysing farmers' responses to a range of agrienvironmental incentive schemes such as Environmentally

Sensitive Areas (ESAs). This work has either emphasized the importance of economic factors (e.g. CEAS, 1987; Brotherton, 1991; Colman *et al.*, 1992) or has investigated in a more descriptive way the farmer and farm family factors that pushed or pulled farmers into environmental and other schemes (see Potter and Gasson, 1988; Gasson and Hill, 1990; Ilbery and Bowler, 1993; Tarrant and Cobb, 1992; Morris 1993). The emphasis here was on analysing the profile of participating farmers and explaining their adoption of schemes in terms of the fit between scheme conditions and the farmer's situation. In explaining participation, the relationship between 'willingness to adopt' (i.e. the attitude of the farmer towards the scheme and also towards other factors like profit) and 'ability to adopt' (i.e. the economic status of the farm and the degree of correspondence with the farming system) was a central theme in many of these studies. Typically, it was the younger farmers, with the largest, more economically buoyant farms who tended to find the schemes attractive, though variations were observed in relation to scheme details and areas of implementation. More recently, there have been attempts to develop deeper, more 'qualitative' assessments of the extent to which farmers are engaging with the objectives of agrienvironmental schemes (e.g. McHenry, 1994; Morris and Potter, 1995). Also, some efforts have been made to approach an understanding of farmers' responses to conservation and the environment in general from the perspective of the farmers themselves (e.g. McEachern, 1992; Clark, 1994). These concerns are reflected, to an extent, in recent work carried out within a political economy framework. Conscious of its neglect of the environment this research has begun to examine how farmers' 'knowledge systems' affect their responses to environmental regulation (Ward and Munton, 1992) and how the relationship between farm family processes and the development of the farm business can inform our understanding of the implementation of farm pollution control measures (Ward and Lowe, 1994).

Whereas there are differences in theory and methodology, a common theme shared by the recent work examining environmental practice on farms and agricultural restructuring is a recognition of the need to develop a greater appreciation of farm level processes of change, while at the same time retaining an awareness of the broader context. As Ward and Munton (1992, pp. 127 and 142) argue, 'the need to understand farm level processes of change ... the abilities of individual farmers and their households to negotiate their own futures ... is increasing because of a series of changes to policies affecting rural areas'. It is at the farm level that broader economic and social changes (e.g. concern for the environment) are interpreted by individuals in a certain context. Significant here is the (often implicit) conclusion in much of the work cited above that the way in which farmers interpret environmental regulation, incentive schemes and conservation advice will be influenced by the particular meanings that they

:t about the environment, nature and conservation. It is suggested se environmental meanings are likely to be a product of values, knowledge systems or 'culture' distinct from and at variance with the policy-makers, conservationists and the general public. This may have important implications for making sense of the responses of farmers to environmental demands placed upon them. In turn, this implies that we should be less interested in what farmers' attitudes towards the environment actually are (a typical pursuit of behavioural research) and instead should be exploring the processes by which environmental attitudes are shaped and formed and how these change with new circumstances. To do this would require a sensitivity, one might say a 'cultural' sensitivity, to the time and place specific processes by which farmers construct their own particular 'version(s)' of the environment. If a cultural approach encompasses a concern for the construction of meanings (Duncan, 1993) and culture is central to human affairs (Cosgrove and Jackson, 1987) then it follows that a cultural understanding may be useful to agrienvironmental research. That agricultural geography should become more sensitive to the role of environmental meanings within farming is a reflection of the interpretative or cultural turn within the social sciences in recent years (Kofman, 1980).

However, understanding the processes by which environmental meanings are constructed within agriculture has to date only received limited research attention and yet, as Shucksmith (1993, p. 477) argues, 'to overcome this reluctance (among farmers) to engage with post-productivist policies will require not only a reformation of these policies ... , but also a *cultural transformation* which redefines the image of "a good farmer" in his own eyes and in those of his peers' (emphasis added).

Towards a 'Culturally Sensitive' Understanding of Agrienvironmental Decision-making

Approaching an understanding of environmental meanings and their influence on environmental practice within farming by adopting a culturally sensitive framework, demands that we have a clear understanding at a theoretical level of what culture is. It is interesting to observe that there has been a move toward considering a set of factors or influences often referred to as 'culture', to help explain farmers' agricultural and agrienvironmental decision-making. Some researchers have adopted 'sociocultural' or 'sociocognitive systems' approaches (Ward and Munton, 1992). Others have engaged in 'actor-oriented research' that investigates a 'cultural as well as a structural account of actions' (McHenry, 1994; Herrman and Shucksmith, 1994) in order to investigate the 'complex web of values and aspirations extending far beyond purely economic rationales' that make up 'agrarian ideology and culture' (Ward and Lowe, 1994, pp. 174 and 183). Similarly, Murdoch and Marsden (1994, p. 135) recognize that 'farmers have increasingly differing attitudes and objectives in relation to their land-holding, these are not simply economic but relate to social and cultural considerations also'.

This small body of work has been significant in introducing a consideration of culture in the farming context. However, the term has often been used in an uncritical way, acknowledged as a potential influence but presented unproblematically, with no clear explanation of the meaning of culture. If culture is a genuine influence on agrienvironmental decision-making then its meaning should be clearly stated and its significance investigated. It has been demonstrated that culture as a concept is very problematic (e.g. Rowntree, 1988; Price and Lewis, 1993; Wagner, 1994). It is often used in explanations to encompass the factors that are left after considering the social, the political and the economic. This has often left it as a very imprecise analytical term. There is also a danger in this approach of reifying culture, i.e. conceiving of it as an observable thing that has independent existence and causal powers, equivalent and alongside the economic and social. Yet clearly culture does not exist in the same way; it cannot be measured and assessed in the same way as the economic and social, nor should it be understood as being distinct and separate from them. As Kofman (1980, p. 54) argues 'cultural elements are not isolated facts, but parts of interrelated systems ... the cultural factor cannot be treated apart from the dominant social and economic processes'. A clear conceptualization of culture is therefore needed if research is to uncover what role it plays in environmental decision-making among farmers.

An important contribution to the debate on the meaning of culture for geographical analysis is made by Mitchell (1995) who suggests a need to focus on the idea or the ideology of culture. He notes that culture cannot have an ontological existence (i.e. an independent, factual existence like the economy) and yet it is often referred to by social actors to justify their actions. It thus becomes important to focus on 'how the very idea of culture has been developed and deployed as a means of attempting to order, control and define' (Mitchell, 1995, p. 104). The idea of culture 'is not what people are doing ... it is the way that people make sense of what they have done. It is the way their activities are reified as culture' (Mitchell, 1995, p. 108). Cultural meanings are therefore socially constructed. What is referred to as 'culture' is in fact 'the mediation of production and consumption within everyday life' (Mitchell, 1995, p. 110). How then does culture 'exist' in the context of agrienvironmental decision-making and practice if it does not have independent existence? Arguably, farming culture exists within individual farmers' minds as a set of meanings, values or attitudes (shared within a particular group with common under-standings) that they use to make sense of their relationship with the environment. It 'exists' in the moments that farmers make reference to it to explain or justify their actions, e.g. appeals to 'traditional farming methods' as 'natural' and unchanging practices. It can also 'exist' and come to have a 'real' force in society in the sense that it can be encapsulated into social products. Through these products the idea of culture is recreated and reproduced. An example of this would be the expression of notions of farmers as 'stewards of the land' in professional farming publications or in media representations of farming. So culture does not exist independently, but neither is it created from nothing. Farmers draw on a variety of experiences within their 'lifeworlds' to construct cultural understandings of the environment: their own socialization and personal experiences (e.g. being brought up in a farming household in a farming community) within a particular locality; their social and professional relationships with other farmers and with other actors in their social and professional networks; and meanings in the public and professional media. The relative significance of each of these influences and the means by which they are translated into cultural understanding of the environment have yet to be properly investigated. As Burgess (1990) notes we know very little about how different social groups consume environmental meanings and use them to make sense of the environment and their relationship to it.

Two examples help to make clear what is involved in this approach. The first concerns how culture is being used with regard to the interaction of farming and the environment in the current row over live animal exports. Under pressure from animal welfare groups and public opinion, which have generated negative images of farmers as exploiters of 'nature', farmers have responded by referring to their own social constructions of their role. To counter public perception farmers have portrayed themselves as 'guardians' or 'stewards' of nature. Why would they be cruel to animals when their role is a nurturing one? Importantly, it is argued, 'people seem to forget that there is no profit in an animal which is not well treated' (Thorley, 1995, p. 32). To gain the support of the state to continue their economic activities, through policing and the law, they portray themselves as responsible business people carrying out a legitimate (i.e. legal) trade who have a right to state protection of their livelihood (and their property). As one professional put it 'as an industry we will take whatever steps are necessary to maintain this legitimate trade. Proper welfare is supported enthusiastically by all right thinking people, especially those in the professional business of caring for animals' (Thorley, 1995, p. 32). An appeal is also made to the notion of the role of farmers in maintaining both rural communities and the countryside the public expects, as to lose sheep farmers will be to 'lose another part of the balance of agriculture and the areas which are currently relatively tidy and reflective of a properly organised society will become barren wastelands' (Thorley, 1995, p. 33). To try and alter the transportation of live sheep within the UK from areas of seasonal grass production to abattoirs 'would be to tamper with nature' (Thorley, 1995, p. 32). Farming 'culture' in this instance comprises socially constructed notions of farmers as 'natural' stewards and nurturers in order to justify a specific set of socioeconomic relationships.

A second example is provided by a study of how farmers in the Yorkshire Dales accommodated the conflict inherent in using land as a source of profit (with environmental consequences) and at the same time fulfilling a nurturing and caring role (McEachern, 1992). The answer lay in the way farmers '*give meaning to* the activity which dominates their lives' (p. 159, emphasis added). The notion of farmers as 'guardians' or 'stewards' of the countryside legitimated farmers' claims that conservation must be undertaken by them without external interference. However, in the 1980s 'the loss of public confidence in farmers meant that the National Farmers' Union and the Country Landowners' Association had to *work hard to resurrect* the image of trusted custodian in order to preserve a voluntary approach which resisted legislated controls over farming' (McEachern, 1992, pp. 160–161; emphasis added). Ideas of what constituted an ideal landscape were constructed by farmers for use in the political struggle over what direction conservation strategies should take (i.e. retaining voluntary adoption). Having a neat, clean agricultural landscape was seen as a sign of a 'good' farmer (although in fact of less value to wildlife), and farmers were held to know 'what is best' because their work and location gave them a 'natural' insight in a way that urban dwellers did not.

These examples raise four points about the way farmers and their professional representatives have constructed a 'farming culture' to legitimate their interaction with the environment. Firstly, they demonstrate that culture is all about meanings and values ascribed to the environment. Secondly, that culture is socially constructed to justify actions and mediate (environmental) problems in production. That is, it does not have a separate 'existence' but is brought into being (made real) as people refer to it to justify their actions. Thirdly, it demonstrates that it is possible to have several ideas of 'farming culture' co-existing, i.e. among farmers themselves ('business' and 'nurture') and between farmers and groups such as Compassion in World Farming. A cultural approach therefore does not imply the idea of a farming culture in relation to the environment, rather that the aim should be to investigate the occurrence of sub-cultures that may vary sectorally between types of farming or geographically between localities. Fourthly, it shows that culture is a contested, fluid thing, i.e. its form is argued over and can change. This is suggested by McEachern (1992, p. 169) who identifies that 'already in 1987, farmers were beginning to accept some of the (National Park) authority's definitions of enhancement and conservation of environment when they entered into management agreements to alter farming practices.... In turn this provided the beginning of new constructions of farming and farmers themselves, though in the late 1980s this was still a highly contested area.'

The Implications of Adopting a Cultural Approach in Agrienvironmental Research

Accepting these ideas about the meaning of culture (the social construction of environmental meanings by farmers) has a number of theoretical and methodological implications for agrienvironmental research. Given the space available, only some of these will be sketched out below.

While implying a continued focus on farm level decision-making, the search to uncover the social construction of environmental meanings among farmers does not imply a narrow approach that emphasizes only individual behaviour and decision-making. By its very nature it is trying to identify any shared meanings in this process. Further, a cultural analysis is explicitly concerned with the broader context in which decisions are made. A cultural analysis looks at how people make sense of everything around them, at all the influences on their agrienvironmental behaviour (the social, the economic, the political) within any given context. This suggests the idea of the farmer as a knowledgeable and active actor capable of 'negotiating' macro-influences or the context of his or her situation (e.g. McHenry, 1994, Murdoch and Marsden, 1994). However, farmers may not always act on off-farm environmental knowledge (e.g. from a conservation agency) because of the influence of local culture in relation to the environment (which may construct an opposing/divergent view of what is good environmental practice).

A cultural analysis therefore demands a more subtle understanding of what socially constructed environmental meanings (i.e. 'culture') farmers actually refer to when making agrienvironmental decisions. We need also to understand how these environmental meanings are constructed, how these constructions can change with new circumstances, and how these constructions affect the interpretation of off-farm agrienvironmental information. What elements of life are drawn on, how do farmers' social and professional networks supply information that is used in these cultural constructions? A range of possible influences can be identified, but one process that may be significant here (at least in terms of challenging existing environmental meanings) is highlighted by Ward and Lowe (1994) who argue that farmers may be being influenced by the environmental meanings held by the growing numbers of incomers into rural areas and also by their own children's education. That this is likely is not in doubt, but at present the geographical nature and importance of their impact in changing farmers' social constructions of the environment is not fully understood or investigated. A further influence may come from conservation agencies involved in the delivery of agrienvironmental policies. The extent to which the meanings awarded to the environment by these agencies concur with the nature of farming culture within any given locality has not received much attention. However, the different use of 'culture' and environmental meaning by farmers and conservationists may have implications for the delivery of agrienvironmental policy.

Clearly, both the social processes involved in the construction of and challenge to farmers' environmental meanings will not take the same form in all areas. A cultural approach to understanding farmers' environmental meanings needs to take into account, therefore, the different material and social conditions and relations of places, e.g. arising from development pressure, the wealth of the farming region, the role of estates, the organization of farming groups and conservation agencies, the presence or absence of conservation designations, and how remote an area is. Together with the local variation in environmental quality, these local conditions are likely to influence how farmers make sense of their environment and thus, in the creation of local farming culture. Within ESAs, for example, it has been shown that features unique to the locality (such as the structure of land-holding and the value of environmental payments relative to the local profitability of farming) played a role in how farmers responded to that scheme (e.g. Brotherton, 1990). An understanding of the use of 'culture' and environmental meaning by farmers and the form this takes in specific localities is likely to raise questions about the policies and practices that will be sensitive to this scale of variability and may add to the case for the local administration of agrienvironmental schemes (e.g. Burchell, 1995).

These points have methodological implications. Importantly, research methods are needed that are appropriate to the investigation of the subtle conception of culture outlined above. Those studies that have suggested a cultural approach have tended to fall back on the more heavily used qualitative research methods such as questionnaires, and then engaged in trying to relate descriptive categories to explanatory schema (see, e.g. Herrman and Shucksmith, 1994; Ward and Lowe, 1994). Such methods do not generate the type of data that allows analysis of meanings, values and cultural constructions in a deeper way. Pile's (1991) work using depth hermeneutics to investigate the political worlds of dairy farmers is a notable exception, as is Gant's (1995) efforts to take seriously ethnographic methods such as oral history in rural geography. Also, the approach advocated by Marsden *et al.* (1993) and Murdoch and Marsden (1994), of literally 'following the actor' as they build their worlds, forge links with others and pursue their interests, allows deeper insights into the processes of rural change within specific localities. The necessary in-depth research methods for a cultural study will involve, for example, semi-structured interviews, participant and non-participant observational techniques, oral histories, focus groups and so on. The work of McHenry (1994, p. 8) using semi-structured interviews demonstrates the importance of research methods designed to 'capture what the farmers thought the questions were', rather than to define the importance of pre-selected questions as is more commonly attempted. The use of these methodologies does not imply a narrow focus on individual understanding, but incorporates understanding of how individuals interpret their particular circumstances.

The method by which farmers are selected for inclusion within a culturally sensitive study is also likely to differ from earlier approaches. The construction of environmental meanings (and how these are changing) is likely to be influenced by, for example, farmers' social and professional networks within and beyond the locality, i.e. factors that transcend simple structural characteristics such as farm type and farm size (typically used to structure a sample in agricultural geography research). This suggests the need to interview a self-defined sample of farmers identified by an initial contact (comprising perhaps key individuals within a farmer's social and professional network within the locality), as opposed to pre-selecting a 'representative' sample or study area on the basis of either farm type or farm size.

All these ideas are currently being developed within research which is examining the social construction of environmental meanings within the farming community in the locality of Cheshire, north-west England.

Conclusion

This chapter has made a case for building on existing agrienvironmental work by integrating some of the ideas and concepts from sociocultural geography to further our understanding of agrienvironmental decision-making. Specifically, it has developed a concept of culture (socially constructed environmental meanings) and explored how this may influence environmental decision-making on farms. Utilizing this approach has a number of implications for theory and methodology. Important here is the use of qualitative methods to uncover what environmental meanings are held by farmers, how these are constructed within specific social and material spaces, and the processes and mechanisms (e.g. rural social change, national incentive schemes and their delivery by local agencies) by which these are being challenged and reconstituted within a changing political economy of farming.

This chapter has focused on the idea of using a concept of culture in agri-environmental research. Adopting such an approach clearly has implications for the development of agri-environmental policy which will need to be thought through and addressed (some of these have been alluded to above). One aspect of this is that large sums of money continue to be concentrated on voluntary grant and incentive schemes targeted at specific areas. There is a vital need to understand the farm level processes which influence how effective these schemes are in practice, but also to investigate how farmers are reacting to environmental pressures in non-designated areas as well, which constitute a large percentage of the land area of the UK. Getting 'inside' the decision-making world of farmers as environmental managers has a large role to play in informing policy makers, and it is hoped that a culturally sensitive approach will be able to contribute to this.

References

Barr, C., Bunce, R.G., Clarke, R.T., Fuller, R.M., Furse, M.T., Gillespie, M.K., Groom, G.B., Hallam, C.J., Hornung, M., Howard, D.C. and Ness, M. (1993) *Countryside Survey 1990: Main Report.* Department of the Environment, London.

Bowler, I.R. (1992) Sustainable agriculture as an alternative path of farm business development. In: Bowler, I.R., Bryant, C.R. and Nellis, M.D. (eds) *Contemporary Rural Systems in Transition: Volume 1 Agriculture and Environment.* CAB International, Wallingford.

Brotherton, I. (1990) Initial participation in UK set-aside and ESA schemes. *Planning Outlook* 33(1), 46–61.

Burchell, M. (1995) Local menus for agri-funds. *Countryside* May/June, 7.

Burgess, J. (1990) The production and consumption of environmental meanings in the mass media: a research agenda for the 1990s. *Transactions of the Institute of British Geographers* 15(2), 139–161.

Carr, S. and Tait, J. (1991) Differences in the attitudes of farmers and conservationists and their implications. *Journal of Environmental Management* 32, 281–294.

C.E.A.S. (1991) *Conservation Advice to Farmers.* CEAS Consultants, Wye.

Clark, J. (1994) Strangers in strange land: farmers and nature conservation on the Pevensey Levels. Paper presented at Rural Economy and Society Study Group Annual Conference. Cheltenham and Gloucester College of H.E., 14–10 September.

Colman, D., Crabtree, B., Froud, J., and O'Carroll, L. (1992) *Comparative Effectiveness of Conservation Mechanisms.* Department of Agricultural Economics, Manchester University.

Cosgrove, D. and Jackson, P. (1987) New directions in cultural geography. *Area* 19, 95–101.

Duncan, J.S. (1993) Landscapes of the self/landscapes of the other(s): cultural geography 1991–92. *Progress in Human Geography* 17(3), 367–377.

Gant, R. (1995) Voices from the past: oral history as a data source for rural geography. Paper presented at the RGS-IBG Rural Geography Study Group, Young Research Workers Conference on 'Practising Rural Geography: Concepts and Methods'. Kingston University, 22–23 May.

Gasson, R. and Hill, P. (1990) *An Economic Evaluation of the Farm Woodland Scheme.* Farm Business Unit Occasional Paper No.17, Wye College.

Herrman, V. and Shucksmith, M. (1994) Habitus and practice of farmers in

Scotland and West Germany. Paper presented at the 35th EAAE Seminar. Aberdeen University, 27–29 June.

Ilbery, B.W. and Bowler, I.R. (1993) The Farm Divesification Grant Scheme: adoption and non-adoption in England and Wales. *Environment and Planning C: Government and Policy* 11, 161–170.

Kofman, E. (1980) Is there a cultural point of view in geography? *Area* 12(1), 54–55.

Marsden, T., Murdoch, J., Lowe, P., Munton, R., and Flynn, A. (1993) *Constructing the Countryside.* UCL Press, London.

McEachern, C. (1992) Farmers and conservation: conflict and accommodation in farming politics. *Journal of Rural Studies* 8(2), 663–676.

McHenry, H. (1994) Farmers' interpretations of their situation: some implications for environmental schemes. Paper presented at the 35th EAAE Seminar. Aberdeen University, 27–29 June.

Ministry of Agriculture, Food and Fisheries (1995) Council agrees regulation on link between market and environmental set-aside. MAFF Press Release, June 1995, London.

Mitchell, D. (1995) There's no such thing as culture: towards a reconceptualization of the idea of culture in geography. *Transactions of the Institute of British Geographers* 20, 102–116.

Morris, C. (1993) Recruiting farmers into conservation: an analysis of participation in agri-environmental schemes in lowland England. PhD Thesis, Wye College, University of London.

Morris, C. and Potter, C.A. (1995) Recruiting the new conservationists: farmers' adoption of agri-environmental schemes in the UK. *Journal of Rural Studies* 11(1), 51–63.

Murdoch, J. and Marsden, T. (1994) *Reconstituting Rurality.* UCL Press, London.

Newby, H., Bell C., Saunders, P., and Rose, D. (1977) Farmers' attitudes to conservation. *Countryside Recreation Review* 2, 23–30.

Pile, S. (1991) 'A load of bloody idiots': Somerset dairy farmers' view of their political world. *Political Geography Quarterly* 10, 405–421.

Potter, C.A. and Gasson, R. (1988) Farmer participation in voluntary land diversion schemes. *Journal of Rural Studies* 4(4), 365–375.

Price, M. and Lewis, M. (1993) The reinvention of cultural geography. *Annals of the Association of American Gegoraphers* 88(1), 1–17.

Rowntree, L.B. (1988) Orthodoxy and new directions: cultural/humanistic geography. *Progress in Human Geography* 12, 575–586

Shucksmith, D.M. (1993) Farm household behaviour and the transition to post-productivism. *Journal of Agricultural Economics* 44(3), 466–478.

Tarrant, J. and Cobb, R. (1992) The convergence of agricultural and environmental policies: the case of extensification in Eastern England. In: Bowler, I.R., Bryant, C.R. and Nellis, M.D. (eds) *Contemporary Rural Systems in Transition*: Volume 1 Agriculture and Environment. CAB International, Wallingford.

Thorley, J. (1995) Transport: an integral part of the sheep industry. *The Sheep Farmer* June, 32–33.

Wagner, P.L. (1994) Culture and geography: thirty years of advance. In: Foote, K.E., Hugill, P., Mathewson, K. and Smith, J. (eds) *Re-Reading Cultural Geography.* University of Texas Press, Texas.

Ward, N. and Lowe, P. (1994) Shifting values in agriculture: the farm family and pollution regulation. *Journal of Rural Studies* 10(2), 173–184.

Ward, N. and Munton, R. (1992) Conceptualizing agriculture–environment relations: combining political economy and socio-cultural approaches to pesticide pollution. *Sociologia Ruralis* 32(1), 127–145.

Westmacott, R. and Worthington, T. (1984) *Agricultural Landscapes: A Second Look.* Countryside Commission, Cheltenham.

7

Community-level Worldviews and the Sustainability of Agriculture

JANEL M. CURRY-ROPER
*Department of Geology, Geography, and Environmental
Studies, Calvin College, Grand Rapids, Michigan 49546, USA*

Introduction

The Midwest United States is a mosaic of ethnic/religious communities on which the prosperous agricultural economy of the region has been built (Borchert, 1987; Egerstrom, 1994). Within agricultural culture of the region, the focus of this study, sociologists have consistently identified community-wide social patterns affecting farming (Parsons and Waples, 1945; Flora and Stitz, 1985; Salamon, 1992). They influence capitalization of the farm enterprise, the extent to which a farm is commercialized, and farmers' risk-reduction strategies (Curry-Roper and Bowles, 1991; Salamon and Davis-Brown, 1986). These patterns suggest that farm community ideologies of the Midwest US, but not limited to the Midwest, may be an important element for understanding social change in agriculture.

Miller (1993, p. 48) has recently argued for the autonomy of religious or ideological belief over mere socioeconomic factors. He contends that beliefs often provide the key to understanding institutional structures and collective social action resulting in measurable social consequences. Winter (1991, p. 206) has likewise criticized social scientists who study rural places for their lack of appreciation of the role of culture and ideology. He also claims that what people actually think and believe need to be included as central concerns in fields like rural sociology. Hummon (1990, p. 11) characterizes these belief systems or perspectives as unconscious and community-wide. He says that, although expressed by individuals, they are public in that they are learned and sustained in the context of relationships with others in a community. These community-wide perspectives interact with larger societal forces, creating regional and community-level responses to change. Gilbert and Akor (1988) recognized the existence of cultural factors, along with others, in their study of divergence in the dairy industry. They found that in one context family dairy farms could persist while elsewhere capitalist–industrial dairying grew to dominate the industry.

This study uses the Midwest as a test case for the exploration of the existence of community-level worldviews. The focus of the study is on those elements of community worldviews that are related to agriculture. One of the key elements is tied to the contrasting philosophies of conventional versus sustainable agriculture – the extent to which rural communities are communitarian in nature as opposed to individualistically oriented. These polar opposites have been associated with other contrasting values and factors as well. In a historical study on New England, Cronon (1983) attempted to show the relationships among communal orientation, property rights and the perception of the environment and use of its resources.

Berkes and Feeny (1990) argue for similar connections. They questioned Hardin's assumption of individual interest and competition in his classical work, '*The Tragedy of the Commons*', and argued instead for the possibility that society is grounded in cooperative, communal action. Their focus of concern was resource management and the often overlooked informal rule-making of communities. They began with the assumption that actions are constrained by the community. They further argued that community-oriented management takes the long view and is more sustainable.

The work of Salamon (1985) points to other factors that may be connected to communal orientation in her typography of Yankee and yeoman farmers. Among communities described as 'Yankee' she found emphases on farms as businesses, and on geographic and economic mobility. Communities of German 'yeoman' farmers emphasized continuity: efforts were made to keep the farm in the family and strong community attachment was evident. Cronon (1983) would argue that this represented a difference in economic orientation. Yankee farmers, on the whole, were capitalistic in orientation, which resulted in the further commodification of land and its resources, with an emphasis on individual, exclusive, property rights. Beus and Dunlap (1990) have linked many of these same elements to the two major agricultural paradigms in the literature. The conventional agricultural paradigm emphasizes commodification of resources, self-interest and farming as a business. The growing alternative agricultural paradigm emphasizes local control, cooperation, community life and imitation of natural ecosystems.

In this study, the hypothesis was tested that distinct and identifiable worldviews relating to sustainable agriculture exist that are grounded in common religious belief systems which, in turn, are often carried by specific cultural and theological traditions. Seven communities (eight social groups) were studied. Using a case study approach, discernible differences among communities were identified by these communities' variations from one another in conceptualization of society along a range from individualistically conceived to communal in emphasis. It was hypothesized that the placement of a community along this scale could predict other value orientations, including: (i) perceived relationship between humans and nature; (ii) prioritization of rights among individuals, communities, and the environment; and (iii) accepted farming paradigms. The findings support the idea that there are differences among communities but the relationships among the variables are much more complex than hypothesized, though the combinations may be explicable through understanding cultural/theological traditions. The findings point in the direction of needing a variety of different intervention strategies to effect change toward a more sustainable agricultural system.

Methodology

This study focused on eight small, farm-based, homogeneous social groups spread across seven communities in Iowa: Wayland (German Mennonite), Stanton (Swedish Lutheran), Wellsburg (German Reformed); Cascade (German and Irish Catholic), Paullina (Anglo and Norwegian Quaker, and German Lutheran), Boyden-Hull (Dutch Christian Reformed) and Lamoni (Anglo-Reorganized Church of the Latter Day Saints) (Fig. 7.1).

Information and data were collected by three different methods. First, within each of the eight groups, two discussion groups were organized (one of each gender) made up of farmers and spouses. In the group discussions, participants were asked to respond to a series of narratives set in farming contexts, that presented situations and/or dilemmas relating to the study variables. Content analysis was used to analyse the transcripts arising out of the discussion groups.

Secondly, participants within each discussion group were asked to fill out a short questionnaire with general population information on themselves and their farms, its size, etc., plus a questionnaire asking them to rank the strength of their agreement or disagreement with 30 value statements discriminating the relevant variables.

In the third part of the study, four individuals or couples from each social group were personally interviewed ($x = 32$). These persons were selected to represent different age groups and farming types. Ninety individuals participated in the study, representing 58 families.

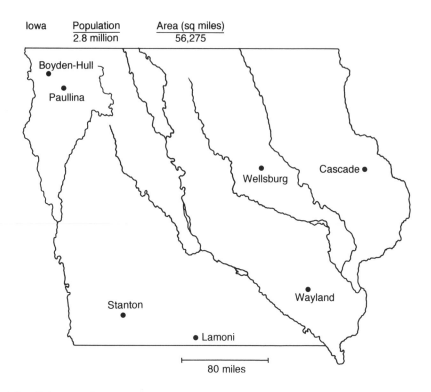

Fig. 7.1. Map of Iowa showing the location of seven communities used in a study of community-level worldviews relating to sustainable agriculture.

Findings

The findings support the idea that there are differences among communities, but the relationships among the variables are more complex than hypothesized. What emerged was a classification of community worldview types. One of the defining characteristics of the types, as originally hypothesized, was whether a group was individualistic or communitarian in its perspective: are issues or dilemmas perceived first from an individual perspective, or are they described first from the stance of a member of a community?

A second important defining characteristic was the expressed relationships between humans and nature: is nature primarily for human use and to enhance human comfort (utilitarian) or does it have a purpose in and of itself? A third characteristic that emerged, but not a defining one, was farm paradigm. Conventional agriculture, as defined by Beus and Dunlap (1990) and measured in this study, included emphasizing farming as a business, farm specialization, world markets, the advantages of technology and science and competition. In contrast, alternative agriculture emphasized farming as a way of life, farm diversification, domestic markets, scepticism toward science and technology, and cooperation among farmers and between farmers and consumers.

One last study variable that cut across all types pertained to priority of rights. What comes first: individual freedoms, the human community as distinct from the needs of the individual, the rights of the government to regulate, or the needs and rights of the non-human world?

Individualistic and utilitarian

A community worldview type that is both individualistic and utilitarian corresponds closely to what Dunlap and Van Liere (1984, p. 1013) describe as the dominant social paradigm of society, and included such values as individualism, materialism, support for limited government and belief in progress. The German Reformed group most clearly reflected this type. On the individual–communal scale the group fell on the individualistic end. In addition, the participants in the group put the most emphasis of all groups on individual rights. They were also the most strongly utilitarian of all groups. The participants emphasized economic growth as the main goal or use for the natural environment, not to be slowed by air and water pollution standards or endangered species.

Dunlap and Van Liere (1984, pp. 1023–1024) similarly found that commitment to the dominant social paradigm led to lower levels of concern for environmental protection, and that some dimensions of the paradigm, particularly commitment to private property rights, economic growth, material abundance and *laissez-faire* government, were more important than others in negatively influencing concern for environmental protection.

The German Reformed group showed strong support for the conventional agricultural paradigm, with the exception of specialization. In spite of their location in a cash grain area, the majority of farm families in the study had livestock. This may be an echo of a different way of life in the past. Though strongly conventional, and viewing farming as a business, many expressed a commitment to farming as a way of life, an alternative agricultural perspective.

This strong commitment to agriculture in the past has been found among other groups who originated, like the German Reformed group, from East Frisia. Salamon (1980, p. 295) characterized a similar group in Illinois as endogamous and ethnically homogeneous. Families tended to remain in the community and were seldom diverted into non-farm occupations. Corner (1928, pp. 23, 27) and Schnucker (1986, p. 263) described similar traits among the East Frisians earlier in this century – attachment to the homestead, pride of land ownership, willingness to do hard work, loyalty to the family, and craving for the security of land ownership. Until recently, these traits were found among the German Reformed group of this study. The last two decades have brought about change as their ethnic solidarity diminished, and their emphasis on education increased along with civic involvement. Farming is no longer seen as an occupational choice to the exclusion of all else.

Individualistic non-utilitarian

Two groups fell on the individualistic end of the individualistic–communal scale, but were not utilitarian in their view of nature – the Quakers, of Norwegian and English ancestry, and the Catholics, of German and Irish descent. Though individualistic in perspective, like the German Reformed, both the Quakers and Catholics put the least emphasis on individual rights of any groups and showed tolerance for the government's role in society. This supported the findings of sociologist Greeley (1989, p. 496) who found that Catholics were more likely to emphasize fairness whereas Protestants were more likely to emphasize freedom and individualism. Catholics were also more likely to support government intervention in the economy, government ownership of industry and equalization of income.

Why are the Quakers so tolerant and why do they put less emphasis on individuals rights? Perhaps their somewhat counter-cultural tradition that emphasizes a non-hierarchical, consensual way of relating to one another, pacifism, rejection of formality in worship, and simplicity and plainness in living offers some explanation (Mott, 1900, p. 271).

Catholics' and Quakers' non-utilitarian view of the relationship between humans and nature may be connected to their lack of emphasis on individual rights, according to Dunlap and Van Liere (1984). The Quaker group was the only one to express a co-equal view of the relationship between humans and nature, which is characterized by nature being seen as an ecological system of interrelated life forms that does best if left alone. Examples of phrases included, 'You shouldn't mess with nature', and 'as soon as we quit messing with it, it will come back.' Several participants from this group also stated that they did not kill spiders – they picked them up and put them outside – and did not like killing deer, referring to them as 'Bambis'. This reverence for life is consistent with their non-combatant stand.

The Catholic group was neither utilitarian nor co-equal. They fell into what could be defined as a stewardly view of nature, believing nature to be created by God, and humans should care for nature for the sake of both humans and nature, not just for human needs. Participants from the Catholic group expressed a sense of awe over the beauty of creation and the process of the creation of life. Some of the phrases that recurred were 'taking care of your surroundings', 'let nature help', 'looking at it from an environmental point of view'. Greeley's (1993, p. 27) explanation for this perspective is that Catholics are more likely to have gracious images of God, which in turn lead to more environmental concern.

Neither of these individualistic, but non-utilitarian, groups was easily categorized when it came to agricultural paradigms. The Quakers were conventional in all areas except their lack of support for farm specialization. They strongly supported the belief that problems can be solved by applying more and better technology, another conventional measure. The strongest alternative agriculture characteristic of the Quakers was their support for the idea that farmers need to encourage cooperative arrangements with consumers and other producers. The involvement of the Quakers in civic affairs was great.

The Catholic participants saw farming as a business and, in contrast to all other groups, did not think farming a more natural occupation than others. Beyond this, however, they showed little support for the conventional paradigm. They also were not clearly within the alternative agricultural paradigm, except for their support for farm diversification.

Communitarian

Three of the groups fell on the communitarian end of the individualistic–communal scale, the German Mennonites, Dutch Reformed, and the Reorganized Church of Latter-Day Saints (RLDS). However, each comes to this position with a unique perspective.

The essence of Anabaptism (Mennonite), a movement that began during the Reformation period, is the submission of individual goals to the collective system (Redekop, 1993, pp. 440–441). Mennonites have traditionally withdrawn from society to the security of the community where life could more easily be modelled after Christ (Nafziger, 1965, p. 188). It was not withdrawal from an evil world that so motivated the RLDS, but rather an effort to bring their vision of Zion, a perfect community, into being. The Dutch Reformed tend to form communities more in line with the RLDS. For Calvin, redemption meant bringing all things, secular and sacred, into proper order (Gingerich, 1985, p. 265). This means building a society where particular rules that govern the conduct of life can be obeyed literally, such as keeping the Sabbath (Bjorklund, 1964, p. 228). This makes them reluctant to leave their community and more willing to pay a considerably higher price for a farm (Van Den Ban, 1960, p. 314). Though communal in perspective, the Dutch Calvinists see the family as the basic unit for the development of the individual (Bjorklund, 1964, p. 230; Van Den Ban, 1960, p. 313).

The Mennonites, true to their long communal history, being less established-minded, and detached from the political–economic status quo (Rushby and Thrush, 1973, p. 23), did not show strong support for individual rights. However, the RLDS and the Dutch Reformed were the strongest supporters of individual rights outside of the German Reformed group. The RLDS have a long history of attempts to live up to their more communal ideal. Though many communistic schemes arose, especially in the nineteenth century, the appeal for private property has generally been too strong for them to succeed (Reneau, 1952, pp. 3–8; Romig and Siebert, 1988, p. 50). Perhaps their communal ideal came too late – in the American context with cultural Americans. The emphasis of the Dutch Reformed can be explained by Calvinistic thought that is supportive of private property and the accumulation of wealth (Gingerich, 1985, p. 249). Calvin insisted that God had not in fact condemned private ownership and was concerned with the preservation of human society, and private property was necessary for that preservation (Gingerich, 1985, pp. 266–267). So while concerned for the order within which the family exists, the Dutch Reformed remained strongly committed to private property and independence. This may also explain the strong commitment of the German Reformed to this position as well. The theological arguments reinforced the East Frisian familistic emphasis on the family farm.

Of the three communal groups, only the Mennonites expressed a utilitarian view of humans' relationship to nature. Although the German Mennonites saw the natural environment as basically for human use, they were less willing to put economic growth above the good of the environment than the German Reformed. An explanation for this perspective may come from the historic emphasis of Mennonites on simplicity of living (Redekop, 1993, pp. 440–441). Two Mennonite scholars confirm the possibility of this utilitarian viewpoint among Mennonites. Redekop (1986, p. 395) and Klassen (1995, pp. 6–7) both stated that there is no reference to the preservation of the earth in Mennonite theology, though practice has tended in that direction.

Both the Dutch Reformed and the RLDS traditions clearly emphasized stewardship – neither utilitarian nor co-equal – and that was evident in this study. They, along with the Catholics, reflected this perspective, talking about 'making animals comfortable', 'looking out for the animal's welfare', 'high regard for the fact that our animals were created by God'. One RLDS participant stated:

> As a whole, most people are coming to realize that even though we have ownership of land ... it's not completely ours to do with only as we want to. In the end it's God's, and we need to look after it the best we can.... It bothers me sometimes to have all these lines of things put into the earth. You have water lines, you have electricity lines ... in Des Moines where there used to be some

really nice fields, it's just paved over with concrete, and will never again see the light of day. I groan, I feel the earth groan; I groan with it, for being covered so . . . it stays all covered, and you know that it'll never be free again.

The RLDS statement of faith includes a strong emphasis on stewardship of resources (RLDS, 1970, para. 10). Humans are seen to suffer not because of a lack of resources but rather because humans fail to see the sacredness of God's creation and manage it for purposes beyond personal self-interest. God is considered the owner of all, while humans are the mere tenants or trustees of resources (RLDS, 1970, p. 154). Likewise, the Calvinistic farmer sees himself as the steward of God on the farm the Lord has given him. Thus the decision to adopt a new farm practice is sacred (Van Den Ban, 1960, p. 316).

All three communal groups were committed to the alternative agricultural paradigm, with the RLDS being the most so. Mennonites were less committed to farming as a business than any other group, and supportive of farming as a way of life and farm diversity. They were also the most supportive of domestic markets in comparison to other groups and were significantly different from several other groups in their scepticism of the idea that most problems can be solved by applying more and better technology. The Dutch Reformed were strongly supportive of family farms, but conventional when it came to farm specialization and world markets. They also strongly supported cooperative arrangements with consumers and other producers and were very leery of technology as a solution. Further reinforcing these findings was the fact that the Mennonite and RLDS groups were the only two groups that had individuals who subscribed to *The New Farm* magazine, an alternative agriculture publication.

Why the communitarian/alternative agriculture connection? What seems to be similar is the emphasis by each of these traditions on the commitment of one's whole life to their religious worldview – a non-dualistic perspective. But why not the German Reformed? German Reformed groups have been influenced less by Kuyperian thought, a strain of Calvinist thought that emphasizes all of life as religion, than Dutch Reformed groups. In addition, the German Reformed group of this study was part of the Reformed Church in American (RCA), while the Dutch Reformed group was part of the Christian Reformed Church (CRC), the more Calvinistic of the two. German Reformed groups, in comparison to Dutch Reformed groups, have traditionally not supported Christian day schools. In addition, those from the RCA similarly have not supported Christian schools, while the CRC has a strong emphasis on them. Thus ethnicity and religious affiliation separate the German Reformed group of this study and the Dutch Reformed group.

The RLDS emphasize the commitment of the whole person to the kind of life and mission as revealed in Christ – everything he/she does from vocation, to community life, to the way one conducts his/herself (RLDS, 1970, p. 157). Christianity, for Mennonites, has meant the idea of responsibility to God in one's life task with no compartmentalization of life into the sacred and secular (Nafziger, 1965, p. 193). For Calvin also, redemption meant not separating church and society but bringing all things into proper order, since society should approximate the order of God (Gingerich, 1985, p. 265).

These groups also have a common emphasis on good works. The RLDS traditions say that humans should 'work without ceasing to produce those conditions in society that bring liberty and freedom to men who are enslaved by ignorance, fear, lustful passions, and hatred' (RLDS, 1970, p. 157).

The Mennonite emphasis on discipleship has led to an emphasis on the place of works and conduct in the Christian life and the regulation of one's total conduct in line with the teachings and life of Christ (Nafziger, 1965, p. 188).

Lutheran

The two Lutheran groups did not clearly fall into the previous categories. The German Lutheran group was part of the Missouri Synod, which began with immigrants from Saxony who settled in Missouri in 1839, and has always emphasized orthodoxy of doctrine (Nelson, 1975, p. 179). The Swedish Lutheran group was part of the Augustana Evangelical Lutheran Church (the Swedish Lutheran denomination), which through several mergers is now part of the Evangelical Lutheran Church in America, the largest Lutheran denomination in the US (Nelson, 1975, p. 507).

The German Lutherans were not clearly individualistic or communal. In fact, it was the only group in which the men and women of the same group fell on opposite extremes – the men were communitarian and the women were individualistic. The Swedish Lutherans tended toward the individualistic end, similar to the Quakers. The closest they got to a communal perspective was the emphasis on their heritage being maintained. The German Lutherans were not especially supportive of individual rights, recognizing the role of government. The Swedish Lutherans, on the other hand, tended toward support of individual rights, but less so than German Reformed and Dutch Reformed.

In the United States one rarely thinks in terms of the rights of a community over the individual. In this study, the nearest exemplar of this perspective was the Swedish Lutheran group. They defended the rights of the community over corporations and outside developers. In addition, this community is known for its civic involvement. Again it raises the question, as with the Quakers, whether a communal outlook is different than cooperative action. Bellah and associates (1985, p. 206), has said as much, pointing out that even political activists can only conceive of 'community' as a voluntary association of individuals.

The Lutherans were also not easily categorized when it came to their relationship to nature. The German Lutheran responses to utilitarian measures were mixed, though there was little support for a co-equal position. They did express an exceptional sense of 'awe' over the beauty and wonder of nature. The comments included the feeling of 'reverence about putting a seed into the ground and watching it sprout and grow and taking care of it and having it produce ... love of plants and flowers ... sense of closeness to your Creator', and 'Doing our best to keep his creation beautiful and fruitful', and a sense that one is closer to God through seeing the process of the creation of life.

The Swedish Lutherans were also mixed in their support for the utilitarian perspective – with more support than the German Lutherans – and also did not support the co-equal position. Nature was viewed for human use, but soil conservation was also emphasized a great deal. The Swedish Lutherans found tremendous pleasure in taking an eroded field and terracing it to bring it back into productive use, thinking only of the long-term benefits and not the short-term costs.

German Lutherans were the most strongly conventional when it came to agriculture paradigms – farming as a business, farm specialization, and world markets. They were somewhat inconsistent, because they also strongly supported farming as a way of life, farm diversification and the idea that farming is a more natural occupation than others.

The Swedish Lutherans tended toward the conventional end, but were also mixed. They supported farming as a business, and world markets, but were not so strong in other measures of the conventional paradigm. They were significantly different from some other groups in their support of technology as an answer to many of today's problems, another measure of conventional farming. The only measure on which they were alternative was support for farm diversity.

Why such a mix among Lutherans? Their theology may give some insight. Luther stressed the powerlessness of humans and their incapacity for doing good by themselves because of an inherent evil nature (Kerston, 1970, p. 22). Lutheranism de-emphasizes works of merit for gaining grace but rather emphasizes salvation by faith alone. This tends to be individual and personal (Kerston, 1970, p. 23). Thus, argued Weber, the Lutheran ethic offered no real total ethical system but just the self-surrender in faith. In fact, Luther did not emphasize behaviour or morality, because it bordered on good works (Kerston, 1970, pp. 30–31). A study of Lutherans confirmed Weber's suspicions and found the Lutheran ethical system to be one of individual morality and piety, with a strong emphasis on religious individualism (Kerston, 1970, p. 31). Furthermore, this emphasis on faith alone leads to a view of social institutions as merely restraint on evil rather than building a new and just society (Kerston, 1970, p. 31). Robert Wuthnow (1981, p. 27) similarly concluded in his study of Lutherans that the Lutheran emphasis on faith alone mitigated the ability of the church to require strict consistency of belief.

The outcomes of these emphases were observed in a Wisconsin study by Van Den Ban (1960, p. 315). She found that though Lutheranism sees humans as stewards for God on this earth, the preaching in church did not stress this point as strongly as the Calvinists, especially the accountability for this stewardship. She was left with the impression that many Lutheran farmers saw no clear connection between their religion and their way of farming.

One commonality among the Lutherans was their expression of agrarian sentiments. Agrarianism, for the purpose of this study, was defined as including any of the following perspectives: (i) farming is the basic occupation on which all other economic pursuits depend; (ii) agricultural life is the natural life of humans – city life is artificial and evil; (iii) farming is a way of life that is based on the independence of the farmer – proud to be so; (iv) hard work is good – suspicious of using head and not hands; and (v) family farms are indissolubly connected with American democracy.

Only the Lutherans repeatedly expressed an agrarian ideology. Among the Swedish Lutherans (Stanton) the belief that the family farm was tied to the survival of democracy in the United States was expressed, whereas among the German Lutherans (Paullina) the emphasis was on farm life as more natural than urban life, with the latter leading to violence and unhappiness. This agrarianism may be explained in at least two ways. First, perhaps the lack of their own unifying religious worldview leaves room for a competing worldview. Secondly, Luther did emphasize a type of agrarian philosophy himself. He thought that Christianity was best served in the context of a self-sufficient agrarian community where individuals could be free to pursue their callings (Miller, 1993, p. 42).

Significance

The findings point to the existence of a range of community-level agricultural worldviews across the state of Iowa, arising from each community's religious and ethnic context. Each community may be unique in its response to economic change and policy change because of corresponding unique worldviews. This may in turn lead to the possibility of different resource management strategies and levels of responses to the philosophy of sustainable agriculture. On the other hand, individualistic policy may undermine the human cooperative resources and values of these 'communities of commitment'. Yet the acceptance of the values that underlie a

sustainable farming philosophy – cooperation, preservation of farm traditions and rural culture, fostering family farms, farming as a way of life,
permanence, aesthetic appreciation – may ultimately be more important to
the long-term viability of family farming and rural communities than the
mere adoption of specific sustainable practices (Lasley, Hoiberg, and
Bultena 1993, p. 137).

Acknowledgement

This project was funded by the Pew Charitable Trusts Evangelical Scholars
Program, The Center for Global and Regional Environmental Change at
the University of Iowa, The Leopold Center for Sustainable Agriculture at
Iowa State University, and the Anne U. White Fund of the Association of
American Geographers.

References

Bellah, R.N., Madsen, R., Sullivan, W.M., Swidler, A. and Tipton, S.M. (1985) *Habits
    of the Heart: Individualism and Commitment in American Life.* University of
    California Press, Berkeley, California.
Berkes, F. and Feeny, D. (1990) Paradigms lost: changing views on the use of
    common property resources. *Alternatives* 17, 48–55.
Beus, C.E. and Dunlap, R.E. (1990) Conventional versus alternative agriculture:
    The paradigmatic roots of the debate. *Rural Sociology* 55, 590–616.
Bjorklund, E.M. (1964) Ideology and culture exemplified in Southwestern Michi-
    gan. *Annals of the Association of American Geographers* 54, 227–241.
Borchert, J.R. (1987) *America's Northern Heartland: An Economic and Historical
    Geography of the Upper Midwest.* University of Minnesota Press, Minneapolis,
    Minnesota.
Corner, F.E. (1928) A non-mobile, cooperative type of community: a study of the
    descendants of an East Friesan Group. *University of Illinois Studies in the Social
    Sciences* 16, 15–80.
Cronon, W. (1983) *Changes in the Land: Indians, Colonists, and the Ecology of New
    England.* Hill and Wang, New York, New York.
Curry-Roper, J.M. and Bowles, J. (1991) Local factors in land tenure change
    patterns. *Geographical Review* 81, 443–456.
Dunlap, R.E. and Van Liere, K.D. (1984) Commitment to the dominant social
    paradigm and concern for environmental quality. *Social Science Quarterly* 65,
    1013–1028.
Egerstrom, L. (1994) *Make No Small Plans: A Cooperative Revival for Rural America.*
    Lone Oak Press, Rochester, Minnesota.
Flora, J.L. and Stitz, J.M. (1985) Ethnicity, persistence, and capitalization of
    agriculture in the Great Plains during the settlement period: wheat production
    and risk avoidance. *Rural Sociology* 50, 341–360.

Gilbert, J. and Akor, R. (1988) Increasing structural divergence in U.S. dairying: California and Wisconsin since 1950. *Rural Sociology* 55, 56–72.

Gingerich, B.N. (1985) Property and the gospel: two reformation perspectives. *Mennonite Quarterly Review* 59, 248–267.

Greeley, A. (1989) Protestant and catholic: is the analogical imagination extinct? *American Sociological Review* 54, 485–502.

Greeley, A. (1993) Religion and attitudes toward the environment. *Journal for the Scientific Study of Religion* 32, 19–28.

Hummon, D.M. (1990) *Common Places: Community Ideology and Identity in American Culture.* SUNY Press, Albany, New York.

Kerston, L.K. (1970) *The Lutheran Ethic: The Impact of Religion on Laymen and Clergy.* Wayne State University Press, Detroit.

Klassen, W. (1995) Pacifism, nonviolence, and the peaceful reign of god. Unpublished manuscript, presented at Creation Summit: Shaping an Anabaptist Theology for Living, 24–25 February, Camp Lake, Wisconsin.

Lasley, P., Hoiberg, E. and Bultena, G. (1993) Is sustainable agriculture an elixir for rural communities? *American Journal of Alternative Agriculture* 8, 133–139.

Miller, J. (1993) Missions, social change and resistance to authority: notes toward an understanding of the relative autonomy of religion. *Journal for the Scientific Study of Religion* 32, 29–50.

Mott, D.C. (1900) The Quakers in Iowa. *Annals of Iowa* 4, 263–276.

Nafziger, E.W. (1965) The Mennonite ethic in the Weberian framework. *Explorations in Entrepreneurial History* 2, 187–204.

Nelson, E.C. (ed.) (1975) *The Lutherans in North America.* Fortress Press, Philadelphia.

Parsons, K.H. and Waples, E.O. (1945) *Keeping the Farm in the Family.* Wisconsin Agricultural Experiment Station Bulletin 157.

Redekop, C. (1986) Toward a Mennonite theology and ethic of creation. *Mennonite Quarterly Review* 60, 387–403.

Redekop, C. (1993) The community of scholars and the essence of anabaptism. *The Mennonite Quarterly Review* 67, 429–450.

Reneau, J.C. (1952) A history of Lamoni, Iowa, 1879–1920. MA Thesis, University of Iowa.

Reorganized Church of Jesus Christ of Latter Day Saints (RLDS) (1970) *Exploring the Faith: A Series of Studies in the Faith of the Church Prepared by a Committee on Basic Beliefs.* Herald Publishing House, Independence, Missouri.

Romig, R.E. and Siebert, J.L. (1988) J.A. Koehler and the stewardship movement at Atherton. *Saints' Heritage: A Journal of the Restoration Trail Foundation*, 44–61.

Rushby, W.F. and Thrush, J.C. (1973) Mennonites and social compassion: the Rokeach hypothesis reconsidered. *Review of Religious Research* 15, 16–28.

Salamon, S. (1980) Ethnic differences in farm family land transfers. *Rural Sociology* 45, 290–308.

Salamon, S. (1985) Ethnic communities and the structure of agriculture. *Rural Sociology* 50, 323–340.

Salamon, S. and Davis-Brown, K. (1986) Middle-range farmers persisting through the agricultural crisis. *Rural Sociology* 51, 503–512.

Salamon, S. (1992) *Prairie Patrimony: Family, Farming, and Community in the Midwest.* University of North Carolina Press, Chapel Hill, North Carolina.

Schnucker, G. (1986) *The East Friesens in America: An Illustrated History of Their Colonies to the Present Time.* Translation of 1917 German edition. Translated by Kenneth De Wall. Jostens, Topeka, Kansas.

Van Den Ban, A.W. (1960) Locality group differences in the adoption of new farm practices. *Rural Sociology* 25, 309–320.

Winter, M. (1991) The sociology of religion and rural sociology. *Sociologia Ruralis* 31, 199–208.

Wuthnow, R. (1981) Two traditions in the study of religion. *Journal for the Scientific Study of Religion* 20, 16–32.

8

Rural Re-regulation and Institutional Sustainability: A Case Study of Alternative Farming Systems in England

GORDON CLARK[1], IAN BOWLER[2], ALASTAIR CROCKETT[3], BRIAN ILBERY[4], ALASTAIR SHAW[2]
[1]Department of Geography, Lancaster University, Lancaster LA1 4YB, UK, [2]Department of Geography, University of Leicester, Leicester LE1 7RH, UK, [3]Department of English Local History, University of Leicester, Leicester LE1 7QR, UK, [4]Department of Geography, University of Coventry, Priory Street, Coventry CV1 5FB, UK

Institutional Sustainability

Sustainable development is a complex concept that has been elaborated primarily in the context of natural resource use and the physical environment (Pearce *et al.*, 1990). In some cases a perceptual and cultural dimension has been included in the discussion of sustainability, notably in some studies of tourism and leisure as land-uses (e.g. Clark *et al.*, 1994). Sustainability can also be considered at a micro-scale in terms of how the viability of individual businesses can be maintained in the face of changing internal and external economic fortunes (Bowler *et al.*, 1996). This chapter extends the scope of the term 'sustainability' to encompass the institutional environment that surrounds the businesses and resources that have been the usual focus for sustainability discourses.

The institutional environment comprises the network of agencies, firms and organizations which, individually and/or collectively, sets the parameters for the freedom of action of individuals and firms. This network can react to global and national forces or it can initiate change. It can act as spokesperson for firms to help further their collective cause, or it can pursue its own agendas. Whatever role it takes, an institutional environment will be structured. It will comprise bodies established by law or common interest; these bodies will have evolved administrative and geographical 'territories'; they will operate among themselves in ways legitimized by law or custom. This structuring will provide a degree of stability both for the institutions and for their clients.

Yet observation of institutions, say in the field of rural development and agriculture, reveals that this environment is not static. Agencies are created, transformed or dissolved; some may become more powerful and some will lose influence; accepted ways of working may be changed. Some of these changes may be consensual – obvious reformulations needed to meet new circumstances. In part, however, they are symptomatic of how individual organizations jostle for power, each seeking by some means to extend its influence, expand its territory and get its agenda accepted by others. It is through this constant conflict that individual institutions seek to survive and the institutional environment sustains and renews itself.

To analyse this process of institutional sustainability, the authors studied how the rural institutional environment coped with the changes in agricultural policy that followed mounting food surpluses and the growth of policies to diversify farmers' incomes by encouraging 'alternative farming systems'. The case study that is reported in this chapter is ideal for our purposes as it concerns a potentially fundamental challenge to established bodies concerned with food production and offers other organizations (concerned with different rural activities) an opportunity to expand their influence. In this chapter we shall examine how the institutional environment reconstructed itself so as to sustain the regulatory environment for farmers in these new 'post-productivist' circumstances.

The Background to Policy Changes

The reforms of the Common Agricultural Policy (CAP) since 1992 and the Uruguay Round of the General Agreement on Tariffs and Trade (GATT) appear to herald a postproductivist era in agriculture in the United Kingdom (UK) in which lower subsidies induce a reduced and cheaper supply of food (CEC, 1993; Robinson and Ilbery, 1993). Yet this intended financial withdrawal has not been accompanied by any lessening of official or public interest in rural areas. Indeed, the need to cope with the potentially negative economic and social consequences of CAP reform has prompted the European Union (EU) and national governments to increase their efforts to find other sources of income for farming families and new uses for farmland (Commission of the European Communities, 1988; Lowe *et al.*, 1993). Additionally new stakeholders have arrived on the scene. Ex-urban social groups living in parts of the British countryside are concerned to protect their country lifestyles as a positional good and as a source of wealth through high houses prices. The environmental movement has also raised the countryside up the political agenda in terms of landscape, habitat, biodiversity and sustainable production. The state is having to respond to all these shifting pressures by changing how (and why) it regulates rural spaces now that food production is being de-emphasized as the dominant *raison d'être* for farmers and the countryside (Cloke and Goodwin, 1992; Marsden *et al.*, 1993). This chapter examines how the institutional environment is sustaining itself in these new economic and political circumstances nationally and at EU level.

The Study of Institutions and Regulation

The regulationist approach, reviewed in geography by Tickell and Peck (1992) and Clark (1992), directs attention towards the importance of the mode of social regulation (including centrally the state and its institutions) in influencing the regime of accumulation; these authors particularly note the approach's potential for modifying the persistent unevenness of capitalist accumulation and development. The state has a key role to play in influencing how different areas respond to common external pressures; this is as true of the local state and the development trajectory of a locality as it is of the guidance which the central state gives to the pathway followed by nations (Gertler, 1995). A regulationist approach focuses on how political action helps create places – in effect the politics of place identities and of their images and histories. The regulationist approach is therefore well suited to this piece of research.

   Yet the regulationist approach is still more of a hypothesis than an axiom and it involves several strands of thought. Nonetheless, what Jessop (1990) calls the 'intermediate concept' of the mode of social regulation (MSR) is a useful one because it invites us to study the ways in which economic systems interact with their social and political domains in order to keep a given accumulation regime stable over a period of time. When the economic system changes, the concept of the MSR prompts us to examine how (at various scales) the institutional environment alters, and is altered by, changes in the productive sphere and vice versa.

   The study of the mode of social regulation (MSR) has principally been developed at the global and national scales (Cloke *et al.* 1990; Le Heron, 1993; Le Heron and Roche, 1995). The nation state obviously provides an assemblage of political organizations, laws, social practices and economic organizations that can effectively create a distinctive national response to capitalist restructuring. Tickell and Peck (1992) note how limited have been the studies of MSR at the local level and how poorly developed is our current appreciation of the links between local and national changes in MSR – Bowler (1994), and Lowe *et al.* (1993) are notable exceptions to this dearth in the rural sphere. This research focuses on these issues in the context of local MSR using a case study of the restructuring of agriculture in northern England through the development of alternative farming systems.

Another complexity, again particularly marked in rural areas, is the richness of the institutional and policy environment. There are many levels of government from parish council through district and regional/county councils to national government and the European Union (EU) and World Trade Organization (WTO). Cutting across this spatial hierarchy are the sectoral agencies for agriculture, forestry, rural development and the environment, for example. Overlain on these are distinctive planning policies for the countryside focusing on how development is controlled and on the protection of landscapes and habitats. This complex multidimensional regulatory regime needs careful study to show which parts of it are the most powerful in any particular geographical and/or historical context.

The mode of social regulation (MSR) reflects the dominant political culture of a country; in agricultural terms that culture is changing rapidly in the UK. The apparatus of rural regulation has recently been subject to a series of external shocks imposed from outside the rural nexus of power. For example, a Conservative ideology of reduced state intervention has placed severe strain on a rural institutional environment constituted to intervene frequently. A free-market ideology, which is opposed to cross-subsidization and intent on making users pay true costs, threatens many subsidies to the rural system. The preference for private-sector bodies over public ones has seen a greater mixing of private and public funding, and a conversion of wholly public departments of state into agencies with some of the features of private-sector operation. A 'value-for-money' approach to evaluating public spending may favour urban investments over their rural counterparts. Another new factor is environmentalism which has prompted further changes in policies, priorities and institutional structures. Finally, farm surpluses and a greater desire for free trade have lessened the political power of a purely productivist agricultural policy. One of the objectives of the research reported here is to examine the two-way relationship between rural regulatory institutions and the need to promote alternative farming systems in an area of traditional upland agriculture. The institutions have the potential to alter how the new farm activities develop, and the task of promoting them may in its turn cause the institutional environment to evolve further. Periods of transition, such as the present one in the countryside, are particularly good for studying the social contestation of rural space between competing institutions and the links between the national and local scales in different places. It is at such times that the sustainability of individual institution's influences, as well as of the whole regulatory system, is under the greatest threat.

Aims and Context

This study builds on previous research on institutions and the state in rural geography (Bowler, 1979; Clark, 1982; Moran and Cocklin, 1989). It examines key features of the transformation of the local modes of social regulation in the context of farmers in the northern Pennines of England diversifying into alternative farming systems (AFS). How was the MSR and especially its institutional structure transformed? How effective can these local structures be in guiding national and international forces for change? How does the institutional environment sustain itself under new conditions? These general aims involve more specific objectives:

1. to determine which organizations influenced the adoption of alternative farming enterprises (AFEs) in order to sustain themselves as policy actors;
2. to describe how institutions in different localities reacted to and coped with their new task of promoting AFEs;
3. to examine the operation of any networks of institutions in terms of interinstitutional effects such as cooperation, conflict and competition;
4. to evaluate the strategies and influence of institutions in different areas on the diffusion of AFEs and to assess the institutions' ability to redress the unevenness of capitalist accumulation in the uplands;
5. to assess the interactions between farmers and institutions on AFE issues.

It is useful to contextualize this research on local MSR by reviewing briefly the state of British upland agriculture and the characteristics of the study area. Upland agriculture in the United Kingdom and throughout the EU has traditionally been supported by programmes to bolster farmers' incomes through special subsidies. The small-scale upland farms are at the start of the food chain and vulnerable to price fluctuations generated by large-scale purchasers. They are prone to poor weather conditions, while a combination of physical limitations and remoteness limits their potential for diversification. However, these subsidies have been inadequate to prevent the traditional responses to the economic fragility of upland farming such as farm amalgamation and out-migration. Consequently there has been renewed interest in off-farm employment (other gainful activities – OGAs) and in the recombination of resources on the farm for both new agricultural and non-agricultural enterprises (Gasson, 1988; MacKinnon *et al.*, 1991; Shucksmith and Smith, 1991).

This chapter focuses on the development of new on-farm enterprises by examining the ways in which farm incomes can be increased through a re-combination of factors (capital, labour or land) within the farm business. This leads to the creation of an alternative farming system (AFS), the word 'alternative' signifying here that the system contains elements that are not traditional in a given area. Some of these elements will be 'alternative farming enterprises' (AFEs), which are new on-farm activities such as new crops, on-farm retailing, woodland or low-intensity agriculture (Ilbery, 1988).

This chapter also needs to be seen in the context of the whole research project that examined at several scales both the social aspects of restructuring (using a model of farmer decision-making) and also its economic dimensions. It emphasized the dynamics of both the internal and external farm environments and the interactions between them. Both the economic and the social aspects can affect, and be affected by, the institutional context of the development of AFS. The results of other sections of the research have been reported elsewhere (Bowler *et al.*, 1996). The research on which this chapter is based was funded by the EU and carried out by geographers and agricultural economists in England, Scotland, Ireland, France and Greece. This chapter reports on only the English results – the international dimension will be explored elsewhere.

The area chosen for detailed study is the northern Pennines of England, which comprise the western parts of the counties of Northumberland and Durham. The former is largely in a National Park, the latter is an Area of Outstanding Natural Beauty and the whole is a Less Favoured Area. In 1993 the area became an Objective 5b region – an EU designation that indicates a rural area with a weak economy, which is therefore eligible to receive grants from the European Structural Funds. The northern Pennines are close to the Tyne and Wear conurbation with its population of 1.1 million and are an area of uplands, low population density and small towns and villages. There is a major forest and parts have been designated for environmental protection (Environmentally Sensitive Areas). Agriculture is dominated by upland farming; sheep and suckler beef cattle dominate the higher land while the lower eastern part has dairying, cattle finishing and some arable land. On the Northumberland uplands the farms tend to be large (up to 1000 ha) whereas in the Durham dales average farm sizes are between 60 and 80 ha.

The problems for this region include the low incomes of the farmers, their vulnerability to market fluctuations in the demand for sheep and beef cattle, and their high dependence on CAP subsidies. Some diversification of the local economy has occurred through forestry, military bases and tourism/leisure sites. However, substantial provision is also made for the protection of landscapes and environmental quality and, especially in the National Park, there is a presumption against inappropriate development. Grouse shooting is a major but contentious source of income for many of the large estates (Wilson, 1992). The study area therefore exhibits a traditional 'rural structured coherence' (Cloke and Goodwin, 1992, derived from Harvey, 1985) of landed interests, a social elite, upland farming, and in different localities, ancillary activities such as tourism, grouse shooting, forestry and, in the past, lead mining.

Bearing in mind the aims of the research, information was collected to determine the scale of the institutional environment surrounding the farmers involved in AFEs in the study area. In total 78 institutions were identified as directly affecting the development of AFEs in this area. A sample of 27 of them was selected for detailed interview. Two interviews were conducted with different departments of Tynedale and Weardale District Councils, bringing to 29 the number of interviews completed. The institutions were selected purposively to represent the main dimensions of the institutional environment and the main agencies which comprise the current and any future local MSR. All the main key agencies such as the Ministry of Agriculture were included and then a sample of the smaller membership-based ones was selected. Hence the sample includes national and local agencies, public and private ones, in agriculture and other sectors with AFEs, and covering the main functions of institutions towards AFEs such as research, grants, training and promotion.

Within each institution a key person (or key people) were identified and interviewed. The person was selected because their position in the organization was sufficiently senior to allow them a synoptic overview of its current activity but also one which still let them appreciate the institution's 'ground-level' activity. The interviewees had to be able to appreciate the context of the institution and how it worked with other bodies; but they also had to know in some detail how the institution interacted with farmers regarding AFEs. We were also seeking people whose length of service with the institution had provided them with an understanding of how its activities had changed.

The Institutional Environment

Diversity and functions

There is a very rich institutional environment that is relevant to farmers in the northern Pennines who wish to develop AFEs; 78 institutions were identified that could affect the adoption or progress of AFEs. It is also a diverse environment since the organizations impinge in so many ways on

AFE activity. Their functions ranged from finance and advice, through marketing and promotion, to regulation, representation and policy-making. They operated at local, regional and national scales; some at a single scale (the district councils, for example), while others covered several scales (a good example being ADAS, the farm advisory service).

Table 8.1 shows that there has been a considerable enrichment of the institutional environment during the last 25 years; 19 of the 27 organizations interviewed had been set up after 1970 and six since 1990, the latter particularly featuring tourism. At least one other, ADAS, had widened its remit to include tourism. This is one indicator of the institutional restructuring related to the shift to a post-productivist agriculture.

Many of the institutions exist in a formally defined hierarchy which combines, as one moves downwards in scale, decreasing spatial responsibility and a narrower range of functions (particularly being less concerned with policy and national representation). The bodies with the narrowest geographical remit (the North Pennines Tourism Partnership and the Durham Rural Dales Centre) are also the least autonomous since they are the creations of 'parent' bodies that finance, advise and guide them.

Few of the institutions are concerned exclusively with AFEs (three out of 27) or are primarily so (five more). For the majority of them AFEs are a minor part of what they do – a function added on to their existing ones. The institutions' functions are set out in Table 8.2.

Lobbying, advice, promotion and research are the major tasks, but the 29 interviewees claimed a total of 124 functions between them (an average of 4.3 each), usually two main functions each and one minor one. Hence it is useful to distinguish between institutional functions and the function of institutions; neither maps exactly onto the other. There are many classificatory criteria that could be used for this set of institutions – size, economic sector, societal sector (public or private) and functions. The composite classification employed here combines function and social sector, to give five broad types of organization namely:

Table 8.1. Date of establishment of AFE-relevant institutions.

Period	Institutions
Pre-1960	6
1960–1969	2
1970–1979	7
1980–1989	6
1990–1992	6

Table 8.2. The functions of AFE institutions.

Functions	Institutions
Lobbying	24
Advice	18
Promote region/product	17
Research	15
Animation/development	12
Regulation	10
Training	8
Grants/loans (admin./source)	8
Provide services/equipment	6
Sell products	3
Government policy input	3

1. those for tourism;
2. membership organizations;
3. those concerned with employment and training;
4. those concerned with advice, research and consultancy;
5. policy and regulatory organizations (either statutory or quangos).

This classification encompasses the national and local scales of regulation; it includes the formerly dominant agricultural organizations and those representing the emerging new sectors; and it covers the public and private sectors and agencies moving between these sectors.

The advisory bodies tend to cover a wide range of AFEs whereas the research and representational functions are focused on usually just one AFE. The training organizations and grant-awarding function are largely in the state sector, and the private membership organizations are mostly national in scale, dealing exclusively with AFEs and concentrating on advice, promotion and lobbying. The research function is found to some extent in many institutions but is the sole activity of just a few of them which concentrate on one or two AFEs. For some institutions involvement with AFEs was integral to their existence (the body had been set up by enthusiasts or by statute to promote an AFE); for others AFEs had been inserted into their original remit (e.g. using AFEs to pursue general tourism or regional development policies).

Institutional Coping Strategies

A second major question concerns the extent to which the institutions have been changed by the new task of promoting AFEs. How did they respond to this in terms of resource allocation? First, only 7 of the interviewees in the 27 institutions knew how many of their staff were involved with AFEs. The rest had none, or did not know, or did not allocate staff specifically to AFEs. Only ADAS (the farm advisory service) had a person specializing in AFEs in the study area, and for the other bodies the number of AFE staff (if any) was small and usually in a national centre far from the study area. Only three out of 29 institutions had increased their staff for AFEs and only three had increased their budget for AFEs. Nearly a third of the institutions had increased the amount of information on AFEs which they produced and 31 per cent now liaised more often with other institutions on AFE issues. Rural tourism is the only AFE that has become more heavily supported and been specifically incorporated into institutional objectives and plans. Apart from the AFE-specific membership organizations, only ADAS has provided training on AFEs for its staff and only two institutions have a formal investment or development plan specifically for AFEs apart from standard tourism. Two institutions had changed their names which reflected their less wholly agricultural remit. In summary, the effect of the AFE mission on institutions depends on how this is measured. Obvious indicators of institutional involvement (staff, training, budget and formal plans) show little response to AFEs, whereas information provision and interagency liaison have increased greatly. The institutions were reactive rather than proactive to the task apart from producing literature and mounting exhibitions.

A second key indicator of institutional involvement is the extent to which the interviewees could articulate an understanding of AFEs and farmers in the study area. Of all the interviewees 27% did not know why farmers should adopt AFEs; those who did, mostly quoted only one reason – raising farm incomes – 31% could think of no reasons why farmers should not or would not adopt AFEs; a further 32% blamed non-adoption on the inadequacies of the farmers (their traditional attitudes or lack of capital or marketing skills); and 17% could see no barriers to anyone adopting AFEs. None of the interviewees knew how many AFE farmers they had helped since the mid-1980s and only four could say how many they were involved with at the time of the interview (and this was no more than the membership organizations quoting their membership figure). Local knowledge – the advantages and disadvantages of the study area for AFEs – was also limited. The tourism organizations were the only ones which knew of a clear advantage which the northern Pennines possessed for their AFE (its landscape and heritage). Awareness of the area's disadvantages concentrated on the poor climate and remoteness but a quarter of the institutions knew of no reason why the area should not develop AFEs. Only three organizations ran workshops which could be a forum through which they might learn from farmers about real local AFE experiences. In summary, it is difficult to argue that the interviewees in the institutions were well informed about either farmers with AFEs, or about AFEs in the study area, or about how AFEs might interact with local farming activities. The understanding of AFEs and farmers exhibited by the key officials did not generally extend much beyond basic received wisdom.

The Influence of the Institutions

The mode of social regulation clearly implies that the effect of institutional action can be to either constrain or enable an activity or do both. The formal aspects of institutional functions that were noted above – advice, grants, lobbying – are all intended to be catalytic. However, our study of the institutions showed that they also acted in various ways to limit AFEs. Four negative aspects of institutional action were noted: physical planning; health, safety and environmental legislation; quality standards; and financial barriers.

The planning departments of the district councils not only promote rural development and enterprise but they also implement building regulations (whose requirements may raise the cost of AFE buildings); they operate landscape protection policies in green belts and National Parks; and they check that proposals are in conformity with approved local and structure plans. Planners' actions may hinder AFE adoption directly through the refusal of planning permission or they may indirectly impede it through delays, legal costs or management effort diverted from running the business. The health, safety and environmental barriers come from a wide range of public bodies and laws such as the Health and Safety at Work Act and the Food Safety Act, fire precautions and pollution controls. These controls have all increased considerably in the last ten years and tend to impede AFEs by raising capital and running costs and requiring more training. The third set of barriers focuses on quality standards and this affects activities as diverse as organic food certification, tourism (site and room standards) and livestock (breed purity). Membership, listing or grading may depend on a producer meeting ever-higher standards. This can raise the entry barriers to the sector and act against novices, the under-capitalized and those wanting to be low-cost producers competing on price. The final set of barriers stems from organizations charging for their services, such as membership and grading. Without this expenditure it may be more difficult to advertise in key publications (e.g. tourist board accommodation lists) or gain premium prices for certified organic or pedigree produce. This need to charge farmers for services has intensified recently as public funding has been restricted and state bodies have been reconstituted as semi-public 'Next Steps' agencies with new goals for business development and revenue generation. On balance the increase in the enabling (pro-AFE) activities of institutions has been matched by other actions which constrain the adoption of AFEs, giving only a modest net effect.

Perhaps the major success of the institutional response to AFEs has been to invest them with a legitimacy they formerly lacked in the eyes of farmers focused only on traditional food production. AFEs, now talked about warmly from local to EU level, have come to seem like a normal part of the farming scene (complete with their own grant schemes) rather than as the last resort of the agriculturally incompetent. This incorporation of AFEs into the mainstream is itself a powerful support for AFE development.

The Institutional Network

Another area of investigation was into the ways in which the numerous institutions interacted. We have noted already how one of the major responses by institutions to AFEs had been an increase in the amount of inter-agency liaison, this focuses attention on networking as an institutional strategy (see Bennett and Krebs, 1994 for a similar approach). Of the 27 institutions 17 had five or more contacts with other institutions regarding AFEs, 86% of the contacts being either frequent or infrequent but regular. The main reasons for the contacts were given as furthering joint ventures, general contacts, training, meeting statutory obligations and marketing or product development. This increase in networking had several benefits for the institutions. One was that they could share expertise: the Northumberland Training and Enterprise Council (TEC) gained experience of farming by liaising with the Agricultural Training Board (Landbase) when helping farmers to reach the BS5750 quality standard; the Northumbria Tourist Board and the RDC helped run the Northumberland Business Centre.

These links, which are based on knowledge, locality and skills, are paralleled by financial and planning contacts. District planning departments consult interested parties when developing land-use plans. The Durham Rural Dales Centre is funded by a series of other institutions. Networking is also a device for integrating policies for small areas using a 'one-stop-shop' approach. The North Pennines Tourism Partnership and the Durham Rural Dales Centre are multi-agency bodies that are designed to avoid inter-agency duplication, and coordinate information flows to farmers on alternative strategies and funding sources. Such local umbrella bodies can also help to integrate specialist skills giving a more rounded assessment by public bodies of individuals' needs. Most of the links involve public or semi-public bodies. The private sector is not yet involved to any degree. Networking was described to us by the institutions as helping to enhance common understanding, reduce conflict and integrate policies.

It is clear, however, that the networking is neither random nor uniform and that a core-and-periphery model best describes the working of the evolving institutional environment. The core was defined by two criteria: (i) by the number of contacts the institutions said they had with other institutions during their AFE activities; (ii) with reference to quantitative and qualitative evidence on the size of the organization; its access to, knowledge of, and standing with farmers; its resources and AFE expertise; its statutory position; and its powers to enable or curtail AFE development. This approach led to the identification of an inner core, an outer core and a periphery of organizations among the 29 that were interviewed (Table 8.3).

The core institutions include national organizations with proven agricultural expertise. Also present are the local authority planning departments that represent the current local state and nexus of political power. There is only one private sector organization (the Farm Holiday Bureau); the rest are all public bodies. It is noticeable that there are more Northumberland than Durham agencies in the core, which illustrates the higher level of support for AFEs there. For example, the Northumberland Economic Development Department helps with the promotion of a county brand-image of quality foods, and there are impressive joint ventures at the county business centre and business school.

Table 8.3. Core and periphery institutions for AFEs.

Inner Core	ADAS (Food, Farming, Land and Leisure)
	Rural Development Commission
	District Planning departments
	Northumbria Tourist Board
Outer core	Northumberland Training and Enterprise Council
	Agricultural Training Board (Landbase)
	North Pennines Tourism Partnership
	Farm Holiday Bureau
Periphery	The other institutions

Conclusions and Implications

This study of the development of alternative farming systems (AFS) in the northern Pennines has shown, at a basic level, that institutions have incorporated their new function of AFS promotion with the minimum of new policies and resource re-allocation (i.e. staff and budgets). They have concentrated either on cheap, controllable, internal measures such as disseminating more information or on rather modest changes such as altering the organization's name. A symptom of the latter was the sub-titling of ADAS as Food, Farming, Land and Leisure since ADAS now includes in its activities non-agricultural enterprises for farmers. Institutions were largely reactive in their behaviour rather than pro-active.

At a deeper level, the local assemblage of institutions has been changed in more subtle ways. There has been a major increase in networking and inter-agency liaison, which cannot be explained simply by reference to the less dominant position of the Ministry of Agriculture in UK rural affairs or by the external stimulus to collaboration provided by EU rural programmes such as LEADER. Networking, as a new way of operating, brings clear benefits to organizations in terms of sharing place- and sector-specific skills, resources and finance. It reduces conflict and helps harmonize institutional planning and thinking. The trend to more diverse economic activities in rural areas makes such collaboration all the more valuable now, given that organizations will find it increasingly difficult to stretch their budgets so that they can be omni-competent and operationally and intellectually self-sufficient in all the relevant aspects of rural development. Networking may also have the advantage of making the policy more effective by raising the quality or credibility of advice and increasing the policy's uptake.

However, networking particularly benefits core institutions that can consolidate their power by orchestrating others to their will, using for leverage their greater resources and, usually, their statutory position. This is clearly a role played by ADAS, the RDC and the district councils; but it is feasible for private bodies (such as the Farm Holiday Bureau) to edge their way into this core using their superior knowledge of a particular sector and their quasi-monopolistic access to key business tools such as listing, grading and advertising. The core institutions remain predominantly public ones, though with the agricultural interest less pre-eminent than before. The general inclusion of private and voluntary bodies (as envisaged in Local Agenda 21) has scarcely started. Networking may also have a role to play in the power politics played out among institutions to secure or expand their 'territory'. It is clear that the decline of agriculture (and the AFE response to this) has helped change how the rural economy is regulated by its institutional environment. The mode of social regulation is showing modest signs of evolving to cope with a new, less agricultural regime of rural accumulation.

At a more fundamental level, there is some evidence from this study of the emergence of a locally differentiated mode of social regulation. Umbrella organizations (such as the Durham Rural Dales Centre and the North Pennines Tourism Partnership) have been set up (in some places but not others) to link the branch offices of agencies and to combine national resources and skills with local knowledge in a manner that has credibility with local entrepreneurs. Further research is needed on the local conditions that govern where such umbrella organizations are established.

Clark (1992) has noted how institutions have some scope to shape how they operate and our work has found evidence of such variation within the study area. There were more examples in Northumberland than Durham of joint ventures and collaboration in county-wide initiatives. It is possible for a region to organize itself in such a way that it can hope to do a little better than the national average whatever the state of the UK economy. Northumberland seems to be constructing a local mode of social regulation that can reasonably aspire to achieve this. Paradoxically this can be viewed as creating another dimension of uneven development along new spatial lines as a consequence of efforts to reduce the traditional upland/lowland and urban/rural welfare gradients. It also marks a new local component in agricultural development which, since World War II, has been largely driven by national objectives and organizations.

A key question is whether these local responses are significant. There is a strong traditional argument for regarding them as minor. There have been relatively few shifts in real resources (staff and budgets) into AFE promotion, which is no doubt explicable by the increasingly constrained position of local government in the UK – it is becoming more limited both financially and legally. Alternatively there is a growing case for saying that this focus on real resources is an outdated view of regional development. Recent work by Saraceno (1994), Commission of the European Communities (1995), Storper (1995) and Herrschel (1995) has argued that regional growth differentials can no longer be explained by the volume of public investment in common goods (e.g. infrastructure), which the private sector will not supply adequately due to market failure. These authors have reverted to notions reminiscent of Marshall's (1919) explanation of industrial districts that posits that the key feature promoting industrial growth today is a *milieu innovateur*. By this they mean regional capacity building and the promotion of local synergy (rather than conflict) through networking, reducing risks, focused activity, and better coordinated planning and information flows. Linking such factors causally to enhanced competitiveness is likely to be even harder statistically than more traditional evaluations of regional policy. Yet if these ideas of enhancing interdependencies in order to amplify rather than stifle individual actions do prove to have value, then it could be argued that the responses we noted in the north Pennines are of great significance. Such measures may have been born of necessity (resource-intensive responses no longer being feasible for so small a sector) but they could prove to be the key method of enhancing local competitiveness. However, at the moment the scale of the response is more a case of coping locally with an externally sourced crisis of agricultural restructuring rather than the formation of a new rural local state or mode of social regulation.

The weakening of the nexus of, on the one hand, the post-war mode of social regulation for agriculture and rural areas and, on the other, the productivist modernization of farming, has required a different form of social regulation in order to achieve the new public goals for the countryside. This chapter has examined, principally at the under-researched local level, the institutional responses to this new challenge. It has shown: how local responses vary; how local initiatives can seek to alter consumption patterns; how networking and inter-agency liaison are the key structural changes in the institutional environment; how this can consolidate the powers of a few core institutions; and how these changes are largely confined to the current and former public sector. Collectively these changes amount to only a modest shift in the local mode of social regulation when viewed from a traditional regional policy stance, but they are of considerably more significance when seen from the perspective of newer ideas on how to reduce regional economic inequalities.

Increasingly institutional sustainability requires networking, cooperation and even formally constituted partnerships, either between the institutions themselves or through local proxy bodies established to provide a single interface with the farmers. Anything less could lead to a body being marginalized from its formerly dominant position, or failing to gain influence over policy in order to secure its objectives. It is clearly better to share power than to lose it or never to enjoy it. It is by these means that the institutional environment is re-regulating the farmers and sustaining itself.

Acknowledgement

This study arises out of a programme of collaborative research by the following: the Departments of Geography at the Universities of Leicester, Caen (France) and Trinity College (Dublin); the Scottish Agricultural College (Aberdeen); CEMAGREF (Clermont-Ferrand); TEAGASC (Dublin) and the Department of Agricultural Economics at the University of Patras (Greece). The research project was funded under the EU's CAMAR Programme.

References

Bennett, R.J. and Krebs, G. (1994) Local economic development partnerships: an analysis of policy networks in EC-LEDA local employment development strategies. *Regional Studies* 28, 119–140.

Bowler, I. (1979) *Government and Agriculture: A Spatial Perspective.* Longman, Harlow, 127pp.

Bowler, I. (1994) The institutional regulation of uneven development: the case of poultry production in the province of Ontario. *Transactions of the Institute of British Geographers* 19, 346–358.

Bowler, I., Clark, G., Crockett, A., Ilbery, B. and Shaw, A. (1996) The development of alternative farm enterprises: a study of family labour farms in the northern Pennines of England. *Journal of Rural Studies* 12, 285–295.

Clark, G. (1982) Institutions and rural development. In: Flowerdew, R. (ed.) *Institutions and Geographical Patterns.* Croom Helm, London, pp. 75–102.

Clark, G., Darrall, J., Grove-White, R., Macnaghten, P. and Urry, J. (1994) *Leisure Landscapes.* CPRE, London, 2 vols, 305pp.

Clark, G.L. (1992) 'Real' regulation: the administrative state. *Environment and Planning A* 24, 615–627.

Cloke, P., and Goodwin, M. (1992) Conceptualizing countryside change: from post-Fordism to rural structured coherence. *Transactions of the Institute of British Geographers* 17, 321–336.

Cloke, P., Le Heron R. and Roche, M. (1990) Towards a geography of political economy perspective on rural change; the example of New Zealand. *Geografiska Annaler* 72B, 13–25.

Commission of the European Communities (1988) *The Future of Rural Society,* COM(88)501. The Commission, Brussels.

Commission of the European Communities (1993) *Support for Farms in Mountain, Hill and Less Favoured Areas,* Green Europe 2/93. Office for Official Publications of the European Communities, Luxembourg.

Commission of the European Communities (1995) *Cohesion and the Development Challenge Facing the Lagging Regions,* Regional Development Studies 24. Office for Official Publications of the European Communities, Luxembourg.

Gasson, R. (1988) Farm diversification and rural development. *Journal of Agricultural Economics* 39, 175–182.

Gertler, M.S. (1995) Debating flexibility: past, present and future. *Scottish Geographical Magazine* 111, 63–64.

Harvey, D. (1985) *The Urbanisation of Capital.* Blackwell, Oxford, 239pp.

Herrschel, T. (1995) Local policy restructuring: a comparative assessment of policy in England and Germany. *Area* 27, 228–241.

Ilbery, B. (1988) Farm diversification and the restructuring of agriculture. *Outlook on Agriculture* 17, 35–39.

Jessop, B. (1990) Regulation theories in retrospect and prospect. *Economy and Society* 19, 153–216.

Le Heron, R. (1993) *Globalized Agriculture: Political Choice.* Pergamon, Oxford, 235pp.

Le Heron, R. and Roche, M. (1995) A 'fresh' place in food's space. *Area* 27, 23–33.

Lowe, P., Murdoch, J., Marsden, T., Munton, R. and Flynn, A. (1993) Regulating the new rural spaces: the uneven development of land. *Journal of Rural Studies* 9, 205–222.

MacKinnon, N., Bryden, J.M., Bell, C., Fuller, A.J. and Spearman, M. (1991) Pluriactivity, structural change and farm household vulnerability in western Europe. *Sociologia Ruralis* 31, 58–71.

Marsden, T., Murdoch, J. and Williams, S. (1992) Regulating agricultures in deregulating economies: emerging trends in the uneven development of agriculture. *Geoforum* 23, 333–345.

Marsden, T., Murdoch, J., Lowe, P., Munton, R. and Flynn, A. (1993) *Constructing the Countryside*, UCL Press, London, 220pp.

Marshall, A. (1919) *Industry and trade.* Macmillan, London, 875pp.

Moran, W. and Cocklin, C. (1989) The state and rural systems. *Journal of Rural Studies* 5, 313–324.

Pearce, D., Barbier, E. and Markandya, A. (1990) *Sustainable development.* Earthscan, London, 217pp.

Pomeroy, A. (1995) Matching New Zealand rural development policy to a changing clientele: the emerging contribution of MAF. *New Zealand Geographer* 51, 49–56.

Robinson, G.M. and Ilbery, B. (1993) Reforming the CAP: beyond MacSharry. In: Gilg, A.W. (ed.) *Progress in Rural Policy and Planning*, Volume 3. Belhaven, London, 246pp.

Saraceno, E. (1994) Recent trends in rural development and their conceptualisation. *Journal of Rural Studies* 10, 321–330.

Shucksmith, M. and Smith, R. (1991) Farm household strategies and pluriactivity in upland Scotland. *Journal of Agricultural Economics* 42, 340–353.

Storper, M. (1995) The resurgence of regional economies. *European Urban and Regional Studies* 2, 191–221.

Tickell, A. and Peck, J. (1992) Accumulation, regulation and the geographies of post-Fordism: missing links in regulationist research. *Progress in Human Geography* 16, 190–218.

Wilson, O. (1992) Landownership and rural development in the North Pennines: a case study. *Journal of Rural Studies* 8, 145–158.

9

On and Off-farm Business Diversification by Farm Households in England

Brian Ilbery[1]; Michael Healey[2] and Julie Higginbottom[1]
[1]Department of Geography, University of Coventry, Priory Street, Coventry CV15FB, UK; [2]Department of Geography and Geology, Cheltenham and Gloucester College of Higher Education, Francis Close Hall, Swindon Road, Cheltenham GL50 4AZ, UK

Agricultural Restructuring

There can be little doubt that agriculture in most developed market economies has entered a post-productivist period (Marsden *et al.*, 1992; Bowler and Ilbery, 1993; Shucksmith, 1993). Stimulated by overproduction and the environmental disbenefits of productivist agriculture, the transition to postproductivism has, at its core, the concept of a 'sustainable' agriculture. Although difficult to define precisely, Brklacich *et al.* (1990) argue that a sustainable agriculture provides simultaneously environmental sustainability, socioeconomic sustainability and productive sustainability. Key attributes of a sustainable agriculture thus include the integration of crop and livestock farming, low energy inputs, and an extensified and diversified agriculture. It is the latter that is the specific concern of this chapter.

The postproductivist transition has been encouraged by a rapidly changing policy situation since the mid-1980s and increasing regulation by the state, especially following the reforms of the Common Agricultural Policy (CAP) in 1992 and the GATT agreement on world agricultural trade in 1993. This in turn has led to an inevitable adjustment of farming to market realities and many farm households are having to restructure their businesses, often in order to survive (Marsden *et al.*, 1989; Ilbery, 1991). A number of restructuring strategies, or '*pathways of farm business development*' are available to farm households (Bowler, 1992; Ilbery and Bowler, 1993). The six different pathways (Box 9.1) can be summarized into three main strategies:

Box 9.1. Pathways of farm business development.

```
I.    Continuation of agricultural production
      1. Extension of industrial model of farming: traditional products
      2. Redeployment of resources into new agricultural products

II.   Diversification of the income base
      3. Redeployment of resources into new non-agricultural products
      4. Redeployment of resources into off-farm OGAs

III.  Marginalization of farming
      5. Traditional farm production with lower income/inputs
      6. Hobby or part-time (semi-retired) farming
```

Source: Based on Bowler (1992).

1. maintaining a full-time, profitable food production element to the farm business (paths 1 and 2);
2. diversifying the income base of the farm business by redirecting resources into non-farm enterprises and occupations (paths 3 and 4);
3. surviving as a marginalized farm business at a low level of income, perhaps supported by investment income, pensions and other direct state payments (paths 5 and 6).

It is possible for farm households to make the transition from pathway 1 through to pathway 6. However, once marginalized it is highly unlikely for a farm business to 'move back' towards pathway 1. Significantly, different localities may be dominated by particular pathways: for example, pathways 1 and 2 in the prosperous agricultural lowlands; pathways 3 and 4 in urban fringe areas; and pathways 3 and 5 in marginal upland areas.

Although more research is needed on the spatial dimensions of agricultural restructuring and adjustment, much has been published on pluriactivity (pathways 3 and 4). Initially, research focused on aspects of either pathway 3 (farm-based diversification) or pathway 4 (other gainful activities (OGAs) off the farm) (See Ilbery, 1988 and 1991 for the former and Gasson, 1986 and 1987 for the latter.) More recently, emphasis has shifted towards examining both on-farm and off-farm 'alternative enterprises' together under the umbrella term pluriactivity (Shucksmith *et al.*, 1989; Fuller, 1990; MacKinnon *et al.*, 1991). Some of the most recent research on pluriactivity has emphasized the importance of socioeconomic and locational influences, especially in the context of marginal upland areas (Shucksmith and Smith, 1991; Edmond *et al.*, 1993; Bateman and Ray, 1994; Edmond and Crabtree, 1994).

Research on pluriactivity has tended to consider all kinds of non-agricultural income-generating activities, either on or off the farm, by all working members of the farm household. It has not distinguished between household members working as paid employees for other people and those who set up and run their own non-agricultural businesses on and/or off the farm. Yet it has been estimated that of the pluriactive farm households in England and Wales, about one half are involved in some form of *business diversification*, either on and/or off the farm (Gasson, 1987). Consequently, this chapter reports on the early stages of a research project on agricultural restructuring and business diversification by farm households in three contrasting areas of England. The aims of the whole research project are to:

1. establish the main characteristics and relative importance of different types of business diversification for farm households;
2. compare the opportunities and constraints for business diversification in three contrasting types of agricultural area;
3. examine the factors both external and internal to the farm business that are influencing the development of business diversification by farm households;
4. explore the impacts of business diversification on the local economy;
5. assess the policy implications of the continued development and uneven uptake of business diversification by farm households.

This particular chapter will focus on some of the conceptual issues surrounding the study of business diversification and report on the initial findings of an extensive survey of nearly 3000 farm businesses in England (aim 1).

Conceptual Issues

Two main conceptual issues need to be outlined briefly here: first, the definition of business diversification by farm households; and secondly, the development of a framework for examining the factors affecting the nature and types of business diversification adopted.

What is business diversification?

Business diversification fits within the wider concept of pluriactivity (Fig. 9.1). Whereas pluriactivity includes all forms of non-agricultural income generation on and off the farm, either as an employer or an employee, business diversification relates specifically to one or both of the following

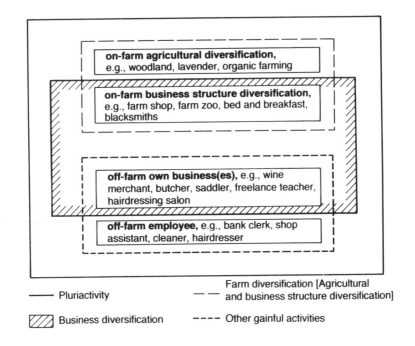

Fig. 9.1. Business diversification and pluriactivity.

activities established and run by any member of the farm household who is also involved in agricultural work on the farm:

1. on-farm, non-agricultural, businesses such as a craft shop or a blacksmith;
2. off-farm businesses such as a saddler, wine merchant or hairdresser.

Thus business diversification incorporates entrepreneurial elements of on-farm diversification and off-farm OGAs. However, in some cases there may be a close relationship between agriculture and the business venture. For example, non-agricultural businesses may, in some way, be involved in either the sale of agricultural produce on the farm (as in a farm shop or PYO scheme) or the use of farm resources off the farm (as in contracting). Although related to agriculture, they are viewed as non-traditional activities and so can be included as types of business diversification.

It needs to be stressed that in this study only those members of the farm household who *work* on the farm are included in this definition of business diversification. This is much more specific than and in contrast to past studies of pluriactivity that have included all members of the farm household who reside on the farm but who do not necessarily work on the farm, i.e. members who have nothing to do with farming but who are coincidentally resident on the farm. Not surprisingly, these studies suggest highly inflated numbers of pluriactive farm households in Britain (Bateman and Ray, 1994; Edmond and Crabtree, 1994).

What factors affect business diversification?

Conceptually, business diversification can be viewed as the outcome of a range of factors working 'externally' and 'internally' to the farm household. The farm family, as an enterprise and a household (Moran *et al.*, 1993), is the main focus of attention, but this is set within the opportunities and constraints created by external structures. Such an approach has been described as 'action in context', where the actions of farm household members are examined within the context of wider (external) structures (Marsden *et al.*, 1992; Ward and Lowe, 1994).

External factors operate at three main levels (Fig. 9.2). The first level constitutes the macro-scale processes recognized in political economy approaches to agricultural change (Marsden *et al.*, 1986). These form an important context within which decisions are made by both external 'agencies' and farm households, but they are not the main focus of attention in this research project. The second level comprises a series of local, regional and national institutions that provide information, advice and finance to farm households. This can take the form of farmers actively seeking such information and advice (demand-led); alternatively, the institutions can be more pro-active and target particular groups of farmers (supply-led). Either way, these *formal* institutions may influence the development of business diversification in different localities and so form an important 'external' focus for the research. The final level relates to a more *informal* external information environment, comprising the exchange of ideas about agriculture and business diversification through, for example, National Farmers' Union meetings, Farm Study Groups, local committees, the media, neighbouring farmers and non-farming family members (Fig. 9.2). Although important, this informal environment has received little attention in the pluriactivity debate.

The farm and farm household creates a set of internal factors that help to determine the type and nature of business diversification, if any, undertaken by the farm business. A contingent relationship may exist between the size, type and tenure of farms and different forms of business diversification. Similarly, research has shown that different farm household members are involved in pluriactivity (and thus almost certainly in business diversification). Consequently, gender relations within the farm family, stages in the life cycle, succession, and educational and occupation experiences of all household members need to be examined as they may

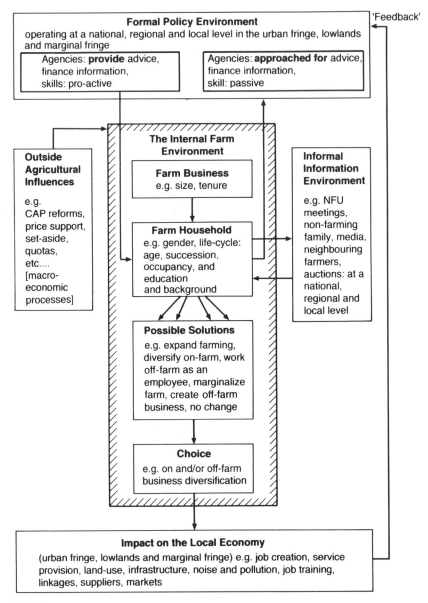

Fig. 9.2. Internal and external factors affecting business diversification: a conceptual framework.

help to provide *indirect* explanations for the development of business diversification. Within the internal environment of the farm household, the conceptual framework allows for more *direct* explanations of business diversification through a search-based model of decision-making (Fig. 9.2). Such a model has been used in a recently conducted examination of alternative farm enterprises in the lagging regions of the European Union (Bowler *et al.*, 1995) and consists of three stages:

1. The perceived *stimulus* to make a decision. This occurs when the *stress tolerance threshold* is exceeded, possibly reflecting the problem of falling farm income and the need to respond in order to 'survive'. Alternatively, the stimulus could be the opportunity to 'accumulate' more capital.
2. The *search* for a satisfactory solution to the problem, necessitating the farm household to interact with elements of the external environment.
3. The *choice* of a final solution from the different pathways of farm business development.

If there is no stimulus to change, the search-based model of decision-making is not activated and there is no change in the behaviour of the farm household. However, once activated there will almost certainly be interaction between 'internal' and 'external' factors. The level of interaction will vary greatly, depending on the type of business and aspirations of household members. A full understanding of such interactions and thus the processes affecting the development of business diversification can best be obtained through in-depth, mainly open-ended *intensive* surveys with a limited number of farm households. The choice of such farm businesses is dependent upon the first main stage of empirical analysis – an *extensive* survey of farm businesses in the three contrasting localities of England.

An Extensive Questionnaire Survey

As a first stage in the research methodology, an extensive survey of approximately 3000 farm businesses was undertaken in 1994. The main objective of such a large survey was to reveal the extent and nature of business diversification in each of three types of study area:

1. the West Midlands urban fringe, in a zone up to 20 km from the built-up area of Birmingham and Coventry;
2. the North Pennines upland marginal fringe (designated as a Less Favoured Area (LFA) for agriculture by the European Union);
3. the prosperous agricultural heartland of Oxfordshire and surrounding lowland areas (Fig. 9.3).

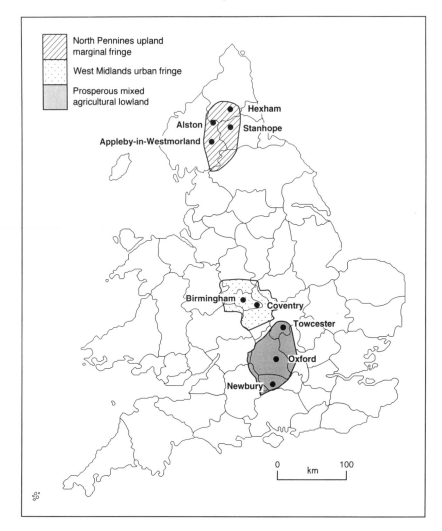

Fig. 9.3. Location of the three study areas in England.

Few comparative surveys of this type have been conducted in England and the incidence of business diversification in different types of area has not been researched before. Yet, different types of opportunity and constraint exist for diversification in different areas, reflecting specific local historical and geographical conditions (Ilbery, 1991). Thus one would not expect all urban fringe areas, for example, to offer the same opportunities and constraints for business diversification as the West Midlands urban fringe.

Rather than using the traditional *Yellow Pages* telephone directories, with their bias towards larger farm businesses (Errington, 1985), as a sampling frame, the *British Telecommunications Business Database* was consulted. This is updated every month and can list farmers (code 3EC0011) by Postcode Area and Postcode District, with the latter being very useful in delimiting the outer boundaries of each study area. Consequently, contiguous postcode districts were selected until the required number of farm businesses (around 1000) had been selected for each area. The lists were purchased on magnetic tape from British Telecommunications and transferred onto two sets of self-adhesive labels ready for the questionnaire survey (and reminder letter). No attempt was made to sample farms from the lists; instead, all farmers were sent a questionnaire.

The questionnaire itself was short (two sides) and structured around three sections: farm characteristics, farm household characteristics, and business diversification. The final response rate was 42.2%, which is quite acceptable for a postal survey of this size; the rates varied from 39.6% in the West Midlands urban fringe to 44.1% in the North Pennines upland marginal fringe. A considerable amount of factual, mainly quantitative information was obtained, which formed the basis of a statistical database (SPSS for Windows). Each farm in the database was classified according to its location (urban fringe, marginal fringe and lowlands), whether or not it had 'adopted' business diversification and, if it had, the main *categories* of business diversification practised (on-farm, off-farm, or on- and off-farm). The different *types* of on-farm and off-farm business diversification were also classified (e.g. accommodation, recreation and leisure, retail, manufacture, services). Using the database, it was then possible to differentiate the adopters and non-adopters of business diversification by the different types of geographical area, farm size, farm type, family composition and so on. Such a methodology permits the search for general trends in the data, but it does not provide direct explanations of them; this must come from more intensive research methodologies.

Extent and Nature of Business Diversification

Survey data indicate that nearly three out of every ten farm households (28%) have some form of business diversification. This is dominated by on-farm businesses, which account for more than the combined total of farms with off-farm and both on- and off-farm businesses (Table 9.1). These results contrast with most studies on pluriactivity, which emphasise the dominance of off-farm OGAs, and suggest that those farm households working off-farm tend to do so as employees rather than as employers. When the results are disaggregated by area, the agricultural heartland of Oxfordshire has the highest incidence of business diversification (33%), followed by the West Midlands urban fringe (29%) and the North Pennines

Table 9.1. The extent of business diversification in the three study areas.

Category	West Midlands		Lowlands		Pennines		Total
	No.	%	No.	%	No.	%	%
Non-adopters	289	71	281	67	329	77	72
Adopters	118	29	140	33	99	23	28
Of which:							
On-farm only	67	17	81	19	58	14	16
Off-farm only	38	9	39	9	29	7	8
Both types	13	3	20	5	12	3	4

Source: Authors' survey.

upland marginal fringe (23%). Yet, the balance between on- and off-farm businesses is remarkably similar in the three regions, with between 41% and 43% of farms with business diversification in each area having just on-farm activities.

The 357 farms with business diversification have a combined total of 512 business enterprises (1.4 per farm household); of these, 347 (68%) occur on-farm. Significantly, a majority of farm households have just one on-farm enterprise (73%) and only 8% have three or more. *On-farm businesses* are dominated by accommodation enterprises (42%), followed by retailing (21%), services (17.5%) and recreation and leisure (13%) (Table 9.2). Retail activities consist mainly of farm gate sales of produce, whereas services are very wide ranging, varying from office and legal services to consultants and livestock services. Some notable differences occur between

Table 9.2. Types of on-farm business diversification.

Type	West Midlands		Lowlands		Pennines		Total	
	No.	%	No.	%	No.	%	No.	%
Accommodation	34	29	47	32	64	78	145	42
Recreation/leisure	18	15	23	16	3	4	44	13
Retail	35	30	30	20	7	9	72	21
Transport/const.	8	7	10	7	3	4	21	6
Manufacturing	2	2	2	1	0	0	4	1
Services	20	17	36	24	5	6	61	18

Source: Authors' survey.

the three areas (Table 9.2). The upland marginal fringe is almost completely dominated by accommodation; indeed, it is the only significant on-farm enterprise and very few farm households here have a second type of on-farm diversification. This contrasts with the wider spread in the lowlands, where accommodation is much better complemented by services, retailing and recreation. Not surprisingly, retailing surpasses accommodation enterprises on the urban fringe and confirms the findings of an earlier study in this area (Ilbery, 1991).

In terms of *off-farm businesses*, most farm households have just one enterprise (89%). These are dominated by services (31.5%), agriculturally related enterprises, especially contracting (27%), and retailing (18%). Again, the services are wide-ranging, from general and office services to personal and education services. Opportunities in the upland marginal fringes are limited and restricted to either agriculturally related enterprises or retailing. Services are of less significance here, which is in marked contrast to their importance in both the lowland and urban fringe areas (Table 9.3). Retailing and agriculturally related enterprises are the second and third types of off-farm business diversification in the lowlands of Oxfordshire, whereas this rank order is reversed for the West Midlands urban fringe.

Quite a clear distinction appears between on- and off-farm enterprises in terms of those household members running the businesses. On-farm businesses have a very strong female presence and 40% of the enterprises are run by farmers' wives, with a further 17% by a husband and wife partnership. The female dominance is most noticeable with accommodation enterprises, where two-thirds are run by farmers' wives. Similar findings were reported by Evans and Ilbery (1992 and 1996) in their national survey of farm-based accommodation. The wife is also the

Table 9.3. Types of off-farm business diversification.

Type	West Midlands		Lowlands		Pennines		Total	
	No.	%	No.	%	No.	%	No.	%
Accommodation	3	6	4	6	2	4	9	5
Recreation/leisure	3	6	0	0	3	6	6	4
Retail	7	13	14	22	9	20	30	18
Transport/const.	4	7	7	11	6	13	17	10
Manufacturing	2	4	3	5	2	4	7	4
Agriculture related	17	31	11	17	16	35	44	27
Services	18	33	26	40	8	17	52	32

Source: Authors' survey.

Table 9.4. Farm size and business diversification.

Category (%)	Farm size (ha)						
	<40	40–79	80–119	120–159	160–199	200–399	>400
Non-adopters	16	24	18	11	7	16	8
Adopters	19	24	15	12	9	12	10
Of which:							
On-farm only	18	26	16	13	11	10	7
Off-farm only	21	24	11	10	9	12	13
Both types	18	15	18	9	7	18	13

Source: Authors' survey.

dominant entrepreneur with off-farm retailing, but joint husband and wife partnerships, the farmer, and son(s) become more important with the third and fourth most frequently occurring on-farm enterprises – services and recreation/leisure activities. Off-farm businesses tend to be run by a more balanced cross-section of household members: 29% by a son, 25% by the wife, and 18% by the farmer. However, this varies slightly according to the type of off-farm business, with the wife dominating services and the son dominating agriculturally related activities, especially contracting.

Farm and Household Characteristics

Some initial indirect explanations of the development of business diversification in the three areas can be gained by looking at certain characteristics that distinguish 'adopters' from 'non-adopters'. Although the extensive survey yielded large amounts of data, just four features of the *internal farm environment* will be examined in this paper: occupancy, farm size, farm type and household composition.

In terms of occupancy, there is little to differentiate the overall sample of adopters from non-adopters. However, there appears to be some association between business diversification and farms of mixed tenure; whereas 27% of the total sample are of that occupancy type, this rises to 30% for those with business diversification and to 32% for those with just off-farm enterprises. In contrast, farm households with both on and off-farm business diversification are over-represented in the owner-occupancy category and extremely underrepresented in the wholly tenanted category. Thus there is a tendency for owner-occupiers to experiment in both on and off-farm types of business diversification, whereas those farm households that rent some of their land are more likely to diversify off-farm. Only 9%

of farms with both on and off-farm business diversification are tenant farms, even though tenanted farms account for nearly 23% of the total sample of farms. Clearly, there are problems associated with the development of business diversification on tenant farms (Ilbery, 1991).

Business diversification tends to favour farms in three different size categories: small (< 40 ha), medium (120–200 ha) and very large (> 400 ha). Clear distinctions are apparent between on-farm and off-farm enterprises, with a tendency for off-farm business diversification to be concentrated on both very small and very large farms and on-farm diversification favouring the medium-sized farms (Table 9.4). This provides further evidence of the U-shaped distribution in the size of part-time farms (Buttel, 1982; Gasson, 1986) and confirms the tendency for on-farm diversification to be biased towards the medium and larger-sized family farms (Ilbery, 1991). Further interesting contrasts emerge when the data are disaggregated according to the type of on-farm business diversification. For example, accommodation enterprises are underrepresented on farms of both under 80 ha and over 400 ha, whereas retailing favours the smaller farms; indeed, 26% of the on-farm retailing enterprises are found on farms of less than 40 ha (only 18% of the entire sample of 1256 farms are smaller than 40 ha). Services follow the general distribution of farm sizes, but leisure and recreation enterprises favour the larger farms (> 400 ha) in particular.

In terms of off-farm business diversification, retailing this time is biased towards both small and very large farms; services are also well over-represented on the largest farms. Yet different again are the agriculturally related activities that tend to be run by households with medium-sized family farms; they are underrepresented on both small (< 40 ha) and very large (> 400) farms. Already, therefore, some reasonably clear trends are emerging between business diversification and farm size. Whereas the medium-sized farms are more likely to diversify into accommodation on-farm and possibly support this with an agriculturally related off-farm activity such as contracting, both the small and large farms are pursuing different business diversification types. Small farms (< 40 ha) favour on- and off-farm retailing, although the latter is also associated with the very large farms, which show a preference for both on-farm recreation and leisure enterprises and off-farm services. Further and more in-depth research is required to explain this link between business diversification and farm size.

Turning to farm type, the results indicate a tendency for business diversification to be associated with farms having cash cropping, and pig and poultry enterprises, and to avoid farms with dairy and beef cattle. In terms of the different categories of business diversification, off-farm activities are over-represented on farms with pigs and on-farm activities, especially retailing, are over-represented on farms with poultry. In contrast, farms with both on and off-farm diversification are biased towards cash cropping. Such broad-scale differences camouflage other relationships within the dataset. For example, whereas adopters of business diversification tend to be underrepresented on farms with beef and dairy cattle, on-farm accommodation enterprises are very much associated with farms that have beef and sheep but not pigs, poultry and cash cropping. Yet on-farm retailing is, not surprisingly, biased towards farms with pigs, poultry and dairy cattle, reflecting the direct marketing of fresh produce to the public. Finally, both on-farm services and recreation are concentrated on farms with cash cropping, but whereas the former also favour farms with dairy cattle, the latter definitely do not. Contrasts also exist in terms of off-farm activities. Whereas off-farm retailing and services are overrepresented on farms with cash cropping and tend to avoid farms with beef, sheep and dairy cattle, agriculturally related enterprises are the opposite and favour farms with beef and dairy cattle at the expense of farms with cash cropping.

Finally, if one looks at the composition of the farm household there appears to be a relationship between business diversification and stages in the family life cycle (Gasson and Potter, 1988). Table 9.5 indicates that, when compared with the overall distribution for the whole sample of farms, adopters of business diversification tend to be couples with children of all ages. Non-adopters are over represented in the 'single' and 'couple with no children' categories. When the adopters are further disaggregated according to the three categories of business diversification, other characteristic features emerge. For example, couples where the eldest child is under 11 are overrepresented on farms that have both on- and off-farm business diversification. In contrast, couples where the eldest child is between 11 and

Table 9.5. Household composition and business diversification.

Category (%)	Single	Couple 1	Couple 2	Couple 3	Couple 4	Single 5
Non-adopters	11	8	11	8	57	5
Adopters	4	5	14	12	61	3
Of which:						
On-farm only	4	5	12	11	64	3
Off-farm only	4	7	15	14	57	3
Both types	2	4	22	11	56	3

1, no children;
2, eldest child under 11;
3, eldest child 11–16;
4, eldest child over 16;
5, children of any age.
Source: Authors' survey.

16 seem to favour off-farm diversification, just as couples where the eldest child is over 16 are biased towards on-farm forms of business diversification. These relationships are quite strong and suggest that when the children are young, there is a need to seek alternative sources of income from wherever possible (on- and off-farm). However, when the children are at secondary school and require less immediate attention, the opportunity exists to develop off-farm enterprises. Yet further into the family life cycle, there is a desire to utilize any surplus labour resources, including children over 16, in on-farm enterprises. Again, further intensive research is required to provide explanations for these trends.

Conclusions

This chapter has conceptualized the development of business diversification by farm households within the wider context of 'pathways of farm business development' and has presented some preliminary findings from an extensive survey of just over 1250 farm businesses in three different areas of England. Variations in the extent and nature of business diversification have been identified between lowland, urban fringe and marginal upland areas, and certain farm and household characteristics are seen to differentiate 'adopters' from 'non-adopters'. These help to provide some initial indirect explanations of the development of business diversification by farm households. However, there are many exceptions to the general trends and the main processes at work have not been identified. This necessitates intensive research methodologies and qualitative analyses of decision-making processes in both the 'internal' environment of selected farm households and the 'external' environment of relevant institutions. These will help to further explain the geography of business diversification. It is already clear, however, that business diversification is an important component of the postproductivist transition and the move towards a more sustainable agriculture in developed market economies.

References

Bateman, D. and Ray, C. (1994) Farm pluriactivity and farm policy: some evidence from Wales. *Journal of Rural Studies* 10, 1–13.
Bowler, I.R. (1992) Sustainable agriculture as an alternative path of farm business development. In: Bowler, I, Bryant, C. and Nellis, D. (eds) *Contemporary Rural Systems in Transition: Agriculture and Environment.* CAB International, Wallingford, pp. 237–253.
Bowler, I.R. and Ilbery, B.W. (1993) Sustaining agriculture in the food supply system. In: Nellis, D. (ed.) *Geographic Perspectives on the Social and Economic*

Restructuring of Rural Areas. Kansas State University Press, Manhattan, Kansas, pp. 4–13.

Bowler, I.R., Clark, G. and Ilbery, B.W. (1995) Sustaining farm business in the less favoured areas of the European Union. In: Sotte, F. (ed.) *The Regional Dimension in Agricultural Economics and Policies.* University of Ancona Press, Ancona, pp. 109–120.

Brklacich, M., Bryant, C. and Smit, B. (1990) Review and appraisal of concepts of sustainable food production systems. *Environmental Management* 15, 1–14.

Buttel, F. (1982) The political economy of part-time farming *GeoJournal* 6, 293–300.

Edmond, H., Corcoran, K. and Crabtree, J. (1993) Modelling locational access to markets for pluriactivity: a study in the Grampian region of Scotland. *Journal of Rural Studies* 9, 339–349.

Edmond, H. and Crabtree, J. (1994) Regional variation in Scottish pluriactivity: the socio-economic context for different types of non-farming activity. *Scottish Geographical Magazine* 110, 76–84.

Errington, A. (1985) Sampling for farm surveys in the UK: some alternatives. *Journal of Agricultural Economics* 36, 251–258.

Evans, N.J. and Ilbery, B.W. (1992) Farm-based accommodation and the restructuring of agriculture: evidence from three English counties. *Journal of Rural Studies* 8, 85–96.

Evans, N.J. and Ilbery, B.W. (1996) Exploring the influence of pluriactivity on gender relations in capitalist agriculture. *Sociologia Ruralis* 36, 74–92.

Fuller, A. (1990) From part-time farming to pluriactivity: a decade of change in rural Europe. *Journal of Rural Studies* 6, 361–371.

Gasson, R. (1986) Part-time farming: strategy for survival? *Sociologia Ruralis* 24, 364–376.

Gasson, R. (1987) The nature and extent of part-time farming in England and Wales. *Journal of Agricultural Economics* 38, 175–182.

Gasson, R. and Potter, C. (1988) Conservation through land diversion: a survey of farmers' attitudes. *Journal of Agricultural Economics* 39, 340–351.

Ilbery, B.W. (1988) Farm diversification and the restructuring of agriculture. *Outlook on Agriculture* 17, 35–39.

Ilbery, B.W. (1991) Farm diversification as an adjustment strategy on the urban fringe of the West Midlands. *Journal of Rural Studies* 7, 207–242.

Ilbery, B.W. and Bowler, I.R. (1993) Land diversion and farm business diversification in EC agriculture. *Nederlandse Geografische Studies* 172, 15–27.

MacKinnon, N., Bryden, J., Bell, C., Fuller, A. and Spearman, M. (1991) Pluriactivity, structural change and farm household vulnerability in Western Europe. *Sociologia Ruralis* 31, 58–71.

Marsden, T., Whatmore, S., Munton, R. and Little J. (1986) The restructuring process and economic centrality in capitalist agriculture. *Journal of Rural Studies* 2, 271–280.

Marsden, T., Munton, R., Whatmore, S. and Little, J. (1989) Strategies for coping in capitalist agriculture: an examination of responses of farm families in British agriculture. *Geoforum* 20, 1–14.

Marsden, T., Murdoch, J. and Williams, S. (1992) Regulating agriculture in deregulating economies: emerging trends in the uneven development of agriculture. *Geoforum* 23, 333–345.

Moran, W., Blunden, G. and Greenwood, J. (1993) The role of family farming in agrarian change. *Progress in Human Geography* 17, 22–42.

Shucksmith, M. (1993) Farm household behaviour and the transition to post-productivism. *Journal of Agricultural Economics* 44, 466–478.

Shucksmith, M., Bryden, J., Rosenthall, P., Short, C. and Winter, M. (1989) Pluriactivity, farm structures and rural change. *Journal of Agricultural Economics* 40, 345–360.

Shucksmith, M. and Smith, R. (1991) Farm household strategies and pluriactivity in upland Scotland. *Journal of Agricultural Economics* 42, 340–353.

Ward, N. and Lowe, P. (1994) Shifting values in agriculture: the farm family and pollution regulation. *Journal of Rural Studies* 10, 173–184.

10 Great Plains Agroecologies: The Continuum from Conventional to Alternative Agriculture in Colorado

Leslie Aileen Duram
Department of Geography, Southern Illinois University,
Carbondale, Illinois 62901-4514, USA

Introduction

This chapter presents an analysis of changes in Great Plains agricultural land use, as evidenced in Colorado farmers who adopt alternative production methods. Although alternative agriculture includes several distinct methods, the focus of this research is organic farming and Holistic Resource Management (HRM) ranching. National standards for organic farming in the US should be established in 1997, and until then numerous state and private agencies have acted to certify organic farmers. Although no universal standards currently exist, organic production usually means that synthetically compounded fertilizers or pesticides are not used for three years prior to certification, nor in production and processing (United States Department of Agriculture, 1980, p. 9). HRM ranching is an alternative range management system based on traditional concepts of rotational grazing and the exclusion of synthetic antibiotic and feed additives (Savory, 1988). All types of farms are reliant on the natural environment, but alternative farms may be to an even greater degree, as no synthetic inputs are used in the organic production process and management of biotic interactions is emphasized (Poincelot, 1986; Stinner and House, 1988).

Numerous studies verify that a key to the adoption of conventional agricultural conservation techniques is in the individual agriculturalist's attitudes and perceptions about the environment and his/her impact on it (Held and Clawson, 1965; Napier and Forster, 1982; Bultena and Hoiberg, 1986). These individual farmer characteristics are also important in explaining the adoption of alternative ranching and organic farming techniques and related land-use choices. A study of conventional and alternative farmers was conducted in Colorado to illuminate variations in agricultural operations, ecological perceptions and agricultural attitudes that play a role in the adoption of alternative agriculture.

Background

The US Great Plains afford a harsh, semi-arid environment that has challenged agricultural development with, for example, the Dust Bowl droughts of the 1930s (Sears, 1959; Worster, 1979). Yet, especially in the last 50 years, humans adapted to and modified the natural conditions of the Plains, and created one of the world's most productive agricultural regions (Spath, 1987). Whereas few deny the productive strengths of the current US agricultural system, its ecological, social and policy dimensions are increasingly criticised (Sampson, 1981; Belden, 1986; Berry, 1986; Hallberg, 1987; Mott and Snyder, 1987; Clancy, 1990; Edwards, 1990; Harl, 1990; King, 1990; Jacobson *et al.*, 1991; Soule and Piper, 1992; Goerring *et al.*, 1993; Anderson, 1995).

Natural resource use in the US West has been the focus of several recent works including Worster (1992) who describes the influence of individualism in land use decisions in 'Under Western Skies: Nature and History in the American West'; and Wilkinson (1992) whose 'Crossing the Next Meridian: Land, Water, and the Future of the West' indicates that traditional extractive uses are increasingly at odds with recent non-consumptive uses. Such broad social critiques coincide with new land-use strategies including the adoption of alternative agriculture by an increasing number of farmers and ranchers, which may be seen as a current form of adaptation in the Great Plains region. This research investigated who adopts organic methods, and what distinguishes them from farmers who practise conventional agriculture.

Study Area

The spatial focus of this research included a core study site in north-eastern Colorado (Fig. 10.1). An ecological gradient near the Central Plains Experimental Range (CPER), north of Nunn, Colorado, was identified by

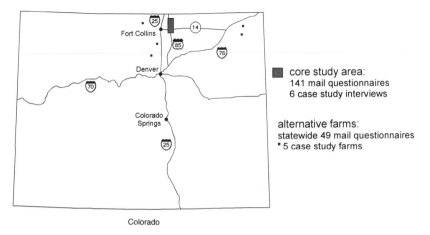

Colorado

Fig. 10.1. Location of surveyed farmers.

researchers from Colorado State University and other institutions. The CPER was established in 1937 to conduct research on grassland management and is administered by the Agricultural Research Service of the United State Department of Agriculture (USDA) (Agricultural Research Service, 1994). Fifty years of ecological research and data collection have been conducted in this area, including a Long Term Ecological Research (LTER) programme (Shoop *et al.*, 1989; Burke *et al.*, 1991).

Precipitation variation has been an important environmental factor, as annual precipitation decreases from the south to the north in the area. A land-use gradient starts with irrigated corn and soybean farms in the south near the town of Ault, shifts into dryland wheat-fallow around Pierce, and extends to pure rangeland grazing near Nunn in the north. This area comprised the core study site for this project because, over a relatively short distance, it provides examples of regional Great Plains agroecological conditions and enterprises.

Whereas conventional farmers were identified in this core study area, the research was expanded to the state level to include alternative agriculturalists, whose overall numbers are lower. Alternative agriculturalists were identified through the Colorado Organic Producers Association (COPA) and the Colorado Branch of Holistic Resource Management (HRM).

Methodology

The survey approach was twofold, drawing on mail questionnaires and in-depth case study interviews. County planning commission maps and

assessor's data were used to identify the names and addresses of farmers
and ranchers in the study site surrounding the towns of Ault, Pierce, and
Nunn. This core study area included approximately 50 irrigated farms, 45
dryland farms and 45 ranches. Mail questionnaires were sent to all of these
agriculturalists in the core study area; and six were interviewed as case study
farms. In addition, all 49 producers listed in the 1993 COPA Marketing
Directory and the four members of HRM located in north-eastern
Colorado were sent mail questionnaires; and five were interviewed for case
study analysis.

Mail questionnaire

Mail questionnaire surveys were developed to investigate operational and
personal farmer characteristics through a ten-page questionnaire in mixed
format including objective, short answer, and essay questions. A return rate
of 22% for the conventional farmers and 53% for the alternative farmers
was obtained, which yielded 57 valid questionnaire responses. Based solely
on membership in these organic/alternative agricultural groups, the
farmers were classified into 'conventional' and 'alternative' categories.
Quantitative analysis revealed variations in operations, demographics,
decision-making, and attitudes between the two categories of farmers. Basic
statistical analyses were performed to find the mean, median, and standard
deviation of each survey question. Cross-tabulation, using Chi Square
analysis, allowed an assessment of the statistical significance of differences
between the groups.

Variation in median age, years in farming, and whether they were raised
on a farm are notable (Table 10.1). Education levels were similar between
the two groups, but the field of study varied, with 50% of conventional and
only 9% of alternative farmers having studied agriculture. Operational and
decision-making factors varied, as conventional farms were likely to be
larger, less diversified, and had experienced substantially fewer operational
changes (Table 10.1). Conventional farmers were also more likely to state
that 'nothing would make me change land uses on my farm'.

Attitudes and experience were different between the groups, as
alternative farmers were more familiar with agroecological terms and their
land use decisions were influenced by personal environmental concerns
(Table 10.2). Conventional farmers, on the other hand, based their
decisions more on the existence of government programmes, attempting to
minimize risk, and the availability of credit. Alternative farmers were more
active in political and conservation organizations, while conventional
farmers were likely in religious organizations.

Based on Beus and Dunlap (1991), attitude scales indicated variation
between the two groups. Alternative farmers believed that farming is first

Table 10.1. Farmer characteristics.

Demographic factors	Conventional	Alternative
Median age	56	42 years
Raised on farm	90	50%
Years farming	31	18 years
Education levels; college graduates	50	50%
Field of study: Agriculture	50	9%
Operational and decision-making factors	**Conventional**	**Alternative**
Farm size	Larger average	Larger variation
Number of crops produced	Few	Many
Recent changes in:		
operation type	10	65%
size	30	50%
diversification	39	62%
use of pesticides	22	42%
use of fertilizer	13	55%
'Nothing would make me change land uses'	48	23%

a way of life, whereas conventional farmers noted that farming was foremost a business. Alternative farmers were also more likely to believe that agriculture causes ecological problems, and that US agriculture is not completely successful because the number of farms is decreasing and average farm size is increasing. Conventional farmers, however, noted that US food prices are low and production yields are high, which they consider an indication of a successful agricultural system. Further, alternative farmers are more likely to use personal experience to guide their operations, while conventional farmers rely on outside sources of scientific information.

Although multivariate statistical analysis is speculative because of the small sample size, factor analysis proved helpful in identifying clusters of responses, such as views of nature or personal values. This assisted in delimiting key factors for a logistic regression equation that indicated three factors are closely associated with alternative farms and farmers: diversification, knowledge of ecological terms, and recent operational change.

Table 10.2. Attitudes and experience.

	Conventional	Alternative
Familiar with agroecological terms	Less	More
Activity, types of organizations	Religious	Political
		Conservation
Influences on land-use decisions	Government	Personal
	programmes	environmental
		concern
	Minimizing risk	
	Credit availability	
Short-term profit maximization	Balance all factors	Environmental
		quality
		Preserve
		long-term
		productivity
		Personal ideals
Attitude scales		
A farm is first . . .	a business	a way of life
Does agriculture cause ecological problems?	No	Yes
Is the US agricultural system successful?	Yes	No
Evidence	High yields	Fewer farms
	Low prices	Large farm size
Information sources	Scientific research	Personal
		experience

Case studies

This survey approach also employed detailed in-depth interviews of 11 farmer participants that provided qualitative data, which were combined with quantitative data for the analysis presented here. Open-ended interviews were conducted with farmers on a variety of types of farms:

Case study farms

1. Dryland organic wheat farm
2. Irrigated part organic farm
3. Small irrigated organic farm
4. Irrigated organic fruit/vegetable farm
5. HRM ranch
6. Dryland minimum-tillage farm
7. Ranch using public lands
8. Dryland wheat/fallow farm
9. Irrigated farm with cattle

10. Irrigated onion farm
11. Ranch with wheat

As well as noting these general operational characteristics, the richness of variations among all farmers surveyed was investigated. Interviews were based on a case study 'Interview Guide', which addressed the following topics: operational diversification and change, water availability, community and personal characteristics, financial factors, conservation affiliation and human/environment relationships.

Results of Questionnaires and Case Studies

There is variation between the two predetermined groups of farmers, but divisive analysis is simplistic. As indicated by essay responses on the mail questionnaire and case study interview data, the agricultural system is composed of a complex spectrum of farmers and farm operations. The agroecological behaviour continuum is a model of farmer attitudes, characteristics, and behaviours, which describes patterns of multiple attributes that vary from farmer to farmer (Fig. 10.2). The crux of this model are the axes that represent farmer characteristics, both agricultural methods and attitudes, that are the key to the adoption of alternative agriculture. The model consists of 12 variables on a scale of 1 to 5 that describe a farmer's position in the agricultural continuum (Table 10.3).

The core where all axes intersect is indicative of conventional farmers, whereas the outer ends of each axis represent alternative farmer characteristics. The term 'Reactive' defines the core and 'Proactive' defines the ends of the farming continuum because these terms are inclusive of personal ideological and operational characteristics. The 'locus of control' is an individual's perception of how much one controls their own destiny versus how much others exert control (Palm, 1990). Reactive farmers have a strong external locus of control, as they perceive outside forces (e.g. government policy, prices, markets) controlling their farm operations. Proactive farmers, on the other hand, have an internal locus of control, because they feel more in charge of their farm. Further, a farmer's decision to adopt organic methods is based on their range of choice (White, 1961). The model describes how many Reactive farmers exclude organic production methods from their practical range of choice. The model also describes that few farmers displayed purely Reactive or Proactive characteristics, rather farmers are at different stages along the continuum between conventional and alternative agriculture.

Farmers exhibit internal conflict because they possess characteristics of both the Reactive and Proactive composites. For example, farmers may believe that diversification is an important goal for farm sustainability, but

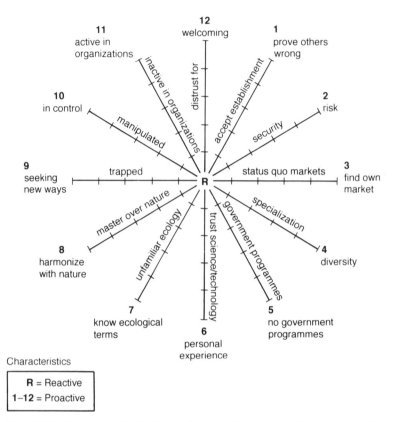

Fig. 10.2. The agroecological behaviour continuum – reactive and proactive characteristics.

they actively seek guidance from scientists and government officials who encourage specialization as a means of increasing crop yields. There is also conflict within the Reactive category, as displayed in axis variable numbers 2 and 9 (security–trapped), 3 and 10 (status quo markets–manipulated), and 11 and 12 (inactive–distrustful). These variables define inconsistent characteristics that are partially responsible for pushing farmers toward alternative agriculture.

There is conflict between farmers' beliefs of how others view them, and how farmers actually view themselves. They revere the security of the current agricultural system (e.g. conventional markets, price supports, CRP), but at the same time they feel trapped and manipulated by policies, regulations, and other outside influences (e.g. environmentalists, urbanites, the government, consumers). They exhibit a circular logic in some cases, such that a farmer perceives negative feelings from outsiders and

Table 10.3. Agro-ecological behaviour continuum variables: axes extremes.

	composite Reactive farmer	transition	composite Proactive farmer
Scale	1　　2	3	4　　5
Locus of control	External		Internal
Axis variables			
1.	Accept establishment		Prove others wrong
2.	Security		Risk
3.	Status quo markets		Find own market
4.	Specialization		Diversity
5.	Government programmes		No government programmes
6.	Trust science/technology		Personal experience
7.	Unfamiliar with ecology		Know ecology terms
8.	Master over Nature		Harmonize with Nature
9.	Feeling trapped		Seeking new ways
10.	Manipulated		In control
11.	Inactive in organizations		Active in organizations
12.	Distrustful		Welcoming

Conflict within Reactive farmer composite: axes 2–9, 3–10, 11–12

feels trapped by those ideas, but at the same time demands the security of maintaining the current system.

There is also conflict between the level of community assimilation that farmers exhibit and their activity in broader social organizations. While Reactive farmers follow their neighbours' agricultural and social behaviours, at the larger community level they are relatively inactive in organizations and activities (e.g. PTA, volunteer fire-fighter, government groups). This may be partially fed by the Reactive farmers' perceptions that outside influences have too much control over agriculture and their individual farms.

Other conflicts involve the use of chemicals. Reactive farmers often believe that synthetic pesticides and fertilizers are needed to 'feed the world' but when this topic is probed further, it is apparent that many distrust these chemicals and fear for their own safety. Yet, they continue to apply the chemicals because of their perception of economic benefit and their overriding faith in science and technology.

Discussion

The continuum is composed of two poles and farmers are distributed between the extremes, as exhibited here by the case study farmers (Fig. 10.3). Farmers possess Reactive attributes to a greater or lesser extent, or Proactive characteristics that are stronger or weaker, or true transitional characteristics that place them in the middle of the continuum. Strong Reactive characteristics are scaled as '1', whereas strong Proactive characteristics are '5' on the scale. Transitional characteristics are complex and fall within the '2', '3' or '4' categories on the scale. An individual farmer is actually composed of numerous variations along the continuum, but the model simplifies these to '2' (somewhat more Reactive), '3' (completely in the middle), and '4' (somewhat more Proactive).

Farmers sometimes display characteristics that are different from the expected profile, and these inconsistencies especially inform one about the

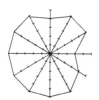

FARMER 1:
dryland organic wheat

FARMER 2:
irrigated part organic

FARMER 3:
small irrigated organic

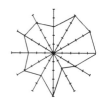

FARMER 4:
organic fruit & vegetable

FARMER 5:
HRM ranch

FARMER 6:
dryland minimum tillage

FARMER 7:
ranch using public lands

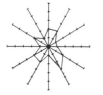

FARMER 8:
dryland wheat

FARMER 9:
irrigated with cattle

FARMER 10:
irrigated onion

FARMER 11:
ranch with wheat

Fig. 10.3. Case study farmers and farm types along the agroecological behaviour continuum.

transitional process. Each farmer possesses a unique chronicle of the transition to alternative methods, which draws upon more Reactive or Proactive characteristics. There are two main routes for this transition to new agricultural techniques. First are people who are not from a farming background who enter agriculture as organic farmers. These farmers are very active in seeking specialized markets, which is the key to successful organic production (Lockeretz and Madden, 1987). These farmers assume an internal locus of control. Second are people from a farm family who were raised in conventional agriculture. Whether they actually decide to take steps toward alternative agriculture depends on their possession of Proactive farmer characteristics and their ability to assume an internal locus of control. They may be initially motivated by the internal conflicts of the Reactive farmer characteristics.

Both routes are valuable in the transition, as the first proves it is possible to step into farming if you desire this lifestyle. Organic farms may provide a viable, socially beneficial option for entrance in the US agricultural system. The second route provides a link between the current and future agricultural system because these farmers prove that alternative agriculture is possible and profitable. Most importantly, such farmers illustrate that organic farming can be a 'normal' means of production for typical farmers.

Conclusion

Quantitative analysis shows basic differences between conventional and alternative farmers, when groups are absolutely defined by COPA/HRM membership. Qualitative analysis shows that rather than two separate types of farmers, a broad range of farm characteristics and farmer attributes exists within US agriculture. The agroecological behaviour continuum model indicates that few farmers completely match a composite Reactive or Proactive farmer, and similarly dichotomous categories of conventional and alternative farmers are deficient. Instead, a continuum best describes the current US agricultural system; there is a spectrum of farm characteristics and farmer attributes between conventional and alternative agriculture, as presently practised in the eastern Colorado Great Plains. The model displays the conflict and complexity found both in individual farmers and among groups of farmers during this transition.

The model describes environmental planning/management at a personal level, directly tied to attitudes about agriculture, perceptions of an individual's farm, and numerous other individual factors. In the near future, however, structural factors will become more influential in the adoption of organic agriculture, as marketing issues and national policies increasingly encompass alternative agriculture. The availability of organic produce is expanding in major supermarket chains (Jolly and Norris, 1991) and sales of organic foods in the USA rose from $174 million in 1980, to $1.25 billion in 1989, to $2.3 billion in 1994 (Jacobson *et al.*, 1991; Mergentime and Emerich, 1995). Local, regional, national and international distribution will play an important role in the future of organic production.

Not only is organic production increasing and gaining acceptance in the US but also organic standards are in transition. The federal government will soon regulate this form of agriculture for the first time. The 1990 Organic Food Protection Act initiated a process for establishing organic certification and production standards, including a National Organic Standards Board (NOSB) within the USDA. Preliminary guidelines were completed in 1994 (United State Department of Agriculture, 1994) and after legal and administrative review, legislation is expected to be implemented in 1997.

Such policy initiatives may indicate acceptance of organic farming within the American agricultural system and may provide a means of 'mainstreaming' this alternative agricultural method; or regulations may further separate organic farmers from the majority of conventional producers in the US farming system. It is still uncertain what effects national certification regulations will have on farmers and Great Plains land use. The transition to alternative agriculture presents an opportunity to geographers and other social scientists to study relationships between people and their local environments in a relevant arena of social change.

References

Agricultural Research Service (ARS) (1994) *Central Plains Experimental Range.* United States Department of Agriculture Pamphlet.

Anderson, M. (1995) The life cycle of alternative agricultural research. *American Journal of Alternative Agriculture* 10(1), 3–9.

Anderson, J., Dillon, J. and Hardaker, B. (1977) *Agricultural Decision Analysis.* Iowa State University, Ames.

Belden, J. (1986) *Dirt Rich, Dirt Poor: America's Food and Farm Crisis.* Routledge, New York.

Berry, W. (1986) *The Unsettling of America: Culture and Agriculture.* Sierra Club, San Francisco.

Beus, C. and Dunlap, R. (1991) Measuring adherence to alternative vs. conventional agricultural paradigms: a proposed scale. *Rural Sociology* 56(3), 432–466.

Bultena, G. and Hoiberg, E. (1986) Sources of information and technical assistance for farmers in controlling soil erosion. In: Lovejoy and Napier (eds) *Conserving Soil: Insights from Socioeconomic Research.* Soil Conservation Society of America, Ankeny, Iowa, pp. 71–82.

Burke, I.C., Kittel, T., Lauenroth, W., Yonker, C. and Snook, P. (1991) Regional analysis of the Central great plains. *Bioscience* 41, 685–692.

Clancy, K. (1990) Agriculture and human health In: Edwards, C.A. (ed.) *Sustainable Agricultural Systems*. Soil and Water Conservation Society of America, Ankeny, Iowa, pp. 655–665.

Colorado Organic Producers Association (1993) *Marketing Directory*. COPA, Denver.

Edwards, C.A. (1990) The importance of integration in sustainable agricultural systems. In: Edwards (ed.) *Sustainable Agricultural Systems*. Soil and Water Conservation Society of America, Ankeny, Iowa, pp. 249–264.

Gilpin, M., Gall, G. and Woodruff, D. (1992) Ecological dynamics and agricultural landscapes. *Agriculture, Ecosystems and Environment* 42, 27–52.

Goering, P., Norberg-Hodge, H. and Page, J. (1993) *From the Ground Up: Rethinking Industrial Agriculture*. Zed Books, London.

Hallberg, G. (1987) Agricultural chemicals in ground water: extent and implications. *American Journal of Alternative Agriculture* 2(1), 3–15.

Harl, N.E. (1990) *The Farm Debt Crisis of the 1980s*. Iowa State University, Ames.

Held, R.B. and Clawson, M. (1965) *Soil Conservation in Perspective*. Johns Hopkins, Baltimore.

Jacobson, M.F., Lefferts, L.Y. and Garland, A.W. (1991) *Safe Food Eating Wisely in a Risky World*. Living Planet Press, Los Angeles.

Jolly, D.A. and Norris, K. (1991) Marketing prospects for organic and pesticide-free produce. *American Journal of Alternative Agriculture* 6(4), 174–179.

Lockeretz, W. and Madden, P. (1987) Midwestern organic farming: a ten-year follow-up. *American Journal of Alternative Agriculture*, 2(3), 57–63.

King, L. (1990) Soil nutrient management in the United States. In: Edwards, (ed.) *Sustainable Agricultural Systems*. Soil and Water Conservation Society of America, Ankeny, Iowa, pp. 89–106.

Mergentime, K. and Emerich, M. (1995) Organic market overview. *Natural Foods Merchandiser*, June.

Mott, L. and Snyder, K. (1987) *Pesticide Alert – a Guide to Pesticides in Fruits and Vegetables*. Sierra Club Books, San Francisco.

Napier, T. and Forster, D. (1982) Farmer attitudes and behavior associated with soil erosion control. In: Halcrow, *et al.* (eds) *Soil Conservation Institutions and Incentives*. Soil Conservation Society of America, Ankeny, Iowa, pp. 137–150.

Palm, R. (1990) *Natural Hazards: An Integrative Framework for Research and Planning*. Johns Hopkins, Baltimore.

Poincelot, R. (1986) *Toward a More Sustainable Agriculture*. AVI, Westport, Connecticut.

Sampson, R.N. (1981) *Farmland or Wasteland: A Time to Choose*. Rodale, Emmaus, Pennsylvania.

Savory, A. (1988) *Holistic Resource Management*. Island Press, Washington DC.

Sears, P.B. (1959) *Deserts on the March*. University of Oklahoma, Norman.

Shoop, M., Kanode, S. and Calvert, M. (1989) Central Plains Experimental Range: 50 Years of Research. *Rangelands* 11(3), 112–117.

Soule, J.D. and Piper, J.K. (1992) *Farming in Nature's Image: An Ecological Approach to Agriculture*. Island, Washington DC.

Spath, H.J.W. (1987) Dryland wheat farming on the Central Great Plains: Sedgwick County, Northeast Colorado. In: Turner, B. and Brush, S. (eds) *Comparative Farming Systems*. Guilford, New York, pp. 313–344.

Stinner, B. and House, G. (1988) Role of ecology in lower-input, sustainable agriculture: an introduction. *American Journal of Alternative Agriculture* 2(4), 146–147.

United States Department of Agriculture (1980) *Report and Recommendations on Organic Farming.* Government Printing Office, Washington DC.

United States Department of Agriculture (1994) *National Organic Standards Board: Final Recommendations for Organic Crop Production Standards.* Adopted 1–4 June. USDA/AMS/TMD, Washington DC.

White, G. (1961) The choice of use in resource management. *Natural Resources Journal* 1, 23–40.

Wilkinson, C. (1992) *Crossing the next meridian: land water and the future of the West.* Island, Washington DC.

Worster, D. (1979) *Dust Bowl: The Southern Great Plains in the 1930s.* Oxford University, New York.

Worster, D. (1992) *Under Western Skies: Nature and History in the American West.* Oxford, New York.

11 Agricultural System Response to Environmental Stress

JOHN SMITHERS AND BARRY SMIT
Department of Geography, University of Guelph, Guelph, Ontario, Canada N1G 2W1

Introduction

Agricultural systems adjust and evolve in response to many types of forces that may be environmental, economic, technological, institutional or social/ideological. Together these parameters establish the context within which farm businesses and agricultural regions develop and function. Changes in any or all of these domains can lead to agricultural change.

Recent concern over the possible agricultural implications of global environmental change has stimulated a renewed, or at least a refocused, interest in how (or if) agricultural systems respond to variations in environmental conditions. Given the importance of weather as a basic resource for agriculture, there is an especially strong interest in the implications of possible changes in climate. Some research has suggested that certain regions of the earth may experience significant changes in climatic conditions and agricultural potential over the next several decades (e.g. Parry *et al.*, 1988; Houghton *et al.*, 1990). Such changes have the potential to alter the ability of farms and regions to produce agricultural commodities, and consequently may threaten the viability and thus the sustainability of agriculture.

Quite apart from the issue of long-term climatic change, there is merit in policy consideration of agricultural adaptation to *current* climatic variation and uncertainty as well. Limits to current levels of adaptation are demonstrated by economic losses and human suffering associated with extreme climatic events (Economic Council of Canada, 1988; Wheaton *et al.*, 1990; Chagnon, 1996). As governments reassess their commitment to public sector-funded stabilization and compensation schemes, the connection between farming and climate or other types of environmental conditions may become more critical in the future regardless of changes in weather.

This chapter explores the issue of agricultural adjustment or adaptation to environmental stress. Concepts from environmental systems management, complex systems theory, agricultural systems research, and the human adaptation paradigm are employed to develop a conceptual model of agricultural adjustment to environmental variability and change. Taken together these concepts establish a framework within which agriculture–environment interactions might be investigated and understood at various scales. Of particular interest is how these ideas might be explored at an aggregate or system level scale of analysis in the context of variation in climatic conditions. For illustrative purposes, some selected results from a recent empirical study conducted at a regional scale of analysis relating past climatic variability and its influence on agricultural structure are provided.

The Building Blocks: Human Adaptation and Agricultural Change

Adaptation

According to Winterhalder (1980) the concept of adaptation has its roots in population biology and evolutionary ecology where applications focus on the survival of species or ecosystems, and not necessarily on individuals within them (Slobodkin and Rapaport, 1974). Application of the concept to human systems has led to a broader meaning of adaptation in the social sciences. An important distinction is that humans possess the ability to plan and even manage adaptation. Thus, whereas the response of biological systems is entirely reactive, the responses of human systems such as agriculture are both reactive and anticipatory, incorporating environmental perception and risk evaluation as important elements of adaptation. Additionally, human systems may adjust in pursuit of goals other than mere survival (Stern *et al.*, 1992).

The concept of adaptation is central to the model of agricultural response to environmental change developed in this chapter. Thus one should begin by elaborating some conceptual dimensions or elements of adaptation that are integral to the analysis of environmentally induced agricultural change. A generic basis for approaching environmental adaptation research and for characterizing adaptation strategies is developed in a recent critique of the 'anatomy of adaptation' by the Canadian Task Force on Adaptation (Smit, 1993). In particular, the task force adopted a descriptive model of adaptation (Fig. 11.1) that summarizes three key dimensions of adaptation (each of which is relevant to adjustments in agriculture).

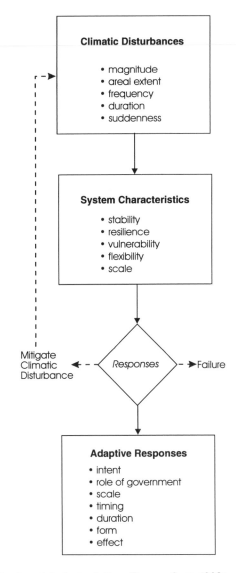

Fig. 11.1. Conceptual model of adaptation. (Source: Smit, 1993.)

The first dimension relates to the characteristics of disturbances or stress. The importance of environmental stresses themselves is a cornerstone of natural hazards and climate impacts research (see Kates *et al.*, 1985; Burton *et al.*, 1993). For agriculture one needs to know something about the attributes of climate that stimulate adaptive responses. What sort of climatic events, trends or regimes, and what *magnitude, frequency, duration* or *areal extent* of climatic conditions do agricultural systems respond to? Are certain types of events more damaging than others for agricultural activities (drought vs. frost, etc.) or is it the conditions under which these events occur that dictate the nature of response (i.e. time of occurrence during the growing season, spatial and temporal crowding of impacts)? In some cases, the timing of a growing season frost may be more important than its absolute severity.

A second dimension of the adaptation question relates to the characteristics of the system that experiences an environmental stress. The fields of environmental systems management and, more recently, complex systems theory have applied certain ecological and general systems concepts to help understand the character of natural systems and how they may be affected by, or respond to, environmental stress (Holling, 1986; Kay and Schneider, 1994). Such concepts have been applied explicitly to agricultural systems or 'agroecosystems' as well (Conway, 1987; Marten, 1988; Wall *et al.*, 1995).

Timmerman (1981) has suggested that the way a system responds to a planned or unexpected disturbance depends on system characteristics such as *stability* and *resiliency*, and that, as change occurs, it is to be expected that regular variability, thresholds of tolerance, and multiple equilibria will exist in natural systems. The concept of multiple equilibria is of particular interest in the context of agricultural response to environmental stress. In natural systems, incremental disturbances may be absorbed until critical thresholds of tolerance or 'absorptive capacity' are passed and the system switches to a different mode of behaviour or organization (what Kay (1991) has termed a 'switch to another attractor'). The same may be true in agricultural systems, but the process is greatly modified by the fact that agricultural systems are not 'natural' – they are heavily managed and continuously modified ecosystems sustained by chemical, economic, and human managerial subsidies. Furthermore, agricultural systems respond to other types of stimuli quite unrelated to environmental signals (Waltner-Toews, 1994).

As suggested in Fig. 11.1, adaptive responses are not the only possible outcome of climatic stress. Mitigation and outright system failure are also possible. The third component of the model of adaptation deals with the characteristics of the responses themselves. Agricultural adaptations can be differentiated and described on several axes including their: *intent, scale, timing, form* and *effect* as well as their sectoral origins (the *role of government*). These distinctions are helpful in understanding the nature of various response options and the connections between certain types of stress and response. Whereas some environmental stresses may result in temporary buffering actions (for example, drought compensation funds) or mitigative responses (for example, decisions on soil amendments or the timing of harvest), others may evoke very different and more profound changes such as a shift to new methods or even new types of production – or perhaps even a move out of farming altogether. These important distinctions apply not only to hypothetical future changes but also to *current* agricultural adaptations to environmental and non-environmental conditions.

Agricultural systems

In order to consider the processes and prospects for agricultural adaptation to environmental stress it is necessary to recognize the multifaceted and hierarchical nature of agricultural systems themselves. In other words, 'what is it that adapts'?

Agricultural systems can be identified as farms, plantations, regional and national agricultures (Spedding, 1979) or as a nested hierarchy of agroecosystems, comprising ecological, economic and human dimensions and ranging from the field or plot to the region and beyond (Conway, 1987; Izak and Swift, 1994, Gallopin, 1994). The level of analysis is particularly important in the study of agricultural systems because the relevant attributes change as consideration progresses from micro (or farm) to macro (or system) scales. For example, an analysis of implications of environmental change for crop production might focus on soil–plant relations at the level of the farm field, but at broader spatial scales the relevant attributes on this same dimension would be quite different, say regional yield levels or national production. At each level there is a variety of components and attributes that are potentially subject to change and a corresponding series of scale dependent indicators (Table 11.1). It is important to select indicators of change that are meaningful and measurable at the scale at which inferences are made (Smit and Smithers, 1994).

For some, the most appropriate level of analysis for agriculture is the farm (e.g. Olmstead, 1970; Altieri, 1987; Dover and Talbot, 1987). This places key importance on the intersection of ecological, economic and human factors at the scale where performance is first assessed and decisions are made regarding intervention and resource allocation. Much recent research examining agricultural adjustment to political and economic change focusing on farmers as decision makers has been conducted at the farm scale (Marsden *et al.*, 1989; Ilbery, 1991; Bryant and Johnston, 1992). In this approach, as at other scales of system definition, farms are seen as functioning within external economic, institutional, technological, and social environments – all with potential to act as forces of change or adjustment or as constraints on responses to environmental stimuli.

These same components, ecological, economic and social, have been used to characterize various regional or national agricultural systems as well. For example, Crosson (1986) has suggested that macro-level food production systems may be characterized along four lines: resources,

Table 11.1. Dimensions and attributes of agricultural change.

Dimension of agriculture	Attributes subject to change		
	Micro		Macro
Crop production	Soil–plant relations, yield per acre	Regional yield levels, production potential	Continental and global level of production
Environmental quality	Soil structure, erosion, moisture, pests	Agroclimatic conditions, water quality and quantity	Climate, soil zones, e.g. desertification
Economic performance	Costs of production, prices, investment, income	Value of production, employment in agriculture, government payments	Trading arrangements, marketing mechanisms
Food production	Adequacy for subsistence, safe for consumption	Regional production ability to meet demands	Global food supply, future reserves, etc.
Rural community	Continuity of family involvement	Stability in farm population, no. of farms, size and function of rural communities	Global poverty, hunger, equity, etc.
Character of production system	Enterprise type, management system, tenure	Mix of enterprises, patterns of production, structural changes	Broad production patterns (continental and large regional)

Source: Smithers (1994).

technology, environment and institutions. Understanding the interaction of these components is necessary in assessing regional agricultural systems in terms of food production, employment, environmental quality and social vitality.

Taken together, the concepts of human adaptation and agricultural system dynamics provide a framework for agricultural adaptation research. Agricultural responses to environmental change are influenced by the nature of the stimulus, the 'ecological' properties of the system in question, the perceptions, motivations, and cognitive abilities of individuals within those systems, and the role played by other institutional, economic, social and technological forces.

To what degree are the concepts articulated above reflected in climate–agriculture impacts research? From existing reviews of this large body of scholarship (Parry, 1990; Smithers, 1994) one can conclude that there has been relatively little attention focused directly on the *process* of agricultural adaptation to environmental change. In many cases the impacts of climatic change are investigated assuming an instantaneous change in average climatic conditions while other factors or potential influences on agriculture are held constant. Yet an essential feature of climate, and most other environmental phenomena, is its variability, and a fundamental characteristic of agriculture, or any resource use system, is that it is subject to a variety of forces including, but not confined to, environmental conditions (Liverman, 1987; Easterling *et al.*, 1992; Kaiser *et al.*, 1993; Rosenberg *et al.*, 1993). In addition, despite the long recognized importance of human response and social adaptation in understanding the interaction of human and natural systems, most analyses of climate – agriculture interaction simply assume adaptive (or non-adaptive) behaviour of one type or another in order to predict initial impacts and subsequent change. Little is really known about the validity of these assumptions.

It is suggested that understanding of agricultural sensitivity and response to environmental stress would be strengthened through better knowledge of the attributes of climate to which agriculture adapts (variability, successive shocks, etc.), of the nature of coping strategies and thresholds of tolerance within agriculture, and of the role played by other non-climatic factors in modifying (or negating) agricultural response to environmental conditions. Further, it is important to understand these processes at a variety of spatial, social and temporal scales in order to estimate the impacts of climatic variability and change, and to develop and promote strategies to encourage successful adaptation by individuals and the agricultural sector.

A Model of Regional Agricultural Adjustment to Climatic Variability

The relationship between agricultural and environmental systems may be seen and explored at a variety of spatial and temporal scales. Analysis at the aggregate, or system, level allows an examination of broad-scale relationships between various types of climatic conditions and general agricultural responses. Research at the level of the agricultural region offers the prospect of extending knowledge of the climatic conditions to which farming is most sensitive generally and gives an indication of the potential future conditions of the system given revealed trends or tendencies.

The largely unknown sensitivity of agricultural systems stems from several factors including the variable nature and intensity of the environmental change, the concurrent influences of other forces of change in agricultural systems, the differing antecedent characteristics of particular agricultural systems (physical, economic, social, etc.), and the possible interventions of individuals or institutions in modifying levels of agricultural sensitivity to environmental conditions. Together, these factors represent the parameters of a theoretical model of agricultural adjustment to climatic variability and change (Fig. 11.2).

The model specifies four measurable *agroclimatic conditions* that are generally held to be of importance for agriculture; growing season length (measured as the frost-free period or growing season start and end), temperature (often expressed in corn heat units or growing degree days), storm events (hail, wind, etc.) and precipitation. These agroclimatic features are depicted in three dimensional form to introduce the notion of cumulative impact to the model. Successive occurrences (e.g. consecutive dry years) or frequent recurrences may evoke different and more substantive adjustments in agriculture than will single and/or short term events, regardless of severity. Further, there may be synergistic effects among these climatic parameters at any point in time (e.g. a single growing season) or over time (e.g. several growing seasons).

Concurrent with the impacts of climatic variability, agricultural systems receive stimuli from numerous other forces. Four such forces are recognized in the model, but clearly others are possible. Agricultural systems are influenced by a wide range of *economic factors* such as changes in the costs of production, and conditions of the commodity markets both domestic and international. Often related to these economic pressures are the *actions of governments* and other institutions in regulating production, stabilizing prices and otherwise supporting various aspects of agriculture. *Sociocultural* pressures are placed on agricultural systems in numerous ways including increasing demand for land for non-agricultural purposes, increasing demand for food necessitating higher production, and changing demands or preferences regarding food quality. Finally, agricultural systems are influenced by *technological innovations* – often designed to respond to the pressures noted above, or simply to improve the efficiency and effectiveness of farming either by enhancing current activities or introducing new opportunities. These factors, along with agroclimatic conditions, lead to a set of *potential constraints* on farming. These constraints may be related to biophysical (e.g. soil capability), personal (e.g. aspirations), micro economic (e.g. debt load) or enterprise (e.g. management) characteristics.

The notion of tolerance *threshold* or coping limit is captured in the model as the point at which the stress or pressure (from climate or any source) is sufficient to evoke an adjustment. Ultimately, adjustment is the result of the surpassing (actual or anticipated) of various thresholds that define the limits of coping or tolerance in agricultural systems. While the prestress character of an agricultural system influences the nature of adjustment that might occur, it is also likely that the future character of the system is significantly influenced by the nature of the adjustments that

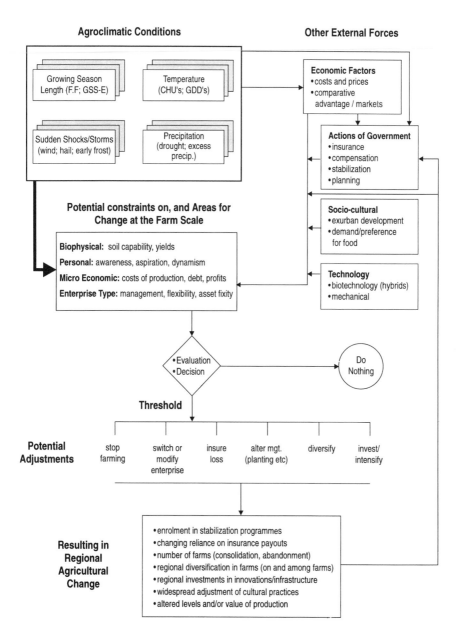

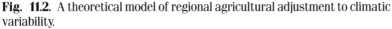

Fig. 11.2. A theoretical model of regional agricultural adjustment to climatic variability.

occur (i.e. an iterative process of adjustment and readjustment).

Several common *potential adjustments* to climatic variability are suggested in the model, reflecting the broad response types developed earlier, i.e. prevention, buffering, change. At the system scale these result in collective structural adjustments in farming, broad-scale changes in land use and management, altered levels and values of production, and the existence of, and participation in, various government stabilization schemes.

Agricultural Responses to Climatic Variation in the Kent-Essex Region of Ontario, Canada

An alternative to assuming some future climate and predicting associated agricultural responses is to examine the past relationship between agriculture and climate over a period of several decades where the actual conditions of climate and agriculture are known. The objective is to isolate any climate-related signatures in the trend of certain key attributes of the agricultural system. The underlying rationale of this approach is that much can be learned about how, or if, agricultural systems tend to respond to climate (which presumably includes periodic shocks) by comparing actual climatic conditions with observable changes in selected attributes or indicators that are thought to be potentially sensitive to climate and reflective of certain types of agricultural response (see for example Glantz, 1988). Rather than assuming adaptations, this 'hindcasting' approach seeks to identify them by exploring spatial and temporal associations between climatic conditions and certain types of adaptation. An important objective of the approach is to identify the thresholds which exist in the agricultural system. Also, there is opportunity for refining understanding of the particular attributes of climate to which agriculture is most sensitive. The study reported in this chapter involves a historical analysis of potential associations between climate and agriculture in the Southwestern Ontario counties of Essex and Kent (Fig. 11.3).

In practical terms this type of analysis requires an accurate and detailed reconstruction of agroclimatic conditions, data on a variety of agricultural attributes, and information on trends in other potential forces of change such as economic conditions (e.g. commodity prices), institutional interventions and incentives (e.g. incentive or stabilization programmes), and technology (e.g. yield trends, major innovations), etc. These elements were assembled over the 33-year period 1960–1992. The region is a comparatively homogeneous agricultural zone by Ontario standards and is dominated by the production of cash crop grain corn and soybeans – hence the analysis focuses on these two crops.

Interannual variations in accumulated heat (corn heat units) and

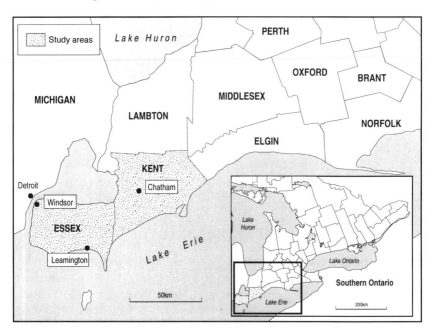

Fig. 11.3. Study area.

moisture conditions (annual moisture deficit and surplus) were compared with variations in three agricultural attributes to check for temporal associations or covariance. In order to address non-climatic stimuli, data reflecting the role of economic conditions and technology were incorporated in the analysis. While the analysis employs both parametric and non-parametric statistical methods (regression and group comparisons) there is no intent or attempt to infer relationships beyond the time period in question. Thus the research is more exploratory than predictive. The specific attributes and measures included in the research are:

1. *Climatic variables*
- Heat conditions: *seasonal corn heat units*
- Moisture conditions: *annual moisture deficit and surplus*
2. *Agricultural attribute variables*
- Crop productivity: *% relative deviation in corn and soybean yield*
- Agricultural land use: *% change in corn and soybean area*
- Agricultural programmes: *% change in corn and soybean area insured*
3. *Non-environmental variables*
- Economic stimulus: *annual soybean/corn price ratio*
- Role of technology: *annual increase in corn and soybean yield 1960–1992*

The role of technology is incorporated in the analysis through an adjustment to yield data. A regression of yield data on a time series reveals a linear, and highly significant, positive relationship. This upward trend provides a credible estimate of the effect of technology on crop productivity over the study period. In the analysis, absolute yields or deviations from mean yield for the period are not examined (both of which would be meaningless for our purposes) but rather deviations from the trend-predicted yield levels – in other words the differences between the expected and attained yield in each year.

Agroclimatic conditions have indeed been variable over the 33-year period, although moisture is much more variable than heat in this region. The coefficients of variation for moisture deficit and surplus are 96 and 132 per cent respectively. When climatic variations are plotted with adjusted yield, it can be seen that corn and soybean producers have experienced significant yield reductions related to variations in moisture deficit conditions over the study period (95% confidence level via regression), with drier conditions resulting in shortfalls in production (Fig. 11.4). Hence there have been 'first order' impacts on farming from variations in weather patterns in the Kent–Essex region.

Given these apparent direct impacts on crop production in the region, one potential response is for adjustment in the amount of land allocated to these crops. Analyses that compared variations in heat and moisture conditions with patterns of corn and soybean production did not provide any clear evidence of an association. However, a regression of crop acres on commodity prices did indicate a significant relationship. Among the obvious explanations is that other factors such as prices simply overwhelm the signals associated with climatic stress. However, it is also possible that the impacts of climatic variations are sufficiently buffered that they are essentially not felt and, in effect, removed from the decision process.

The notion of 'buffering' in the system was explored via an analysis of farmer participation in the provincial crop insurance programme (the 'agricultural programmes' variable). Crop insurance may be conceived as an exogenous factor which mediates the impacts of climatic variations, and may also be viewed as a response to climatic stress in its own right. To investigate the former role an attempt was made to examine land use patterns in light of the possible mediating influence of crop insurance.

By the end of the study period, the Kent–Essex agricultural system was heavily insured against weather-related crop damage. However, because the programme was only initiated part way through the study period (1968), and because of low participation rates in the first few years of the programme, it is possible to distinguish between a period (1960–1979) when the system was relatively uninsured (less than 25% of the crop) and one (1980–1992) during which insurance was more pervasive (over 80% insured by 1992). Differences in the direction and magnitude of area

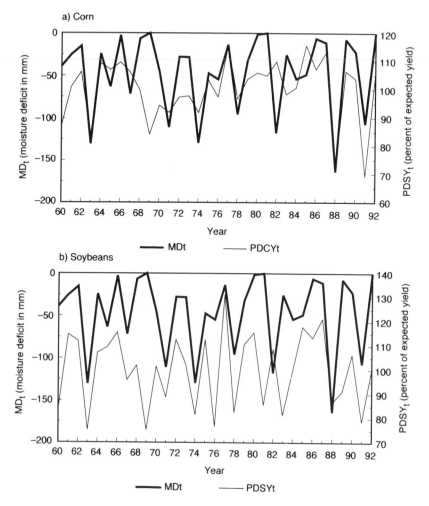

Fig. 11.4. Variation in moisture deficit and adjusted corn and soybean yields.

change between time periods were checked via a comparison of mean area changes following years when climatic conditions were poor versus mean area changes in other years. In the relatively uninsured period, extreme moisture conditions are followed, on average, by a retreat in soybean area (−2%), while climatically favourable years are followed by a sizable average increase (+6.5%), and this difference is statistically significant (see Table 11.2). In the later period, during which crop insurance is more widely used, there is an increase in soybean area after both poor and good years (+1.8% and +2.9% respectively) and the difference in response between favourable and unfavourable years is no longer statistically significant. Based on this

Table 11.2. Mean area change following moisture extremes: differences between periods.

Period	Moisture extremes	Change in soybean area
1960–79	Extreme	$\bar{x} = -1.9$
(insurance low)	Not Extreme	$\bar{x} = \;\;\;6.5$
1980–92	Extreme	$\bar{x} = \;\;\;1.8$
(insurance high)	Not extreme	$\bar{x} = \;\;\;2.9$

simple analysis, there is some evidence that producers' land-use decisions are more strongly affected by climatic conditions before the widespread adoption of crop insurance than they are after its arrival, suggesting the possible mediating role of crop insurance in land-use responses to climatic stress.

This empirical investigation provides certain insights into the relationship between climatic and agricultural systems, and, more generally, a test of some of the elements and issues highlighted in the conceptual framework. Regional crop production has been affected by variability in agroclimatic conditions over the study period. Yet these impacts seem to be less influential on cropping decisions than might be expected. There are indications that economic conditions (commodity prices) and institutional mechanisms (crop insurance) outweigh environmental factors. Hence, in this case at least, the relationship between climate and regional agricultural response is mediated by non-climatic factors.

Conclusion

While the model of agricultural adjustment to climatic variability and change was interpreted in a regional context for this chapter, the concepts embedded within it are relevant at other scales of analysis. The conceptual contributions of the model derive from the linking of several interrelated issues: the notion of cumulative impact from climate, the pathways of climatic impact, the role of economic and institutional forces and the nature of system level adaptation to environmental change as characterized by the types of adjustments that are undertaken.

The methodological approach employed in the Southwestern Ontario case study suggests some of the insights that are possible from hindcast analyses at this scale. Research at an aggregate level permits the use of a longer period of climatic record than would be practical in behavioural research, and thus there are improved opportunities to contrast the impacts and responses to different types and combinations of events, and to explore notions of cumulative impact. Further, this method and scale of analysis offers some understanding of the general role of other factors and forces, either as direct influences on agriculture or as mediators of climatic impacts.

The method has significant limitations as well. First, some types of adjustment are simply not identifiable at the regional scale with the data available (e.g. modified soil management practices). Hence, the analysis is insensitive to the suite of routine or tactical adjustments that might be undertaken in response to climatic stress. This limitation is not just because of an absence of data, but because of the scale of analysis itself. Second, the interpretation of results is problematic. At this scale, interpretations, of necessity, are based on temporal associations for which causal connections must be inferred. While the theoretical model is of some assistance, the establishment of cause and effect relationships is beyond the ability of the approach. Knowledge of producer motivations and perceptions is relevant, but these are typically investigated at the level of the farmer, not the farming region.

A consideration of agricultural sustainability requires a knowledge of the characteristics of structural change brought about by adaptation. Further progress in understanding the complex nature and process of agricultural adaptation to environmental change requires continued effort on both theoretical and methodological fronts, and empirical cases are required to test concepts and methods at different spatial scales and locations. Clearly there are a variety of theoretical perspectives and methodological routes possible in environmental adaptation research. The task is to recognize the potential that each of them has to inform about agriculture and environment interactions and to select conceptual and methodological approaches, and spatial scales of inquiry, that are appropriate to the research questions one wishes to ask.

Acknowledgement

This research was supported by the Great Lakes – St. Lawrence Basin Project of Environment Canada; the Social Science and Humanities Research Council of Canada; the Ontario Ministry of Agriculture, Food and Rural Affairs; and the Eco-Research Program of the Tri-Council of Canada, via the Agroecosystem Health Project at the University of Guelph. The authors are grateful for the editorial advice and assistance of Johanna Wandel.

References

Altieri, M. (1987) *Agroecology: The Scientific Basis for Alternative Agriculture.* Westview Press, Boulder, Colorado.

Bryant, C.R. and Johnston, T.R.R. (1992) *Agriculture in the City's Countryside.* University of Toronto Press, Toronto.

Burton, I., Kates, R.W. and White, G.F. (1993) *The Environment as Hazard.* Guilford Publications, New York.

Chagnon, S.A. (ed.) (1996) *The Great Flood of 1993: Causes, Impacts and Responses.* Westview Press, Boulder, Colorado.

Conway, G. (1987) The properties of agroecosystems. *Agricultural Systems* 24, 95–117.

Crosson, P. (1983) A schematic view of resources, technology and environment in agricultural development. *Agriculture, Ecosystems and Environment* 9(4), 339–357.

Dover, M. and Talbot, L.M. (1987) *To Feed the Earth: Agroecology for Sustainable Development.* World Resources Institute, Washington DC.

Easterling, W.E., Rosenberg, N.J., McKenney, M.S. and Jones, C.A. (1992) Preparing the erosion productivity impact calculator (EPIC) model to simulate crop responses to climate change and the direct effects of CO_2. *Agricultural and Forest Meteorology* 59(1), 17–34.

Economic Council of Canada (1988) *Handling the Risks: A Report on the Prairie Grain Economy.* Queen's Printer, Ottawa.

Gallopin, G. (1994) *Impoverishment and Sustainable Development: A Systems Approach.* International Institute for Sustainable Development, Winnipeg.

Glantz, M. (ed.) (1988) *Societal Responses to Regional Climatic Change: Forecasting by Analogy.* Westview Press, Boulder, Colorado.

Holling, C.S. (1986) The resilience of terrestrial ecosystems: local surprise and global change. In: Clark, W.C. and Munn, R.E. (eds) *Sustainable Development of the Biosphere.* Cambridge University Press, Cambridge, pp. 297–317.

Houghton, J.T., Jenkins, G.J. and Ephraums, J.J. (eds) (1990) *Climate Change: the I.P.C.C. Scientific Assessment.* Cambridge University Press, Cambridge.

Ilbery, B.W. (1991) Farm diversification as an adjustment strategy on the urban fringe of the West Midlands. *Journal of Rural Studies* 7(3), 207–218.

Izak, A.M. and Swift, M. (1994) On agricultural sustainability in small-scale farming in sub-Saharan Africa. *Ecological Economics* 11, 205–225.

Kaiser, H.M., Riha, S.J., Wilks, D.S. and Sampath, R. (1993) Adaptation to global climate change at the farm level. In: Kaiser, H.M. and Drennen, T.E. (eds) *Agricultural Dimensions of Global Climate Change.* St. Lucie Press, Delray Beach, Florida.

Kates, R.W. (1985) The interaction of climate and society. In: Kates, K.W., Ausubel, J.H. and Berberian, M. (eds) *Climate Impact Assessment: Studies in the interaction of climate and society.* John Wiley, Chichester, pp. 3–36.

Kay, J.J. (1991) A nonequilibrium thermodynamic framework for discussing ecosystem integrity. *Journal of Environmental Management* 15, 483–495.

Kay, J.J. and Schneider, E. (1994) Embracing complexity: the challenge of the ecosystem approach. *Alternatives* 20(3), 33–39.

Liverman, D.M. (1987) Forecasting the impact of climate on food systems: model testing and linkage. *Climatic Change* 11:267–285.

Marsden, T., Munton, R., Whatmore, S. and Little, J. (1989) Strategies for coping in capitalist agriculture: an examination of the response of farm families in British agriculture. *Geoforum* 20:1–14.

Marten, G.G. (1988) Productivity, stability, sustainability, equitability and autonomy as properties for agro-ecosystem assessment. *Agricultural Systems* 26:291–316.

Olmstead, C.W. (1970) The phenomena, functioning units, and systems of agriculture. In: Kostrowicki, J. and Tyszkiewicz, W. (eds) *Essays on Agricultural Typology and Land Utilization. Geographica Polanica* 19. Polish Academy of Sciences, Warsaw.

Parry, M.L. (1990) *Climate Change and World Agriculture.* Earthscan Publications, London.

Parry, M.L., Carter, T.R. and Konijn, N.T. (eds) (1988) *The Impacts of Climate Variations on Agriculture,* Volume 1. Reidel Publications, Dordrecht.

Rosenberg, N.J., Crosson, P.R., Frederick, K.D., Easterling, W.E., McKenney, M.S., Bowes, M.D., Sedjo, R.A., Darmstadter, J., Katz, L.A. and Lemon, K.M. (1993) The MINK methodology: background and baseline. *Climatic Change* 24(1), 7–22.

Slobodkin, L.A. and Rapaport, A. (1974) An optimal strategy of evolution. *The Quarterly Review of Biology* 49(3), 181–200.

Smit, B. and Smithers, J. (1994) Sustainable agriculture and agroecosystem health. *Agroecosystem Health: Proceedings of an International Workshop.* University of Guelph, Guelph.

Smit, B. (ed.) (1993) *Adaptation to Climatic Variability and Change: Report of the Task Force on Climate Adaptation.* Prepared for the Canadian Climate Program, Department of Geography Occasional Paper No. 19. University of Guelph, Guelph.

Smithers, J. (1994) *Agricultural Adjustment to Climatic Variability.* Unpublished PhD thesis, Department of Geography, University of Guelph.

Spedding, C.R.W. (1979) *An Introduction to Agricultural Systems.* Applied Science Publishers, Barking, Essex.

Stern, P.C., Young, O.R. and Druckman, D. (eds) (1992) *Global Environmental Change: Understanding the Human Dimensions.* National Academy Press, Washington, DC.

Timmerman, P. (1981) *Vulnerability, Resilience, and the Collapse of Society,* Monograph 1. Institute of Environmental Studies, University of Toronto, Toronto.

Wall, E., Smithers, J. and Wichert, G. (1995) *A Framework for the Integration of Concepts and Research in Agroecosystem Health.* Agroecosystem health discussion paper No. 18. Faculty of Environmental Sciences, University of Guelph, Guelph.

Wheaton, E., Arthur, L., Wittrock, V., Whiting, J. and Shewchuk, S. (1990) *The Drought of 1988: Executive Summary of Some Environmental and Economic Impacts of the 1988 Drought: Saskatchewan and Manitoba.* Publication No. E-2330-4-E90. Saskatchewan Research Council, Saskatoon.

Winterhalder, B. (1980) Environmental analysis in human evolution and adaptation research. *Human Ecology* 8(2), 135–170.

Adaptability of Agriculture Systems to Global Climate Change: A Renfrew County, Ontario, Canada Pilot Study

MICHAEL BRKLACICH[1], DAVID MCNABB[1], CHRIS BRYANT[2] AND JULIAN DUMANSKI[3]

[1]Department of Geography, Carleton University, 1125 Colonel By Drive, Ottawa, Ontario, Canada, K1S 5B6.

[2]Département de Géographie, Université de Montréal, C.P. 6128, Succursale Centre-ville, Montréal, Québec, Canada H3C 3J7.

[3]Research Branch, Agriculture and Agri-Food Canada, Central Experimental Farm, Building #74, Ottawa, Ontario, Canada K1A 0C6.

Conventional Climatic Impact Assessment Framework

Atmospheric scientists and climatologists have for over a decade been presenting compelling evidence that human activities are and will continue to modify the chemistry of the earth's atmosphere, and that these changes will contribute to unprecedented alteration of climate systems on a global scale (Houghton *et al.*, 1990). The prospect of an unprecedented global climate change has stimulated a wealth of studies into the possible impacts of a wide range of scenarios for global climatic change on agriculture (Parry, 1990).

Research frameworks and methods for assessing the potential agricultural impacts of global climatic change have been evolving during this period. Figure 12.1 outlines the basic structure underpinning many of these studies. The specification of multiple scenarios for global climatic change (box 1) is the usual starting point. The use of multiple scenarios reflects the current uncertainty associated with the estimates of the future

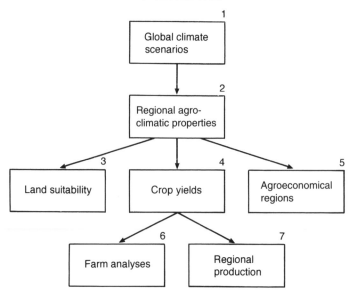

Fig. 12.1. Assessing the agricultural impacts of global climatic change.

evolution of climate regimes (Carter *et al.*, 1994). These global scenarios are often derived from General Circulation Models (GCMs) which have been used to forecast the effects of an altered atmosphere (e.g. a $2 \times CO_2$ atmosphere) on macro-scale climatic properties. Other approaches for specifying climatic change scenarios are based upon historical or spatial analogues (Easterling *et al.*, 1992) and incremental changes to the observed weather record (Bootsma *et al.*, 1984).

The specification of regional agroclimatic properties (box 2) from global scenarios is based upon two processes. One is the spatial interpolation of macro-scale data to a regional scale, the other is the conversion of basic climate parameters (e.g. maximum daily temperature) into agroclimatic properties (e.g. growing-degree days). There are several approaches for conducting these tasks and as a result this has led to methodological inconsistencies among studies. A systematic assessment of the implications of these various methods on the resultant estimates of regional agroclimatic properties has not been undertaken. Thus it is not currently possible to estimate the extent to which subsequent impact analyses reflect the specified climatic changes rather than variations in the interpolation methods used to convert global climatic change scenarios into regional agroclimatic properties (Cohen, 1990).

Many impact studies focus on land suitability (box 3) and agroecological assessments (box 4), which typically employ resource rating schemes to assign land parcels to broadly defined suitability or agroecological classes

(Carter *et al.*, 1990; Brklacich and Curran, 1994). The rating schemes are usually based on comparisons of basic climatic requirements (e.g. growing period, moisture supply) for broad categories of agricultural production (e.g. cereal grains, pulses) in relation to specified shifts in selected climatic properties.

Crop yield analyses (box 5) (McKenney *et al.*, 1992; Baethgen and Magrin, 1995) are also based upon comparisons of the availability of production inputs to production requirements, but differ from land suitability and agroecologic assessments in three ways. First, yield analyses are conducted for specific crops (e.g. winter wheat) rather than categories of agricultural production. Second, crop development and yield models are analytically more sophisticated than resource rating schemes. The current generation of crop models consider interactions among crop growth factors whereas resource rating schemes have limited capability in this area. Third, crop models generate estimates of output per land unit (e.g. kg ha^{-1}) rather than providing general assessments of land potential (e.g. high capability).

Two other approaches for assessing the agricultural impacts of global climatic change have utilized the outputs from crop yield analyses. Farm level analyses (box 6) have employed whole farm models to estimate impacts of yield changes stemming from global climatic change on cash flow and vulnerability of different farm types (Mooney and Arthur, 1990). At the broad scale (box 7), regional production and macroeconomic models have been utilized to estimate the effects of global climatic change on regional production potential (Adams *et al.*, 1990) and international trade of agricultural products (Fischer *et al.*, 1995).

Contributions and Limitations of Conventional Frameworks

The framework outlined in Fig. 12.1 resembles standard scientific practices that have been applied to field level agricultural research. With this approach, a control plot is used to represent baseline conditions and performance indicators such as biomass and grain yield would be observed. Simultaneously, other plots are subjected to altered conditions or treatments such as alternative application rates for fertilizer or irrigation, and changes in the performance indicators would be attributed to the treatment.

In the context of research into the agricultural impacts of climatic change, the current climatic regime represents baseline conditions and the scenarios for global climatic change are conceptually equivalent to treatments. Overall, this approach has relied heavily on land resource and crop modelling methods, and has been instrumental in isolating the sensitivity of particular attributes of agricultural production systems to specified climatic perturbations.

However, this approach has led to the irresistible temptation to extrapolate these sensitivity assessments and to predict responses for entire agricultural systems to climatic change. Assumptions that underpin these extrapolations are often left unstated, and include:

1. climate is the only condition that will vary;
2. farmers will perceive the change in climate;
3. agricultural systems are vulnerable to the changed climate; and
4. farmers will therefore choose to adapt to the altered climate.

The Blasing and Solomon (1983) study is typical of this type of extension, but it is by no means the only study that has extrapolated results in this way[1]. Their assessment, which investigated the impacts of a climatic change scenario on agroclimatic properties that influence grain corn production, delineated the area within the continental USA with an agroclimatic potential for corn under the current and an altered climate. The conclusion drawn from the study was that a 1°C temperature increase would move the Corn Belt in the United States 175 km to the north and northeast of its present location.

While the future spatial displacement of conditions that are physically suitable for the production of a particular agricultural activity will potentially impact the prospects for agriculture, this sort of information by itself provides little insight into the vulnerability of agricultural systems to changing conditions and the capacity of agriculture to adapt to change (Smit, 1993, Carter *et al.*, 1994; HDP, 1994, Chiotti and Johnston, 1995).

In order to investigate the adaptive capacity of agricultural systems to potential changes in climatic and other conditions that influence agriculture, there is a need to revise the conventional research framework and it is argued here that it is essential to develop frameworks which place climatic change research into a broader context of agricultural decision-making. This would facilitate a new generation of climatic impact assessments which could address questions relating to:

1. Do farmers perceive a change (in climate and/or other conditions)?
2. What role does climate play in agricultural decision-making relative to other influences including other environmental, economic, political and sociocultural factors?
3. Is the farm in any way vulnerable to the changing conditions?
4. If the farm is vulnerable, what is the perceived range of adaptive responses?
5. Which of these adaptive responses could be implemented?
6. Which of the feasible adaptive responses comes closest to meeting the

broader goals for farming, such as desired production and income levels, consumer satisfaction, and so on?

Addressing these questions suggests the need for a different research framework than the one outlined in Fig. 12.1. This alternative approach would need to be based on agricultural decision-making, and climatic change by itself would not be the departure point for the research. Rather the revised research framework would need to consider explicitly the role of climate and climatic change *vis-à-vis* other factors that influence agricultural decision-making.

Renfrew County Pilot Study

Purpose and overview

The Renfrew County pilot study represents an attempt to expand the context of research into the impacts of global climatic change on agricultural systems. Its overall purpose is to gain insight into the adaptability of agriculture in Renfrew County to changing conditions, and to examine the importance of climate and climate change relative to other factors influencing agricultural decision-making at the farm level. This is accomplished by examining previous farm-level decisions, and investigating responses to perceived and hypothetical changes to climatic change.

The research framework used in the Renfrew County pilot study has three major components:

Component 1: focus group meeting I

The pilot project commenced with a focus group meeting with local area farmers, and was arranged in conjunction with the Ontario Federation of Agriculture and held in Renfrew County in December 1994. Background information describing the purpose of the project and workshop was made available to the participants about one week in advance of the meeting. The meeting served several important functions including:

1. establishing the basis for collaboration with the local farming community;
2. identifying factors that have prompted agricultural adaptation over the past decade;
3. developing an inventory of adaptive responses; and
4. providing a foundation for the development of an agricultural adaptation survey.

Component II: farmer interviews

The information gathered during the focus group meeting was used to develop a questionnaire probing farm level decision-making processes. The survey gathered information on:

1. background information on the farm's history, present production and farm structure;
2. farm level decision-making practices and farm changes since 1984; and
3. climate, climatic change and agriculture.

The structure of the questionnaire purposefully commenced with questions soliciting information on broad issues regarding agricultural adaptation over the recent past and then proceeds to investigate the importance of climate and potential climatic change in the decision-making process. This approach was employed in order to unravel the importance of climate and a potential climatic change relative to other factors in agricultural decision-making at the farm level.

A pretest of the questionnaire was completed in March 1995 and minor adjustments were made to the survey instrument. In April 1995, personal interviews were administered to 30 randomly selected farmers in Renfrew County.

Component 3: focus group meeting II

Participants from the first focus group meeting were reassembled early in 1996. At this second meeting, participants had the opportunity to review findings and to evaluate the methods employed in the Pilot Study. This provided an opportunity to ensure that the collected data were interpreted accurately and also provided a basis for enhancing the research methods.

Renfrew County agricultural profile

Renfrew County is located in the northeast portion of eastern Ontario, adjacent to the Ottawa River (Fig. 12.2). The frost-free season in the agricultural regions of Renfrew County is in the 125 to 135 day range, and begins in mid-May. Summer temperatures are relatively cool, with long-term normals for average daily temperatures in June, July and August being 18°C to 19°C. Accumulated corn heat units (CHUs) range from 2300 and 2500 CHUs per year. Average annual precipitation is about 76 cm and is distributed fairly uniformly throughout the year. Surface water is abundant throughout Renfrew County (Gillespie *et al.*; 1964, Keddie and Mage, 1985).

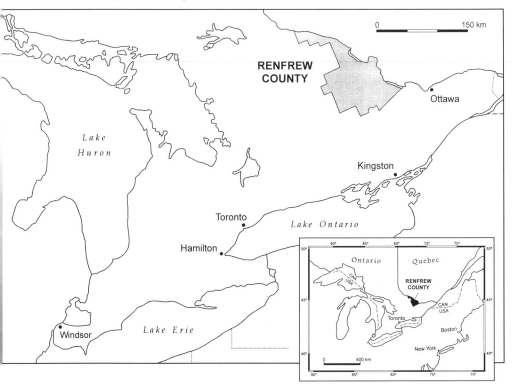

Fig. 12.2. Location of Renfrew County, Ontario.

More than half of the County's 780,000 ha is Precambrian Shield and without potential for agriculture (Gillespie *et al.*, 1964). Land with a physical potential for agriculture occurs within the Ottawa Valley lowlands. Land with no significant limitations for agriculture (e.g. Canada Land Inventory Class 1 for Agriculture) do not occur in Renfrew County. The majority of the land used for agriculture in Renfrew County has moderate to severe limitations for arable agriculture, and land rated as CLI 2 and 3 accounts for about 25% of the County's land base (Hoffman and Noble, 1975).

Similar to trends throughout the Province of Ontario, the number of farms and farmland area in Renfrew County continues to decline (OMAF, 1992). For the period 1976 to 1991, the total number of farms decreased from 1842 to 1505, while total farmland was reduced from 200,000 ha to 165,000 ha. Approximately one-third of the farmland is in crop production, and used primarily to support a beef and dairy based livestock sector. Hay is the dominant crop, accounting for about two-thirds of Renfrew County's cropland. The remaining cropland is used primarily for the production of spring-seeded small grains, with grain and fodder corn being of minor importance.

The decision-making environment

The Renfrew County pilot study investigated the decision-making environment, with specific questions probing factors that influenced farm level decision-making and information sources. This included questions about decision-making structures, the relative importance of a range of farmer identified factors influencing decisions, and information sources.

Not surprisingly, the decision-making environment was characterized by a complex set of interrelated factors (Table 12.1). Important factors identifed by farmers affecting agricultural decisions included economic factors, environmental conditions, changes in agricultural technology, off-farm factors, evolution and restructuring of the rural community and

Table 12.1. Importance of factors influencing farm management decisions.

Factors	No. of respondents indicating important or very important
Economic	
Commodity prices	23
Production costs	22
Off-farm work	16
Labour costs	11
Environmental	
Climate	18
Soil	15
Water	11
New technologies	
Cost	14
Availability	9
Extra-agricultural	
Public perception	10
Public demands	14
Rural community	
Rural population change	11
Peer pressure	4
Public programmes	
Insurance premiums	11
Incentives	6
Regulations	6

Sample size = 30.

agricultural policies and programmes.

Farmers indicated economic conditions were the most important factors influencing agricultural decision-making. Agricultural decisions were based heavily upon commodity prices and production costs, as well as labour costs. The availability of off-farm work to supplement agricultural incomes was also seen as an important factor influencing agricultural decision-making.

With respect to overall importance, a second set of factors involving environment, technology and extra-agricultural concerns emerged. Agricultural decisions were to a large extent shaped by the availability and quality of climate, soil and water resources. The cost and availability of emerging agricultural technologies as well as urban perceptions of agriculture were identified as frequently as environmental concerns as factors influencing agricultural decisions.

The evolution of the rural community and public programmes were also important factors. However, these factors were clearly of lesser importance than economic, environmental, technological and extra-agricultural factors. This suggests that the rural community and public programmes act as tempering agents rather than primary factors influencing agricultural decisions.

Even though the role of the family farm within agricultural systems has diminished considerably and it is now widely accepted that agricultural decision-making is a function of an ever widening range of factors beyond the farm gate, family members and other farmers in the local community provided the most important sources of information influencing on-farm decisions in Renfrew County (Table 12.2). This does not suggest that farmers in Renfrew County failed to consider non-local conditions in their decision-making, but rather it indicates that the farm family and the local

Table 12.2. Influence of information sources on farm-level decision-making.

	No. of respondents indicating:	
Information source	Most important	First, second or third most important
Family members	19	25
Other farmers	5	15
Government	4	12
Farm organizations	1	3
Media	0	3
Employees	0	2

Sample size = 30.

farming community continue to play a pivotal role in the provision and weighing of information.

Overall, the agricultural decision-making environment is characterized by a complex interplay among many factors. Economic, environmental, technological and extra-agricultural factors were identified by farmers as factors exerting the greatest influence, and the farm family and local agricultural community remain instrumental in agricultural decision-making.

Recent farm level changes

The last decade has been characterized by substantial changes to agricultural systems in Renfrew County. The agricultural sector has been far from static, with several farms reporting changes to cropping systems, livestock systems, land resources and farm structure (Table 12.3). The changes were often in response to several factors, representing complex adjustments involving multiple aspects of the farm operation and implemented over a period of time ranging up to several years.

The majority of the modifications to cropping and livestock production systems were relatively conservative changes involving adjustments to the mix of agricultural commodities produced on the farm. This sort of behaviour is as expected, as on-farm changes usually occur in small increments and do not typically involve wholesale change. More radical shifts involving the introduction of new crops or new types of livestock were undertaken by considerably fewer farmers.

The overwhelming factor prompting changes to cropping and livestock systems was to enhance economic and labour efficiency. The overall trend was to make better use of labour saving agricultural technologies and to decrease reliance on hired labour. In addition to reducing production costs, decisions to reduce hired labour were also tied to protecting investments in expensive technology. Superior management skills are required to operate modern agricultural machinery, and several farm operators expressed a reluctance to rely on non-family members. Modifications to cropping and livestock systems were also influenced to a large extent by market changes, including increasing demand for agricultural commodities and enhanced demand for premium quality livestock products such as leaner beef products. These sorts of market forces, coupled with the need to improve production efficiency, also encouraged several farmers to improve the overall quality of their livestock herd.

Environmental conditions in general, and more specifically climatic conditions, were relatively unimportant factors influencing agricultural decisions. Decisions to implement soil conservation practices were directly associated with improving soil quality, but reasons underpinning adjustments to production and management systems, insurance programmes,

Table 12.3. Farm level changes since 1984.

Cropping	Livestock	Land	Farm structure
Production (16)	Production (26)	Land area (15)	Hired labour (13)
Change	Change	Change	Change
Crop mix (11)	Livestock mix (21)	Increase (11)	Decrease (11)
New crops (3)	New livestock (6)	Decrease (4)	Increase (2)
Reasons	Reasons	Reasons	Reasons
Economic/labour	Economic/labour	Demand (10)	Family (5)
efficiency (11)	efficiency (11)	Family (1)	Technology (5)
Demand/quality (4)	Demand/quality (9)	Economic/labour	
Environmental (2)	Family (5)	efficiency (1)	
	Risk reduction (4)		
	Environment (1)		
Management (19)	Management (18)	Management (15)	Capital investment (22)
Change	Change	Change	Change
Agricultural	Improve herd (13)	Improve land (9)	Technology
chemicals (9)	Feed rations (3)	Conservation (7)	increase (22)
Harvesting (10)			More quota (9)
Reasons	Reasons	Reasons	Reasons
Economic/labour	Economic/labour	Economic/labour	Economic/labour
efficiency (14)	efficiency (8)	efficiency (14)	efficiency (22)
Demand/quality (1)	Demand/quality (9)	Environmental (7)	Demand/quality (2)
Environmental (2)			
Insurance (11)			
Change			
Increase (8)			
Decrease (3)			
Reasons			
Risk reduction (4)			
Too expensive (2)			
Environmental (2)			

Note: Response frequencies in parentheses. Sample size equals 30.

and farm structure were in most cases not based on reasons relating to environmental concerns and issues.

Table 12.1 indicates environmental factors exert considerable influence over farm management decisions, whereas many of the recent farm level changes summarized in Table 12.3 suggest agricultural decisions are relatively independent of environmental concerns. Whereas it is only possible to speculate on this apparent contradiction in the evidence compiled under this pilot study, it is possible that environmental factors are critical in establishing the boundary conditions which determine in the broadest sense which agricultural commodities are produced. The majority of recent changes to crop and livestock production systems were previously characterized as fine-tuning of an existing system, and more substantive adjustments were comparatively rare.

One interpretation of the evidence presented in Tables 12.1 and 12.3 is that the majority of the farms in the pilot study were able to cope with the range of environmental factors experienced over the recent past. For example, some farmers indicated that round balers had reduced the vulnerability of haying to weather conditions. Wet conditions no longer resulted in delays and a decline in crop quality. The capacity to ensile the round bales in plastic bags reduced the risk and made haying operations less sensitive to weather conditions. Overall, this one example provides additional evidence that industrialized agricultural systems are becoming less reliant on natural resource inputs and becoming increasingly dependent upon managerial skills and agricultural technology.

Climatic change perception

The vast majority of the farmers interviewed during the Renfrew County pilot study believed climate had changed over the past two decades (Fig. 12.3). The consensus opinion was that precipitation had decreased and climate was becoming less predictable. There was no apparent consensus regarding changes to temperature.

Adaptive Responses to Climatic Change

Perceived climatic changes over the past 20 years

The 27 farmers who perceived climate had changed over the past two decades were asked to indicate what, if any, adjustments were made to their farm operations in response to the altered climate. Explicit responses to a perceived climatic change was limited to 20% of this subset of 27 farmers. The adaptations included changing crop varieties and types, adoption of alternative harvesting methods and infrastructure additions.

Overall, these findings support the earlier assertion that an altered climate does not automatically imply a response in agricultural operations. One interpretation of these findings is that the perceived climatic changes were not sufficient to warrant a direct or explicit response for the majority of the farmers in this pilot study. Whereas the reasons underpinning the decision not to respond are complex and cannot be isolated from other factors prompting agricultural adaptation in this pilot study, it would appear

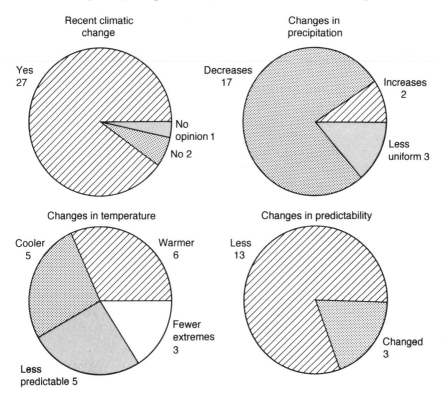

Fig. 12.3. Perception of recent climatic changes.

the lack of response to the perceived changes in climate was attributed to:

1. many of the farm operations were already adapted to operate under a range of conditions, including the perceived changes in climate over the past two decades, and hence a specific response was not warranted; and
2. the importance of weather and climate *vis-à-vis* other factors in farm decisions was declining.

Hypothetical climate changes in the future

The future climatic change scenario used in the pilot study was derived from the Goddard Institute for Space Studies (GISS) research into the effects of a $2 \times CO_2$ atmosphere on climate patterns (Hansen *et al.*, 1983). The scenario portrayed a future climate that would increase the frost-free season by 25 days, increase accumulated CHUs by 55%, and increase precipitation by 25%. Under this altered climate, grain corn, cereal grain and soybean yields would be expected to increase by 50%, 15% and 40% respectively. The information included in the survey did not specify a time frame for these changes other than some time in the future.

The majority of the respondents (70%) indicated these changes to climate would prompt a response. The main responses suggested farmers would employ crop varieties that could benefit from longer growing periods or that they would consider growing crops that were not suited to the current climate in Renfrew County. The climatic change scenario was not perceived as having a direct impact on livestock operations, but only indirectly through the availability of feed crops. The proposed responses were supported by a desire to increase economic efficiency, expand production and enhance profits.

Whereas the majority of respondents indicated a climatic change equivalent to the GISS scenario would prompt adaptations at the farm level, many of the proposed adaptations were vague and non-specific. These sorts of responses are common to farmers when confronted with hypothetical situations. Furthermore, some respondents indicated that farm decisions were at least as sensitive to the timing of weather events, extreme climatic events and climatic variability as they were to changes in average conditions. This information is unavailable from GCMs and this in turn suggests there is a need to re-evaluate the use of GCM output in the context of assessing agricultural adaptation to global climate change.

Summary

The findings from the Renfrew County pilot study indicate that current agricultural systems have a certain degree of built-in elasticity and that not all changes in socioeconomic and biophysical conditions call for a farm level adjustment. The extent to which agricultural systems can respond is highly dependent upon the overall context governing agriculture in a particular region. Agriculture in Renfrew County is currently characterized by a relatively cool climate, abundant water and diversified grass-based livestock systems. These types of diversified agricultural enterprises are often robust and have the internal capacity to function effectively under a relatively wide range of conditions, and thereby reduce vulnerability to changes in climate and other factors that shape agriculture.

Agricultural decision-making is a complex process, and perceived changes to the prevailing climatic regime do not always result in farm level adjustments. It is unreasonable to consider climate and climatic change as the primary forces underpinning agricultural decision-making. There is an urgent need to reframe the research context for investigating relationships among climate, climate change and agriculture, and to examine more carefully agricultural decision-making and adaptation processes.

Acknowledgements

This research was supported by funding from Agriculture and Agri-food Canada and Environment Canada, and C. Merminod assisted with the workshops, survey design and interviewing. Collaboration with the Ontario Federation of Agriculture, the Canadian Federation of Agriculture and farmers throughout Renfrew County was an essential component of this research and was greatly appreciated.

Note

1. For other examples, see Rosenzweig (1985); Carter *et al.* (1990).

References

Adams, R., Rosenzweig, C., Peart, R., Ritchie, J., McCarl, B., Glyer, D., Curry, R., Jones, J., Boote, B. and Allen, L. (1990) Global climate change and US agriculture. *Nature* 345, 219–224.

Baethgen, W. and Magrin, G. (1995) Assessing the impacts of winter crop production in Uruguay and Argentina using crop simulation models. In: Rosenzweig, C., Allen, L., Harper, L., Hollinger, S. and Jones, J. (eds) *Climate Change and Agriculture: Analysis of Potential International Impacts*, ASA Special Publication Number 59. American Society of Agronomy, Madison, pp. 207–228.

Blasing, T. and Solomon, A. (1983) *Responses of North American Corn Belt to Climatic Warming.* DOE/N88-004, US Department of Energy, Carbon Dioxide Research Division, Washington.

Bootsma, A., Blackburn, W., Stewart, R., Muma, R. and Dumanski, J. (1984) *Possible Effects of Climatic Change on Estimated Crop Yields in Canada*, LRRI Contribution No. 83–64, Research Branch, Agriculture Canada, Ottawa.

Brklacich, M. and Curran, P. (1994) Climate change and agricultural potential in the Mackenzie Basin. In S. Cohen (ed.) *Mackenzie Basin Impact Study Interim Report #2: Proceedings of the Sixth Biennial AES/DIAND Meeting on Northern Climate Environment Canada.* Downsview, pp. 459–464.

Carter, T., Parry, M. and Porter, J. (1990) Climatic change and future agroclimatic potential in Europe. *International Journal of Climatology* 11, 251–269.

Carter, T., Parry, M., Harasawa, H. and Nishioka, S. (1994) *IPCC Technical Guidelines for Assessing Climate Change Impacts and Adaptations.* Department of Geography, University College, London and Centre for Global Environmental Research, National Institute for Environmental Studies, Japan.

Chiotti, Q. and Johnston, T. (1995) Extending the boundaries of climate change research: a discussion on agriculture. *Journal of Rural Studies* 11, 335–350.

Cohen, S. (1990) Methodological issues in regional impacts research. In: Wall, G. and Sanderson, M. (eds) *Climate Change: Implications for Water and Ecological*

Resources. Department of Geography, Occasional Paper No. 11, University of Waterloo, Waterloo, pp. 31–43.

Easterling, W., Rosenberg, N., McKenney, M., and Jones, C. (1992) An introduction to the methodology, the region of the study, and a historical analog of climate change. *Agricultural and Forest Meteorology* 56, 3–15.

Fischer, G., Frohberg, K., Parry, M. and Rosenzweig, C. (1995) Climate change and world food supply, demand and trade. In: Rosenzweig, C., Allen, L., Harper, L., Hollinger, S. and Jones, J. (eds) *Climate Change and Agriculture: Analysis of Potential International Impacts,* ASA Special Publication Number 59. American Society of Agronomy, Madison, pp. 341–382.

Gillespie, J., Wicklund, R. and Matthews, B. (1964) *Soil Survey of Renfrew County,* Report No. 37. Ontario Soil Survey, Research Branch, Agriculture Canada, Ottawa.

Hansen, J., Russell, G., Rind, D., Stone, P. Lacis, A., Lebedeff, S., Ruedy, R. and Travis, L. (1983) Efficient three-dimensional global models for climate studies: models I and II. *Monthly Weather Review* 4, 609–662.

HDP (1994) *Human Dimensions of Global Environmental Change Programme Work Plan 1994–1995,* Occasional Paper Number 6. International Social Science Council, Geneva.

Hoffman, D. and Noble, H. (1975) *Acreages of Soil Capability Classes for Agriculture in Ontario.* Ministry of Agriculture and Food, Toronto.

Houghton, J., Jenkins, G. and Ephramus, J. (eds) (1990) *Climatic Change: The IPCC Scientific Assessment.* Cambridge University Press, Cambridge.

Keddie, P. and Mage, J. (1985) *Southern Ontario Atlas of Agriculture: Contemporary Patterns and Recent Changes.* Department of Geography, University of Guelph, Guelph.

McKenney, M., Easterling, W. and Rosenberg, N. (1992) Simulation of crop productivity and responses to climate change in the year 2030: the role of future technologies, adjustments and adaptations. *Agricultural and Forest Meteorology* 56, 103–127.

Mooney, S. and Arthur, L. (1990) The impacts of climate change on agriculture in Manitoba. *Canadian Journal of Agricultural Economics* 38, 685–694.

OMAF (1992) *1991 Agricultural Statistics for Ontario.* Ontario Ministry of Agriculture and Food, Toronto.

Parry, M. (1990) *Climate Change and World Agriculture.* Earthscan Publications, London.

Rosenzweig, C. (1985) Potential CO_2-induced climate effects on North American wheat-producing regions. *Climatic Change* 7, 367–389.

Smit, B. (ed.) (1993) *Adaptation to Climatic Variability and Change.* Department of Geography, University of Guelph, Guelph.

13 Agricultural Response to Climate Change: A Preliminary Investigation of Farm-level Adaptation in Southern Alberta

QUENTIN CHIOTTI[1], TOM JOHNSTON[2], BARRY SMIT[3] AND BERND EBEL[4]
[1]*Environmental Adaptation Research Group, Institute for Environmental Studies, University of Toronto, 33 Willcocks Street, Toronto, Ontario, Canada M5S 3E8;* [2]*Department of Geography, The University of Lethbridge, 4401 University Drive, Lethbridge, Alberta, Canada T1K 3M4;* [3]*Department of Geography, University of Guelph, Guelph, Ontario, Canada N1G 2W1;* [4]*Department of Economics, The University of Lethbridge, 4401 University Drive, Lethbridge, Alberta, Canada T1K 3M4*

Introduction

Growing concern over increasing atmospheric concentrations of carbon dioxide and other 'greenhouse' gases has stimulated a plethora of interdisciplinary environmental change research (Rabb, 1983; Henderson-Sellers, 1991a). Much of this work has been directed towards developing baseline scenarios of future climates and associated biophysical conditions, including assessments of effects and impacts upon human systems. In recent years there has been growing interest in the adaptability of human systems, particularly in the area of agriculture (Easterling *et al.*, 1989). Agriculture has attracted special attention due to a number of factors, including the need to ensure that food production is not threatened (IPCC,

1995), the vulnerability of agriculture (especially in developing countries) to climate change and the influence that relatively simple land management adjustments can have in reducing the adverse effects from climate change (Easterling *et al.*, 1993; Rosenzweig and Parry, 1993). A wide knowledge gap, however, exists between modelling broad parameters of climate change and an understanding of regional impacts and agricultural adaptations. In part, this gap can be attributed to the theoretical perspectives and methodological frameworks that conventional approaches employ, especially their assumptions regarding adaptive behaviour and farm-level decision-making.

In this chapter the adaptive strategies adopted by farm households are examined within the context of agricultural restructuring and the environmental stresses arising from recent variations in climate. The primary objective is to begin the process of linking empirical evidence with theoretical constructs developed in the agricultural restructuring literature. As in Chapters 11 and 12, the primary focus of the investigation is on the regional and farm-level scale, and land management responses in particular. First, an outline of the conventional climate change impact assessment approach and the knowledge gaps pertaining to farm-level decision-making and adaptation is given. This chapter then draws upon the agricultural restructuring literature to develop an integrative framework that offers a more critical understanding of how land management decisions can be influenced by a complex of external and internal forces. Due to space constraints, the outline and development of the framework is brief, condensing a more thorough discussion articulated elsewhere (Chiotti and Johnston, 1995). In the next section a regional description of agriculture in southern Alberta is presented, providing the broader social and environmental context within which farmers operate. The results from an investigation of three farming systems in southern Alberta, which was carried out in the spring and summer of 1994, are then analysed. For each farming system aspects of climatic variability to which farmers are sensitive are identified, the importance of climate *vis-à-vis* other societal forces in the agricultural adaptation process is appraised, and land management responses documented. This chapter concludes by discussing how the concept of locality can be used to develop a more fully informed appreciation of climate change impacts, as well as addressing the significance of the findings in terms of sustainable agriculture.

Climate Change and Agriculture

During the past decade concern over global climatic change has become a major research area (Handel and Risbey, 1992). A voluminous body of literature now exists that addresses the physical science of atmospheric processes and change, as well as the environmental and socioeconomic impacts of possible changes in the climatic regime (Houghton *et al.*, 1990). In the latter context, the scenario-based impact assessment approach has emerged since the late 1970s as the most widely used research protocol in assessing the possible implications of environmental change (Gates, 1987). Typically, a set of future climates is postulated, sometimes from regional or temporal analogues, or from General Circulation Models (GCMs), which are then superimposed upon a study area, and the effects are estimated for whatever system is the subject of inquiry.

Given the nutritional, economic and social importance of food and fibre production, it is not surprising that agriculture has received considerable attention from climatic change impact assessment analysts (Parry, 1990). Agricultural impact studies have been conducted at various spatial scales, ranging from the regional (Cohen, 1994) to the national/ continental (Rotmans, 1990; United Kingdom Climate Change Impacts Review Group, 1991), to the global (Rosenzweig and Parry, 1994), as well as for major crop (Smit *et al.*, 1989; Easterling *et al.*, 1993), and livestock production systems (Baker *et al.*, 1993). Many of these studies employ quite sophisticated conceptual frameworks of analysis, often taking into consideration first-order impacts, higher-order ones, and in some cases even impacts between different economic sectors (e.g. agriculture and water resources). However, despite much analytical and conceptual progress in scenario development, researchers interested in assessing the impacts of climate change face several noteworthy challenges. Many concerns are already a matter of record (Cohen, 1988; Henderson-Sellers, 1991b; Easterling *et al.*, 1993), ranging from computational limits imposed by existing computing technology, to cascading error when assessing higher-order effects. As a result of these limitations, tremendous uncertainty surrounds the results that conventional climate change impact assessment studies generate, especially at the regional scale of analysis.

Despite recent efforts to advance an understanding of agricultural adaptation (e.g. Easterling *et al.*, 1993; Smit, 1993), critical gaps remain within much of the literature on the potential impacts of climate change on agriculture, including the narrow treatment of non-climatic forces, and particularly the implicit assumption that climatic influences control human decisions *vis-à-vis* agriculture. Indeed, there is a need to understand better the relationship between *present* climate and agriculture, before estimating future impacts (Wheaton, 1994). Much of the conventional research on climate change impacts assessment is grounded within the neoclassical economic paradigm, which assumes that the 'invisible hand' of the market place will encourage or discourage various adjustments. Under such theoretical assumptions, the land resource will be devoted to the so-called highest and best use (Mendelsohn *et al.*, 1994), as 'smart-farmers' access available technologies and successfully adjust farming practices and operations to suit a changing and variable climate (Easterling *et al.*, 1993).

Whereas such assumptions may be useful for modeling purposes, especially under an 'Everything Else Remains Equal' scenario, they fail to appreciate three important aspects of modern commercial agriculture that could significantly influence land management responses. First, the complexities of social and political–economic conditions tend to be overly simplified, especially the multivariate nature of the forces which drive agricultural adjustments. The role of the market place and government policy (e.g. agricultural support, crop insurance and monetary exchange rates), for example, is known to influence the decision-making process (Warrick and Riebsame, 1983; Maunder, 1989; Smit, 1993), but are rarely incorporated into impact assessment models. Second, an understanding of the complex adaptive process is also obscured by the limited range of adjustments and policy responses that are typically addressed in the conventional approach. As Mendelsohn *et al.* (1994) note, the climate change literature needs to consider a broader range of adjustments available to farmers, such as 'the 1001 other productive uses of land in a modern postindustrial society' (p. 754). Third, conventional approaches tend to underemphasize what Dow (1992) refers to as the processes creating exposures and shaping coping abilities, thereby neglecting the social and economic reasons why farmers may be reluctant or unable to implement adaptation measures (Rosenzweig and Parry, 1994).

In terms of redressing these knowledge gaps, researchers interested in climate change impact assessment do not have to venture into theoretically barren territory. There is a growing body of literature in the fields of natural hazards and underdevelopment which adopts a neo-Marxist perspective emphasizing the importance of understanding how and why farming systems are vulnerable to climate change. This literature tends to address what Parry (1986) refers to as the constraints on choice, and specifically the social–economic and political factors contributing to vulnerability. Much of this literature, however, is focused on the relationship between drought and famine, almost exclusively in a developing world context (Garcia and Escudero, 1981; Susman *et al.*, 1983; Watts, 1983; Blaikie and Brookfield, 1987), and only occasionally in the context of climate change (Liverman, 1989, 1991). Consequently, whereas this literature provides useful insights regarding the relationship between society, environment and land-based resources, there is a need to look elsewhere for a framework more applicable to modern commercial agriculture and climate change.

Adaptation and Agricultural Restructuring

The field of agricultural geography, and particularly agricultural restructuring, provides a rich body of scholarship that examines agricultural landscapes in developed countries at various spatial scales. In the past decade, a political-economy perspective has emerged, advocating a broader conceptual and more critical theoretical framework to advance the understanding of changing agricultural landscapes (Marsden *et al.*, 1986; Bowler and Ilbery, 1987; Bowler, 1989). One thrust has been directed towards the broad structural forces shaping agricultural development (Buttel, 1989; Friedmann and McMichael, 1989; Goodman and Redclift, 1989), and another emphasizing the importance of locality and human agency in determining the postindustrial landscape (Moran, 1988; Marsden *et al.*, 1989; Bryant and Johnston, 1992; Ilbery 1992). Still others have drawn upon the concept of uneven development (Smith, 1984), arguing that agricultural landscapes are created through the interaction of dialectical processes played out in localities (Roberts, 1992; Munton *et al.*, 1992), where, among other things, state intervention and human agency shape the environment within which farmers operate.

Drawing from these bodies of scholarship, an integrative framework can be envisaged (Fig. 13.1) which provides for a more comprehensive understanding of farm-level decision-making and adaptation, recognizing that land management decisions are socially constructed, environmentally influenced and historically contingent (Blaikie and Brookfield, 1987). At the farm-level of analysis, decision-making can be influenced by the relations internal to the family household, such as gender and intergenerational considerations, and external relations between farm businesses and finance capitals. Farm-level adaptations, such as adjustments in land management, are then viewed as responses to a complex of motivating factors, incorporating several stimuli (Box 13.1). Land management adjustments, however, are merely one form of numerous adaptations undertaken by farmers. Responding to the international farm crisis, farm businesses have been found to follow different 'pathways of development', adopting a variety of adjustment (or survival) strategies which redeploy human and financial capital in various ways on and off the farm (Whatmore *et al.*, 1987; Bowler, 1992). One can also expect that the processes influencing decision-making and adaptive responses will be spatially variable, reflecting an uneven pattern of capital penetration and technological control over nature, yet shaped by the specific locality within which farmers, and agricultural systems, operate (Lobao, 1990). Consequently, the interplay of societal forces and farm-level adjustments vary spatially and temporally, conditioned by the availability of, and access to, marketing opportunities (Bryant and Johnston, 1992), credit (Marsden *et al.*, 1989), and irrigation (Roberts, 1992).

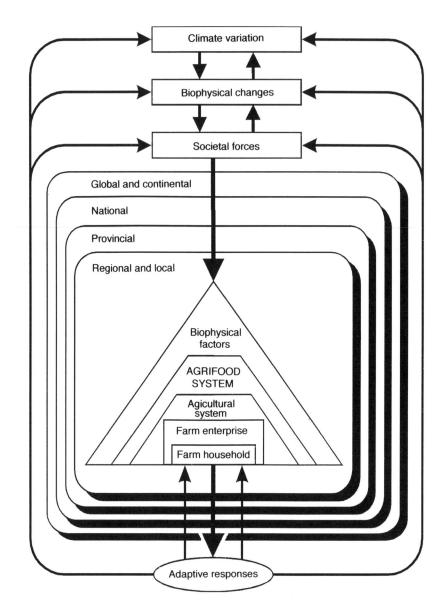

Fig. 13.1. The climate change – farm-level decision-making nexus.

Whereas this integrated approach provides a more critical framework from which to assess farm-level adaptation within the broader context of agricultural restructuring, it can also be extended to include climate change (Chiotti and Johnston, 1995). When extended, and applied to the

Box 13.1. Motivating factors and farm level responses in an integrated approach.

Climatic factors
  Temperature
  Precipitation
  Sub-soil moisture
  Available heat units
  Length of growing season
  Frequency, severity and duration of extreme climatic events (such as heavy rains, hail storms, heat waves, cold spells, drought)

Societal forces
  Economic conditions
  Family and demographic considerations
  Government policies
  Market conditions
  Technological considerations

Adjustments (changes in)
  Production systems and land management practices
  Soil and water conservation practices
  Capital investments
  Credit arrangements
  Crop and livestock production
  Farm size and tenure arrangements
  Labour inputs
  Legal status
  Marketing channels and contracts
  Non-agricultural investment patterns
  Off-farm employment and pluriactivity

Source: Chiotti and Johnston (1995).

knowledge gaps noted above, various questions can be raised regarding the complex interactions between climate and modern commercial agriculture. For example, how important is climate relative to other societal forces confronting farmers? To which aspects of climate are farm operations sensitive? In response to climate, what types of adaptations do farmers engage in and what land management responses are sustainable under an altered natural and social environment? How important are the local social and political–economic conditions in ensuring sustainable farm-level adjustments? Answers to these types of questions are fundamentally important to develop public policy that empowers farmers to adapt in a sustainable manner to climate change.

An Overview of Agriculture in (Irrigated) Southern Alberta

Agriculture and related activities play an important role in the regional economy of southern Alberta. Indeed, when production and processing of food and fibre are combined, the agricultural sector is the single most important in the region. To illustrate, according to a recent study (Russell *et al.*, 1993), more than 20% of the region's workforce is employed in the production and processing of agricultural products, whereas by contrast only 8% of the provincial labour force is employed in this manner. And in the City of Lethbridge, the largest in the region, it has been estimated that more than half of those employed in the City's manufacturing sector work for firms linked in some way (i.e. forward or backward linkages) to agriculture (Johnston and Sundstrom, 1995).

Owing in large part to the presence of irrigation – the region contains more than 480,000 ha of irrigated land, representing about 95% of the provincial total – southern Alberta boasts a remarkably diverse agricultural structure. Cropland in the region is dominated by wheat and barley production, which together in 1990 accounted for approximately 1.78 million ha, but canola (87,000 ha in 1990), and oats (43,000 ha in 1990) are also important crops. A range of speciality crops are also produced in the region, including sugar beets, mustard, sunflower, legumes and tame hay, much of which is dehydrated and cubed for export to Japan. With regards to livestock, the region is home to one of the highest concentrations of feeder cattle in the country. The County of Lethbridge, for instance, is estimated to have a feedlot throughput capacity of 1 million cattle, and is the location of the largest single owner of cattle in Canada (100,000 head in four feedlots). This concentration of cattle has been a relatively new phenomenon, for as recently as 1980, the cattle finishing and processing industries were market-oriented, and hence based primarily in southern Ontario and southern Quebec.

Until 1 August, 1995, and for nearly 100 years, the Canadian government subsidized the cost of transporting western Canadian grain to export terminals, thus reducing the locational disadvantage experienced by Prairie grain farmers. An unanticipated consequence of this policy was that it created a shortage of feed grain in the region, as most farmers chose to grow for the more lucrative export market. In the early 1980s, the provincial government in Alberta established the Crow Rate Off-set programme, a payment made directly to farmers for grain which was used for cattle feed. This policy, which was copied briefly by both Saskatchewan and Manitoba, encouraged a shift in the feeder-cattle industry away from its traditional base in central Canada. During this same period, the province of Alberta provided financial incentives for the expansion of cattle-processing capacity in the region. For example, Cargill, an American-based company, took advantage of this policy to build a state of the art slaughter and meat packing plant in High River, located just south of Calgary. Southern Alberta enjoys several other locational advantages with respect to feeder cattle production, which derive from its proximity to the USA market, its relatively mild agroclimatic regime, and especially a secure supply of cheap water as the region's water-management system supports the needs of cattle producers as well as irrigated crop farmers. Numerous processing activities have also been established in the region, engaged in the manufacture of canola oil, pasta, mustard seed and many other value-added agrifood commodities.

Whereas agriculture in southern Alberta has been relatively prosperous during the past decade, it has also been subject to a variety of broader forces shaping Canadian agriculture. For example, grain farmers have endured a decade of chronically low prices, at a time of rising input costs. Canola has recently emerged as a profitable alternative to wheat in many parts of the prairies, whereas the cattle industry has enjoyed a relatively buoyant market during the early 1990s. Trade has also become more liberalized, through negotiations of bi-, tri-, and multilateral trade agreements (e.g. CUSTA, NAFTA and GATT); consequently, the agrifood system is becoming more market driven, with exports increasingly being directed to the USA and the Pacific Rim. Many of the agricultural policies introduced in the 1980s (e.g. NISA, GRIP, crop insurance and the WGTA) are currently being either revamped or dismantled, to meet trade commitments, or in response to a growing fiscal crisis of the state. The 1980s were also a period of frequently occurring droughts across the prairies, especially in the portion of southern Saskatchewan and Alberta known as Palliser's Triangle (Koshida, 1991). The effects of the 1988 drought were particularly harsh, with a combination of low spring run off, record high spring and summer temperatures, and precipitation less than 50% normal, causing an estimated direct production loss of $1.8 billion (Wheaton and Arthur, 1989), and prompting the Federal Government to pay out over $1.3 billion through crop insurance and special drought assistance (Statistics Canada, 1989). The decade between 1985 and 1994, then, was an extremely volatile period for agriculture in the Canadian prairies, and agriculture in southern Alberta was not isolated from these dynamics.

Farm-level Response to Climatic Variation in Southern Alberta

It is widely recognized that weather in southern Alberta can be extremely variable. This has surely been true since the early 1980s, when there were at least 5 years of drought (1981, 1984/85, 1987/88), followed by three relatively wet years (1991–1993). Perhaps reflecting the onset of global warming, there is no doubt that climate in southern Alberta has experienced considerable variation during this period. Indeed, in 1988 the weather station at Lethbridge recorded significantly lower levels of precipitation, higher temperatures, and a greater number of growing degree days compared with 'normal' years (1950–1981) (Environment Canada, 1988). In contrast, 1993 was a particularly cool and wet year, resulting in a smaller number of growing degree days (Environment Canada, 1993). It is within this climatic and agricultural context that the researchers undertaking the Nat Christie Farm Adaptation and Sustainability Project interviewed 38 farmers in southern Alberta, representing three farming systems (e.g. arable dryland, irrigation, beef feedlot).

The Farm Adaptation and Sustainability Project is one component of the Nat Christie Foundation Programme: The Impacts of Climate Variability on Agricultural Sustainability in Alberta (McNaughton, 1993); a 5-year joint study undertaken by the University of Lethbridge, and Agriculture and Agri-Food Canada, Lethbridge Research Centre, Lethbridge, Alberta. Tape recorded interviews were carried out with 38 farmers, selected through the assistance of local informants, from the County of Lethbridge, the County of Vulcan, and the Municipal District of Taber. An additional 192 detailed questionnaires were mailed out to a random selection of farmers, of which 50 were returned. This discussion is based on a qualitative assessment of the 38 in-depth interviews. Adopting a variation of the life history approach used elsewhere in the literature (e.g. Thompson, 1982; Pile, 1990), each farmer was asked a series of questions pertaining to their views on recent and future challenges and changes to Alberta agriculture, from which it was possible to derive some sense of the relative importance of climate *vis-à-vis* societal forces, and also the particular aspects of climate to which they were sensitive. More specific inquiries regarding farm-level decision-making were also made, including the types of land management changes undertaken during the past decade.

Most farmers, regardless of their farming system, identified societal forces (e.g. government policy, marketing, the cost-price squeeze, transportation and farm management) as the greatest challenges experienced over the past ten years. Climate, including references to drought, was rarely mentioned. Similarly, when asked to identify the major challenges facing Alberta agriculture over the next ten years, farmers ignored climate entirely, instead expressing concern over issues more attributable to the forces of agricultural restructuring (e.g. government policy, the cost-price squeeze, the need for value added processing, and the importance of marketing and diversification). Environmental issues were also mentioned, although exclusively by cattlemen who were concerned about the possibility of constraining regulations being imposed upon them by an ill-informed urban-based society. Although not conclusive, the relative absence of climate as an important challenge in both the recent past and immediate future suggests that farmers are not particularly concerned with climate, perhaps illustrating a high degree of confidence in their ability to adapt, or reflecting a preoccupation with other societal influences such as trade or government policy. For example, a grain farmer engaged primarily in dryland production stated:

> ...the biggest attribution has been government involvement world wide. I think our government involvement has come in just as a reaction to try and keep us on somewhat of a level playing field with the subsidy war that has gone on. Climate has always had its up's and down's, dry years, good years, and if you can't weather those ... I don't think that's really ever changed in history. I remember going with my Dad and cutting very poor crops and very good crops, that's just something you live with and hopefully in the good years you can put some away, but it's sure in the government policy that's stifled our industry.

Moreover, this may also reflect a certain malaise that farmers have towards climate and weather, resigned to the fact that this facet of 'nature' is an uncontrollable aspect of farming.

Despite the absence of climate as a major challenge, most farmers mentioned various parameters of climate to which they have been sensitive, either during the life history of the farm, or in terms of the land management adjustments that have taken place over the past decade. Dryland farmers are particularly sensitive to drought effects, notably soil and wind erosion, due to a deficiency in available subsoil moisture. In contrast, irrigated farms were generally insulated from the effects of drought, and many of these farmers mentioned that they actually benefited relative to their dryland competitors during such conditions. Neither dryland nor irrigated farms, however, are immune to the effects of extreme weather events, particularly hail storms and frost. In the latter case, an early autumn frost and snowfall in October, 1993, had an adverse effect on crop quality, reducing many crops to feed grade status, and consequently severely decreasing farm income. Further, as irrigation is normally not applied until May, both dryland and irrigation crop producers cited the importance of the snow pack, and the depth of spring soil moisture as a necessary condition for a successful crop year.

Feedlot operators presented a different, if not unique, perspective. They seemed to be more reticent to discuss sensitivities to climate, instead opting to interpret the question as one regarding the importance of climate and other 'natural advantages' to cattle production, rather than identifying particular aspects of climate. Echoing the sentiments of many feedlot operators, one cattle producer responded:

> We have in this irrigated area expensive forages, but we have an abundance of grain – the barleys and wheats.... We also have a substantial amount of

guaranteed water because we've got irrigation, so we all have surplus water that we use for feeding our cattle. Number two, we have good markets. We have the Cargill plant within 100 miles, we have IBP in Washington, we have Lakeside, we have Calgary XL, we have the Miller brothers in Utah. We're right in almost the centre of the hub of all the markets for our finished cattle.... We have everything right here. The availability of water, the availability of the market and our climate. We don't have harsh winters, we don't have really harsh summers. We don't have a lot of rainfall, our annual rainfall on an average is not great, that's why we're irrigated. And so that lends itself to feeding cattle.

Moreover, cattlemen often expressed interest in how climate change would affect their competitive advantage, in relation to production in the USA. When prompted further, most cattlemen were eventually able to identify some aspects of climate to which their operations were sensitive, but these were very different from farmers engaged in crop production. These included the sensitivity of cattle to extremes in temperature (e.g. summer heat and winter cold), as well as to blizzards and excessive rain. Low precipitation and the drying effects of the Chinook winds tend to promote ideal conditions for growth rates and effluent management, whereas excessive rain (and associated muck!) have been known to reduce efficiencies in productivity.

Farmers within each farming system also responded quite differently in their land management adaptations to changes in climate and societal forces (Table 13.1). Overall, most farmers were shifting or expanding

Table 13.1. Climate sensitivities and adaptive responses.

Farming system	Climate sensitivity	Adaptive response
Dryland	drought	minimum tillage
	soil moisture	trash/stubble
	wind erosion	chemical fallow
	frost (early autumn)	half/half rotation
	hail	crop share
		crop insurance
Irrigation	wind erosion	pivots
	frost (early autumn)	less tillage
	hail	some trash/stubble
	heat units	chemical fallow
Feedlot	summer heat	sprinklers
	winter cold	barns
	blizzards	rotational grazing
	chinooks	seeding grass
		livestock breeds

production into higher value crops (e.g. canola or speciality crops) and cattle, perhaps reflecting a strategic adjustment to ensure economic viability, rather than an adaptive response to climatic variability. Dryland farmers were often following a variety of soil and water conservation practices (e.g. minimum tillage, leaving trash or stubble on the land and utilizing chemical fallow instead of the traditional summer fallow). They were also more likely to engage in such practices compared with their irrigation counterparts. Some speciality crops (e.g. legumes and sugar beets), for example, require the entire removal of the plant, thereby making them less suitable for minimum tillage. Feedlot operators typically used sprinklers in their pens to cool livestock during excessive summer heat, and those with range cattle tended to rotate their grazing land or seed land that was previously used for more productive crops into grass.

Crop insurance continued to be a viable response option against crop failure, and yet the future of this adaptation is uncertain. Farmers' attitudes towards crop insurance tended to be somewhat mixed, and often difficult to evaluate. Some farmers espoused a defiant disposition towards crop insurance, as with most other forms of government assistance. In some cases it was uncertain if farmers were opting for crop insurance as a buffer against climatic perturbations, or in response to pressures from the government (as a condition of GRIP) or lending institutions. As one dryland grain grower lamented:

> I don't believe in government programmes, but there they are. Actually, the first few years of drought that we had, we weren't even involved in crop insurance, because I didn't believe in government assistance. But when the banks finally said, you know we won't lend you any more money unless you get involved in crop insurance, the next day we just kept going down hill.

Most farmers were also in the position whereby premiums into GRIP had become greater than the support payments; subsequently, many expressed their intention of opting out of this programme. The future of GRIP itself, however, is also in doubt, likely to be revamped or revoked in the near future, in compliance with the trade agreements. Further, crop insurance is also currently under reform, with the share of producer premiums expected to increase relative to shrinking contributions from the Federal and Provincial governments. Hence, it is unclear how long crop insurance will remain a viable option in defence against climate change.

Discussion and Conclusions

Agriculture in southern Alberta represents a unique example of a system that is characterized by an uneven pattern and degree of capital penetration, created through the interaction of dialectical processes. The development of extensive irrigation facilities, the expansion of the beef-feed grain complex, and considerable state intervention in the form of income supports and transportation subsidies within this locality, have contributed to the development of a dynamic and interdependent agrifood system, which has facilitated the deployment of various adaptive strategies in response to the forces of agricultural restructuring and climate. On an aggregate basis, it seems that agriculture is coping quite well with recent variations in climate, as farmers and farming systems are seemingly engaged in a symbiotic relationship. Indeed, unfavourable growing conditions for one farming system may in fact benefit another. For example, irrigated farms may benefit relative to dryland producers under drought conditions, and feedlot operators may be the beneficiaries via an abundance of cheap feed when barley producers are adversely affected by an unexpected early autumn frost.

At the disaggregate level, however, a somewhat different picture emerges. Each of the three farming systems is sensitive to somewhat different parameters of climate, and in response to these stimuli each deploy distinct land management adjustments. Depending upon the climatic conditions, some farming systems may be more vulnerable than others, and one can expect that farmers within farming systems will also exhibit differential degrees of vulnerability. It is inconclusive, however, as to the degree that the regional agrifood system, the respective farming systems, and/or individual farms, are in fact vulnerable to future perturbations in climate. Considering the evidence that suggests that farmers are more concerned with economic viability and societal forces, rather than changes in climate, more research is needed to determine if their adjustments will be sustainable under an altered climate. Are farmers, for example, more at risk by introducing new crops that may be initially more profitable, yet are proven to be more vulnerable to the vagaries of climate change?

In this context, it is also worth considering the response by a farmer engaged in a multi-household feedlot/irrigation operation. When asked if their operation would be able to adapt if climate returned to the drought conditions of the 1980s, he replied:

> . . . as far as the norm for last year, or the last couple of years, I would say the extreme to extreme has happened. The extreme drought to the extreme wet, and somewhat in-between would be what I call the norm. . . . definitely not 7 years in a row would be the norm for drought. . . . So I hope the norm is somewhere in-between that we will see some day.

It is quite possible that climate is becoming more variable, and that wider swings in events may become the norm rather than the exception. Coupled with a political climate featuring re-regulation and less state intervention, farmers will inevitably have to become more self-reliant when responding to these challenges. How successful southern Alberta agriculture is in adapting remains to be seen.

Acknowledgements

The authors would like to acknowledge the financial support from the Nat Christie Foundation, Calgary, Alberta, Canada. Research for this paper was conducted while Quentin Chiotti was a Nat Christie postdoctoral fellow at The University of Lethbridge.

References

Baker, B.B., Hanson, J.D., Bourdon, R.M. and Eckert, J.B. (1993) The potential effects of climate change on ecosystem pressures and cattle production on US rangelands. *Climatic Change* 25, 97–177.

Blaikie, P. and Brookfield, H. (1987) *Land Degradation and Society*. Methuen, London, 296pp.

Bowler, I. (1989) Revising the research agenda on agricultural policy in developed market economies. *Journal of Rural Studies* 5, 385–394.

Bowler, I. (1992) 'Sustainable agriculture' as an alternative path of farm business development. In: Bowler, I.R., Bryant, C.R. and Nellis, M.D. (eds) *Contemporary Rural Systems in Transition, Volume 1: Agriculture and the Environment*, CAB International, Wallingford, pp. 237–253.

Bowler, I. and Ilbery, B. (1987) Redefining agricultural geography. *Area* 19, 327–332.

Bryant, C.R. and Johnston, T.J. (1992) *Agriculture in the City's Countryside*. University of Toronto Press, Toronto, 233 pp.

Buttel, F.H. (1989) The U.S. farm crisis and the restructuring of American agriculture: domestic and international dimensions. In: Goodman, D. and Redclift, M. (eds) *The International Farm Crisis*. St. Martin's Press, New York, pp. 46–83.

Chiotti, Q.P. and Johnston, T.R. (1995) Extending the boundaries of climate change research: a discussion on agriculture. *Journal of Rural Studies* 11(3), 335–350.

Cohen, S.J. (1988) Application of climate model output to impact studies: water resources. In: Magill, B.L. and Geddes, F. (eds) *The Impact of Climate Variability and Change on the Canadian Prairies: Symposium/Workshop Proceedings*, Alberta Department of the Environment, Edmonton, pp. 63–89.

Cohen, S.J. (ed.) (1994) *Mackenzie Basin Impact Study, Interim Report #2*. Environment Canada, Ottawa, 485pp.

Dow, K. (1992) Exploring differences in our common future(s): the meaning of vulnerability to global environmental change. *Geoforum* 23, 417–436.

Easterling, W.E., Parry, M.L. and Crosson, P.R. (1989) Adapting future agriculture to changes in climate. In: Rosenberg, N.J., Easterling, W.E., Crosson, P.R. and Darmstadter, J. (eds) *Greenhouse Warming: Abatement and Adaptation.* Resources for the future, Washington, DC, pp. 91–104.

Easterling, W.E., Crosson, P.R., Rosenberg, N.J., McKenney, M.S., Katz, L.A. and Lemon, K.M. (1993) Agricultural impacts of and responses to climate change in the Missouri–Iowa–Nebraska–Kansas (MINK) Region. *Climatic Change* 24, 23–61.

Environment Canada (1988) *Annual Meteorological Summary: Lethbridge 1988.* Lethbridge Weather Office, Atmospheric Environmental Service, Lethbridge.

Environment Canada (1993) *Annual Meteorological Summary: Lethbridge 1993.* Lethbridge Weather Office, Atmospheric Environmental Service, Lethbridge.

Friedmann, H. and McMichael, P. (1989) Agriculture and the state system. *Sociologia Ruralis* 29, 93–117.

Garcia, R.V. and Escudero, J.C. (1981) *Drought and Man: The Constant Catastrophe, Volume 2.* Pergamon Press, New York, 204 pp.

Gates, W.L. (1987) The use of General Circulation Models in the analysis of the ecosystem impacts of climate change. *Climate Change* 7, 267–284.

Goodman, D. and Redclift, M. (1989) Introduction: the international farm crisis. In: Goodman, D. and Redclift, M. (eds) *The International Farm Crisis.* St. Martin's Press, New York, pp. 1–22.

Handel, M.D. and Risbey, J.S. (1992) An annotated bibliography on the greenhouse effect and climatic change. *Climatic Change* 21, 97–255.

Henderson-Sellers, A. (1991a) Climatic impacts: the Cinderella of global climate change research funding. *Climatic Change* 19, 267–270.

Henderson-Sellers, A. (1991b) Policy advice on greenhouse-induced climatic change: the scientist's dielmma. *Progress in Physical Geography* 15, 53–79.

Houghton, J.T., Jenkins, G.J. and Ephraums, J.J. (eds) (1990) *Climate Change: The IPCC Scientific Assessment.* Cambridge University Press, Cambridge, 364 pp.

Ilbery, B.W. (1992) State assisted farm diversification in the United Kingdom. In: Bowler, I.R., Bryant, C.R. and Nellis, M.D. (eds) *Contemporary Rural Systems in Transition, Volume 1: Agriculture and the Environment.* CAB International, Wallingford, pp. 100–116.

Intergovernmental Panel on Climate Change (IPCC) (1995) Summary for policy makers: impacts, adaptation and mitigation options, *IPCC Working Group II Second Assessment Report,* Intergovernmental Panel on Climate Change Working Group II, Technical Support Unit, Montreal, 26 pp.

Johnston, T.J. and Sundstrom, M. (1995) Irrigation agriculture and local economic development: the case of Lethbridge, Alberta. In: Bryant, C.R. and Marois, C. (eds) *The Sustainability of Rural Systems: Proceedings of the 1st Meeting of the IGU Study Group on the Sustainability of Rural Systems, Montreal, 1993.* Department de Geography, Universite de Montreal, Montreal, pp. 290–303.

Koshida, G. (1991) *About Drought in Canada.* Atmospheric Environment Service, Environment Canada, CLI-I-92, 41 pp.

Liverman, D.M. (1989) Vulnerability to global environmental change. In: Kasperson, R.E., Dow, K., Golding, D. and Kasperson, J.X. (eds) *Understanding Global Environmental Change: The Contributions of Risk Analysis and Management,*

A report on an international workshop, Clark University, Worchester, pp. 27–44.

Liverman, D.M. (1991) The regional impact of global warming in Mexico: uncertainty, vulnerability and response. In: Schmandt, J. and Clarkson, J. (eds) *The Regions and Global Warming: Impacts and Response Strategies.* Oxford University Press, New York, pp. 44–68.

Lobao, L.M. (1990) *Locality and Inequality: Farm and Industry Structure and Socioeconomic Conditions.* State University of New York Press, Albany, 291 pp.

Marsden, T.K., Munton, R.J.C., Whatmore, S.J. and Little, J.K. (1989) Strategies for coping in capitalist agriculture: an examination of the response of farm families in British agriculture. *Geoforum* 20, 1–14.

Maunder, W.J. (1989) *The Human Impact of Climate Uncertainty: Weather Information, Economic Planning and Business Management.* Methuen, London, 178 pp.

McNaughton, R. (1993) *The Impacts of Climate Variability on Agricultural Sustainability in Alberta, 1993,* Annual Report. The University of Lethbridge, and Agriculture Canada Research Station, Lethbridge, Alberta.

Mendelsohn, R., Nordhaus, W.D. and Shaw, D. (1994) The impact of global warming on agriculture: a Ricardian analysis. *The American Economic Review* 84, 753–771.

Moran, W. (1988) The farm equity cycle and enterprise choice. *Geographical Analysis* 20, 84–91.

Munton, R., Marsden, T. and Ward, N. (1992) Uneven agrarian development and the social relations of farm households. In: Bowler, I.R., Bryant, C.R. and Nellis, M.D. (eds) *Contemporary Rural Systems in Transition, Volume 1: Agriculture and the Environment.* CAB International, Wallingford, pp. 61–73.

Parry, M.L. (1986) Some implications of climatic change for human development. In: Clark, W.C. and Munn, R.E. (eds) *Sustainable Development of the Biosphere.* Cambridge University Press, Cambridge, pp. 382–407.

Parry, M.L. (1990) *Climate Change and World Agriculture.* Earthscan Publications, London, 157 pp.

Pile, S. (1990) *The Private Farmer: Transformation and Legitimation in Advanced Capitalist Agriculture.* Aldershot, Dartmouth, 205 pp.

Rabb, T.K. (1983) Bibliography. In: Chen, R.S., Boulding, E. and Schneider, S.H. (eds) *Social Science Research and Climate Change: An Interdisciplinary Appraisal.* Reidel, Dordrecht, pp. 77–114.

Roberts, R. (1992) Nature, uneven development and the agricultural landscape. In: Bowler, I.R., Bryant, C.R. and Nellis, M.D. (eds) *Contemporary Rural Systems in Transition, Volume 1: Agriculture and the Environment.* CAB International, Wallingford, pp. 119–130.

Rosenzweig, C. and Parry, M.L. (1993) Potential impacts of climate change on world food supply: a summary of a recent international study. In: Kaiser, H.M., Drennen, T.E. (eds) *Agricultural Dimensions of Global Climate Change.* St. Lucie Press, Delray Beach, Florida, pp. 87–116.

Rosenzweig, C. and Parry, M.L. (1994) Potential impact of climate change on world food supply. *Nature* 367, 133–138.

Rotmans, J. (1990) *IMAGE: An Integrated Model to Assess the Greenhouse Effect.* Kluwer Academic Pub., Dordrecht, 289 pp.

Russel, K.D., Craig, K.R. and Kulshreshtha, S.N. (1993) *Irrigation Impact Study,*

Volume 5: Regional Development. UMA Engineering and the Alberta Projects Association, Lethbridge, 55 pp.

Smit, B. (ed.) (1993) *Adaptation to Climatic Variability and Change: Report of the Task Force on Climate Adaptation.* Guelph, Ontario: Department of Geography, University of Guelph, Guelph, Ontario, 53 pp.

Smit, B., Brklacich, M., Stewart, R.B., McBride, R., Brown, M. and Bond, D. (1989) Sensitivity of crop yields and land resource potential to climate change in Ontario, Canada. *Climatic Change* 14,153–174.

Smith, N. (1984) *Uneven Development: Nature, Capital and the Production of Space.* Basil Blackwell, Oxford, 198 pp.

Susman, P., O'Keefe, P. and Wisner, B. (1983) Global disasters, a radical interpretation. In: Hewitt, K. (ed.) *Interpretations of Calamity.* Allen & Unwin, Winchester, Mass., pp. 263–283.

Statistics Canada (1989) *Farm Income and Prices Section.* Government of Canada, Ottawa.

Thompson, P. (ed.) (1982) *Our Common History.* Pluto Press, London, 334 pp.

United Kingdom Climate Change Impacts Review Group (1991) *The Potential Effects of Climate change in the United Kingdom.* IIMSO, London, 124 pp.

Warrick, R. and Riebsame, W.E. (1983) Societal response to CO_2-induced climate change: opportunities. In: Chen, W.E., Boulding, E. and Schneider, S.H. (eds) *Social Science Research and Climate Change: An Interdisciplinary Appraisal.* Reidel, Dordrecht, pp. 20–60.

Watts, M. (1983) *Silent Violence: Food, Famine and Peasantry in Northern Nigeria.* University of California Press, Berkeley, 687 pp.

Whatmore, S.J.R., Munton, R.C., Marsden, T.K. and Little, J.K. (1987) Towards a typology of farm businesses in advanced capitalist agriculture. *Sociologia Ruralis* 27, 21–37.

Wheaton, E.E. (1994) *Impacts of a Variable and Changing Climate on the Canadian Prairie Provinces: A Preliminary Integration and Annotated Bibliography.* SRC Publication No. E-2900-7-E-93. Saskatchewan Research Council, Saskatoon, 140 pp.

Wheaton, E.E. and Arthur, L.M. (eds) (1989) *Environmental and Economic Impacts of the 1988 Drought: with Emphasis on Saskatchewan and Manitoba, Volume 1.* SRC Publication No. E-2330-4-E-89. Saskatchewan Research Council, Saskatoon, 362 pp.

14 Policy, Sustainability and Scale: The US Conservation Reserve Program

M. Duane Nellis, Lisa M.B. Harrington and Jason Sheeley*
Department of Geography, Kansas State University, Manhattan, Kansas 66506, USA

Introduction

Sustainability of human use of the environment in the United States, including sustainability of both the resource base and the economic systems that natural resources support, has become a major concern and focus of research and policy decisions in recent years. The prevention of accelerated soil erosion serves to protect an essential resource for agricultural production, and also helps to provide protection of other resources that may be damaged by the deposition of soil eroded from agricultural fields. 'Soil conservation was the first major, national recognition of the need to manage the natural environment of private lands' (Steiner, 1990, p. 7). The need for federal programmes to control the national problem of soil loss (and consequent sedimentation) has been evident since the 1930s. For example, on-going soil concerns led to evolution of the Soil Conservation Service, now the Natural Resources Conservation Service. One of the important means used to address soil management and sustain the productivity of natural resources in recent years has been the US Conservation Reserve Program.

* J. Sheeley is now working on a graduate degree at the University of Iowa, Iowa City, Iowa, USA.

History and Purposes

The US Conservation Reserve Program (CRP) is part of a line of land and soil conservation measures dating back to the Dust Bowl. Actions have repeatedly been taken to help control soil loss and the use of marginal or 'submarginal' lands. Control of soil erosion quickly came to be tied to a number of other environmental and social goals, including natural resource protection, flood control, reservoir protection, navigation, and public health (Steiner, 1990, pp. 5–6). Withdrawal of land in crop production serves two *major* purposes for most 'retirement' programmes: reduction of soil loss, thereby supporting sustainability of the resource; and reduction of crop surpluses, supporting economic sustainability of farming. Whether these are equal purposes of the CRP, or if conservation of the natural resource is primary, is not always clear (Zinn, 1988).

The first major effort to set aside lands in order to protect them from erosion involved US federal government purchase and management of 'submarginal' lands from the mid-1930s to mid-1950s; the acquired land areas were referred to as Land Utilization (LU) Projects. Much of the LU lands of the Great Plains region eventually became National Grasslands, under US Forest Service administration (Laycock, 1991; Duram, 1995).

A second major precursor to the US CRP was the Soil Bank. The Soil Bank programme was established in the mid-1950s, following the redistribution of LU lands. The Soil Bank Act of 1956 was meant 'to divert land from crop production' (primarily), and 'to establish and maintain protective cover' on land withdrawn from cropping (Laycock, 1988, 1991). The Soil Bank programme was set up so that farmers could voluntarily contract to withdraw cropland from production for a period of 3 to 10 years, with rental payments and cost sharing by the government. The programme peaked in 1960/61, with about 306,000 farms and 11.6 million ha involved; soil banking ended in 1969, with expiration of the last contracts. At the height of the programme, about 5.7 million ha in the Great Plains were set aside, with most of this land planted to perennial grasses (Laycock, 1988). The total federal cost for the Soil Bank programme included about US$2.5 billion in rental payments and US$162 million in cost sharing (Laycock, 1991). Congress ended the programme due to costs (Bedenbaugh, 1988).

Laycock (1988, 1991) questioned the success of the Soil Bank Program, citing a lack of clear reduction in wheat acreage across the Great Plains. In the early–mid-1970s wheat acreage rapidly expanded across the Great Plains in response to price increases. Although the Soil Bank was optimistically seen by some to represent a permanent shift from production of annual crops to other types of land cover/land use, after it ended 'nearly all Soil Bank land reverted to crop production' (Osborn *et al.*, 1994). Laycock's (1988) conclusion was that 'the Soil Bank Program was successful as a conservation measure only during the life of the contracts'.

US 'harvested cropland' increased from 135 million ha in 1972 to 158 million ha in 1982 (Chapman, 1988), resulting in increased erosion and concern about farmland management. The Payment-In-Kind programme begun in 1983 resulted in the conversion of 31.6 million ha to 'conserving uses' (Bedenbaugh, 1988). More action was seen as necessary, however, and with the passage of the Food Security Act of 1985 (FSA), a new soil protection programme was established, the Conservation Reserve Program. One section of this act (sometimes referred to as the 1985 Farm Bill), the Conservation Title, created the CRP. The FSA represents 'the first time eligibility for farm program benefits have been made contingent upon proper soil stewardship' (Bedenbaugh, 1988).

Myers (1988) credited the development of conservation provisions in the 1985 legislation to several factors, including the following:

1. development of studies and policies related to natural resources, including a US Department of Agriculture plan that 'sought to ensure the sustained productivity of soil, water, and related resources';
2. public interest/environmental concerns;
3. a coalition of organizations – the Conservation Coalition – that together were powerful in their support of the Conservation Title of the Bill; and
4. Congressional awareness of public perception of widespread soil erosion problems.

The goal of the CRP was to remove 16 to 18 million ha of highly erodible cropland from production, with conversion of land cover to grass and trees for at least 10 years. The CRP was modified somewhat through the 1990 'Farm Bill' – the Food, Agriculture, Conservation and Trade Act. By 1990, 13.7 million ha had been enroled in CRP, with an average rental cost of about US$49/acre (Bjerke, 1991; Heimlich and Kula, 1991). This legislation provided expanded eligibility criteria, and also sought to increase the acreage enroled (Bjerke, 1991). Another 1 million ha were added under the provisions of the 1990 Farm Bill (Osborn *et al.*, 1994). CRP enrolments will begin expiring in 1996, ten years after the first sign-up period.

Issues and Concerns: The Impact of CRP

With the potential expiration of contracts, a number of key variables associated with the CRP must be addressed. These include such general concerns as the future use of CRP lands; the environmental and commodity market implications, particularly as they relate to sustainability; and the response of various levels of government, as well as land managers.

Although CRP has gained the support of conservationists and a number of agriculturalists, concerns have been raised. The scale of concern is often local, although issues often are treated regionally or nationally, as well. Most of the drawbacks of CRP have been economic. At the local level, concerns relate to farmers and/or rural communities. At the national level, economic objections are most closely related to current US Congressional attitudes and emphasis on federal budget-cutting. Overall, objections to CRP may arise from the following concerns.

Federal economic costs

CRP costs have reached more than US$1.4 billion/year (Steiner, 1990). As with the Soil Bank programme, Congressional cost-cutting may terminate the programme.

Impacts on local economies

Declines in agribusiness across the Great Plains have been projected (Devino *et al.*, 1988), but with increased spending in other sectors compensating for such declines (Saltiel, 1994). Greater problems may arise in areas that lack retail businesses and have relatively large quantities of land enrolled in CRP (Saltiel, 1994).

Increase in land prices due to decrease in availability

With only the most productive, sustainable lands available for agricultural production, these lands will be priced at a higher cost (Saltiel, 1994).

Increase in weeds and pests

CRP lands are perceived as a weed source during the time it takes to establish 'permanent' plant cover (Saltiel, 1994). Grassed reserves of the US Great Plains, for example, became infested with wheat aphids, at great cost to farmers (Steiner, 1990).

Competition with livestock production versus lack of agricultural use

The possibility for grazing use of, or livestock feed production from, CRP lands has concerned stock growers (Saltiel, 1994). Although initially not allowed, by mid-1989 selective haying and grazing was permitted in certain counties (Steiner, 1990).

Politics and farmer autonomy

Larson (1988) stated that the CRP 'is a program warmly received by resource managers and well accepted by landowners'. From the viewpoint of the farmer, however, the tendency to change laws and regulations from one Congress to the next (Kleckner, 1988) signifies a difficult environment for farm-level administration and decision-making. As described by Steiner (1990), 'farmers are concerned about the government's long-term commitment to the program'. In addition, there is often a lack of confidence in federal legislators and bureaucracy, and a feeling that such interference is unfounded. Farmers generally believe that farmers are best attuned to their land, and know best how to manage for sustainability of agricultural resources.

Beneficial aspects of the CRP also are attached to some controversy. Benefits are linked to its primary purposes and include decreased erosion and increased farm income and long-term sustainability. In addition, CRP eligibility criteria 'evolved' and came to include other environmentally beneficial concerns, such as wetland protection and improvement of water quality (Nellis, 1992; Young *et al.*, 1994).

Decreased erosion

The estimated average reduction in erosion for enrolled land is 7 metric tons/ha/year (Lindstrom *et al.*, 1994). Butz (1993) estimated total reduction in soil erosion of 'nearly 636 million metric tons annually'. However, the CRP in its earlier (1985–1989) form was 'criticized for its cost ineffectiveness in achieving soil conservation goals' (Young *et al.*, 1994). Inefficiency was blamed on early emphasis on target enrolments, rather than on erosion; subsequent changes in the programme seem to have resulted in greater participation in highly erodible counties. Other criticisms of the CRP regarding resource protection relate to its lack of permanence and voluntary nature (Steiner, 1990).

Increase in farm income

Farm income was expected to increase for those with CRP contracts, due to federal rental payments and a decrease in production costs (Saltiel, 1994). Actual benefits vary with location: in a study of highly erodible southeast Washington farmland, it was estimated that farmers in areas that normally are lower yielding benefited up to $3.6/ha/year by enrolling in the 1985 version of the CRP, but farmers in higher yielding locations could lose $3.2/ha/year. The 1990 revisions to administration of the CRP seem to have improved the cost-effectiveness of the programme (Young *et al.*, 1994).

CRP also should represent an increase in farm income for all producers, *if* it has been successful in reducing output and surplus of particular crops (Saltiel, 1994).

Associated environmental/economic benefits

Based on approximately 14.7 million ha in CRP, Butz (1993) reported that pesticide use had been reduced by an estimated 27.7 million kg/year and fertilizer use had been reduced by about 2.2 million metric tons/year. Water quality also has benefited by a reduction of more than 191 million metric tons/year of sediment loadings. Benefits that may be associated with the CRP include improvements in surface water quality, air quality and wildlife habitat (and an associated increase in recreational opportunities and activities), and a decrease in groundwater withdrawals. In 1989, the estimated economic value of these benefits, with eventual attainment of 18.1 million ha enrolled, was about US$10 billion (Ribaudo *et al.*, 1989). Other estimates placed the *net* economic benefits of 18 million ha in the CRP at US$3.4 to US$11 billion (Young and Osborn, 1990).

Landowner perceptions are the most important factor in determining conservation actions, as reported by the Government Accounting Office (GAO) (Steiner, 1990). Similar findings were obtained by a 1990 study of Montana farmers – 'farmers who thought CRP would have a strong impact on reducing soil erosion were about 55% more likely to favour continuation of the program than those who thought no effect would be seen' (Saltiel, 1994). Additionally, younger farmers and those with higher levels of education were more likely to support the CRP.

Surveys of farmers with land enrolled prior to the 1990 Farm Bill rules indicate that, as contract expirations approach, farmers plan to remove more land from current CRP protective cover (Osborn *et al.*, 1994). According to the 1990 survey, postcontract plans for CRP land would have 53% of the land converted back to crop use; a 1993 survey indicated plans for conversion of 63% of the land. However, the 1993 survey also found that, at 100% of the current CRP rent, farmers were likely to extend CRP contracts (without haying or grazing) for an additional ten years on 63% of the current CRP land. For the Northern Plains states 67%, and for the Southern Plains states 72%, were likely to be re-enroled. However, even at rental payments of 135% of current rates, farmers in the Northeastern states responded that they were only likely to re-enrol 36% of CRP land.

Spatial Distribution and Scale Perspectives

Early estimates by the US Soil Conservation Service determined potential land areas eligible for the CRP at 40.9 million ha (Padgitt, 1989). As mentioned earlier, a little over one-third of this estimate actually is included in CRP lands. These enrolled lands present a variety of interesting spatial and scale dimensions that are important to analyse. First, the distribution of CRP lands should show where marginal lands have been used historically for crop production. Secondly, such analysis may show areas where opportunities for greater sustainability in the cropland rural land use system are more attainable. Finally, such an analysis may provide some indication as to the regional impacts of Federal relaxation of the Farm Bill initiatives through reductions in this effort starting in 1995.

Figure 14.1 shows the percentage of cropland enrolled in CRP by county for the United States. Data were provided by the US Soil Conservation Service and entered into an ARC-INFO GIS system to create this map. The major CRP enrolment regions include the Great Plains, the Corn Belt, and the historic 'Cotton Belt'.

Nearly 60% of the US CRP-enrolled lands are in the Great Plains States (Texas, New Mexico, Oklahoma, Kansas, Colorado, Nebraska, Wyoming, North Dakota, South Dakota, and Montana) (Fig. 14.1). This includes the historic 'Dust Bowl' region of the 1930s. More than 20 of the counties in the 'Dust Bowl' region have greater than 25% of their cropland enrolled in CRP. Even more telling, counties have to petition the USDA on an individual basis to enrol more than 25% of their total acreage (Butz, 1993). These lands tend to be in water deficit zones heavily dependent on irrigated agriculture. Without irrigation, the lands in this region used for cropland are often fallowed, and only of marginal cropland productivity. The southeastern region of the US, the historic 'Cotton Belt' area, also shows a clear concentration of CRP lands. Many of these formerly intensively used and highly eroded lands are being converted to other uses, such as pine plantations as part of the southeast emergence as a focal point for the timber industry.

The Corn Belt of the Midwest also is a prime area for CRP lands. This is consistent with the high levels of soil erosion associated with these productive agricultural lands in the Northwest. Irrigated areas in the Columbia River Basin and Snake River Basin also show up as prime CRP concentrations, and reflect interesting owner economic sustainability decisions on whether or not to irrigate.

A closer look at the state and local scale of CRP implementation reveals the importance of factors other than soil sustainability. Kansas ranks second in the total number of CRP contracts (Iowa is first) and the area enrolled totals 1.2 million ha (Frazee, 1995). Although the thin soils of the Kansas Flint Hills, in the eastern part of the state, show up as one prime CRP area,

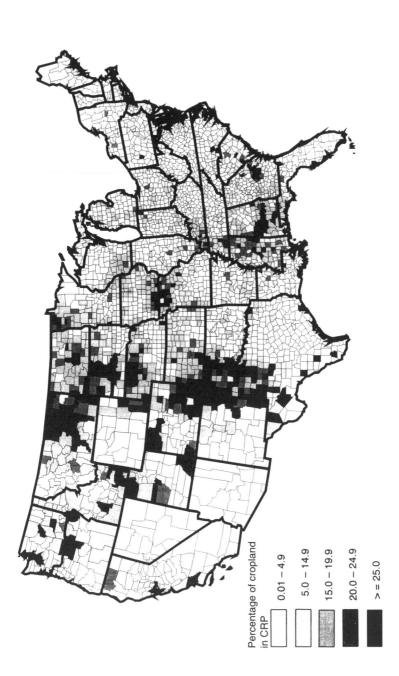

Percentage of cropland
in CRP

0.01 – 4.9

5.0 – 14.9

15.0 – 19.9

20.0 – 24.9

> = 25.0

Fig. 14.1. Percentage of cropland in the Conservation Reserve Program by county, 1992. (Source: Soil Conservation Service, 1992.)

most of the CRP lands in Kansas are in the southwestern portion of the state (60%), again the old 'Dust Bowl' area (Fig. 14.2). The former 'Dust Bowl' region includes contrasts between intensively developed irrigated areas, where farmers have access to the Ogallala Aquifer, and areas of marginal farming systems where farmers lack access to water. The owners with more marginal lands found it more profitable to enrol lands in CRP. The margin of this participation, however, is limited. If the programme is not extended, 34% of southwest Kansas CRP participants questioned said they will return land to crops; 26% will keep land in grass for livestock grazing, and the remaining are undecided (Diebel *et al.*, 1993).

   In Finney County of Southwest Kansas, for example, groundwater access, not necessarily soil quality, was a controlling factor in farmers' decisions to enrol in CRP. Satellite and geographic information systems analysis of Finney County in 1992 revealed that approximately 23,600 ha were enrolled in CRP (Egbert *et al.*, 1995). The majority of these lands are in the county's panhandle region, where no usable groundwater is available, and in the northern reaches, where severe groundwater declines have occurred since 1940 (Fig. 14.2). Further, by overlaying the distribution of CRP lands with soil properties in Finney County, it is clear that lands with the most marginal soils are not necessarily those being set aside in CRP. In 1992, only 4.6% of CRP lands had soils with a surface horizon of less than 10 cm, for example, whereas 32.5% of the land planted to corn was in areas

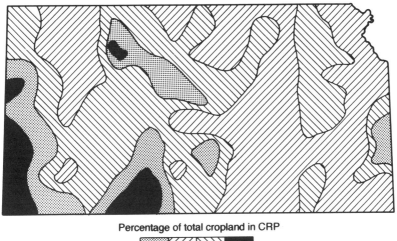

Percentage of total cropland in CRP

1.4   7.3   14.1   20.4   30.0

Fig. 14.2. Percentage of total cropland in the Conservation Reserve Program in the state of Kansas, 1992. (Source: Soil Conservation Service, 1992.)

of less than 10 cm soil surface horizon. Soil texture was also more suitable for crop growth on the CRP lands versus planted cropland. Such enrolment patterns at the county level reflect some of the broader issues already addressed in this chapter.

In this county level case, the scale of concern should be local, although programme guidelines are treated regionally or nationally. At the local level, Finney County farmers are interested in maximizing their economic gain, even if this includes utilizing more marginal lands, with underlying groundwater supplies, as cropland rather than placing the land in longer term conservation reserve. At the same time, farmers with moderate quality soil conditions, but limited or non-existent groundwater reserves, find their marginal rate of return maximized through participation in the CRP initiative. Although the benefits of decreased erosion are attained at the farm level, regional and national level sustainability objectives are not necessarily being addressed in the most appropriate manner.

Prospects for the Future and Conclusions

At the national level, the CRP and associated environmental aspects of past agricultural legislation have wide support among environmentalists and conservationists. For example, both the Audubon Society and the National Wildlife Federation have recently urged members to support CRP in the 1995 Farm Bill (National Audubon Society, 1995; National Wildlife Federation, 1995). In a series of regional forums held by the Soil and Water Conservation Society (SWCS), the general consensus of participants included the idea that 'American agriculture is not sustainable', as well as support for continuation of the CRP (SWCS, April 1995). A 1994/95 Gallup poll of the general public indicated that 51% of the respondents 'wanted to see federal spending on agricultural natural resource conservation increase a little or a lot, and 16% wanted spending to decrease a little or a lot' (SWCS, May 1995).

According to the SWCS (1994), the 'CRP has proved to be popular and, in most cases, a sound natural resource conservation program'. The official SWCS position is that CRP:

> needs to be extended well beyond the present 10-year contract period. Moreover, it should become a conservation program, first and foremost, not an agricultural commodity supply management program.... In acknowledging budget concerns, some economic use of less environmentally sensitive land now enrolled in the CRP, haying and grazing for example, should be allowed in return for reduced rental payments.

Although widely supported, the future of the CRP is currently unknown. In early 1995, the Secretary of Agriculture 'used his authority under the 1990 farm law to announce an extension of the CRP for 10 years' (McBeth, 1995). In response, the Congressional Budget Office prepared a budget 'forecast' that included cuts of CRP enrolments by about 2 million ha/year beginning in 1998. The actual fate of CRP funding is still in question. The general atmosphere in the US Congress is strongly anti-spending; the CRP may be seen as an 'easy' cut in federal spending for agriculture. Some agricultural-state Congress persons are in relative positions of power, however, and there has been some indication of support for CRP. The major problem will be in determining where, in the 1995 Farm Bill, and in the Department of Agriculture budget, monies will be spent and cuts will be made. (Note: it is rumoured that the current version (H.R. 2854) will not continue CRP.)

Past history indicates that disappearance of the CRP is likely to mean renewed cultivation of highly erodible lands over most or all of the current grassed and/or wooded reserves. By the repeated institution of soil reserve programmes, history also indicates a need for continued land reservation programmes to sustain healthy natural resource conditions.

References

Bedenbaugh, E.J. (1988) History of cropland set aside programs in the Great Plains. In: Mitchell, J.E. (ed.) *Impacts of the Conservation Reserve Program in the Great Plains.* GTR RM-158. USDA Forest Service, Rocky Mountain Forest and Range Experiment Station, Fort Collins, Colorado, pp. 14–17.

Bjerke, K. (1991) An overview of the Agricultural Resources Conservation Program. In: Joyce, L.A., Mitchell, J.E. and Skold, M.D. (eds) *The Conservation Reserve – Yesterday, Today and Tomorrow.* GTR RM-203. USDA Forest Servcice, Rocky Mountain Forest and Range Experiment Station, Fort Collins, Colorado, pp. 7–10.

Butz, D. (1993) Future of CRP lands: an SCS perspective. In: *Proceedings of the Mid-Continent CRP Conference.* Kansas Wildlife and Parks, Manhattan, Kansas, pp. 18–30.

Chapman, E.W. (1988) Rationale and legislation for the creation of the Conservation Reserve Program. In: Mitchell, J.E. (ed.) *Impacts of the Conservation Reserve Program in the Great Plains: Symposium Proceedings.* USDA Forest Service, Fort Collins, Colorado, pp. 9–13.

Devino, G., Van Dyne, D. and Braschler, C. (1988) Agribusiness and the CRP. *Journal of Soil and Water Conservation* 43(5), 379–380.

Diebel, P., Cable, T. and Cook, P. (1993) *The Future of the Conservation Reserve Program Land in Kansas The Landowner's View,* Report of Progress 690. Agricultural Experiment Station, Kansas State University, Manhattan, Kansas.

Duram, L.A. (1995) The national grasslands: past, present and future land management issues. *Rangelands* 17(2), 36–42.

Egbert, S., Price, K., Nellis, M.D. and Lee, K. (1995) Developing a land use cover modeling protocol for the high plains using multi-seasonal thematic mapper imagery. *Proceedings of the American Society of Photogrammetry and Remote Semnsing* (3), 836–845.

Frazee, B. (1995) CRP land is budget battlefield: wildlife managers fight to protect prime habitat. *Kansas City Star,* 26 Feb., p. C-17.

Heimlich, R.E. and Kula, O.E. (1991) Economics of livestock and crop production on post-CRP lands. In: Joyce, L.A., Mitchell, J.E. and Skold, M.D. (eds) *The Conservation Reserve – Yesterday, Today and Tomorrow,* GTR RM-203. USDA Forest Service, Rocky Mountain Forest and Range Experiment Station, Fort Collins, Colorado, pp. 11–23.

Kleckner, D. (1988) Implementing CRP: a private perspective. *Journal of Soil and Water Conservation* 43(1), 18–20.

Larson, G. (1988) Implementing CRP: a state/local perspective. *Journal of Soil and Water Conservation* 43(1), 16–18.

Laycock, W.A. (1988) History of grassland plowing and grass planting on the Great Plains. In: Mitchell, J.E. (ed.) *Impacts of the Conservation Reserve Program in the Great Plains,* GTR RM-158. USDA Forest Service, Rocky Mountain Forest and Range Experiment Station, Fort Colllins, Colorado, pp. 3–8.

Laycock, W.A. (1991) The conservation reserve program – how did we get where we are and where do we go from here? In: Joyce, L.A., Mitchell, J.E. and Skold, M.D. (eds) *The Conservation Reserve – Yesterday, Today and Tomorrow,* GTR RM-203. USDA Forest Service, Rocky Mountain Forest and Range Experiment Station, Fort Collins, Colorado, pp. 1–6.

Lindstrom, M.J., Schumacher, T.E. and Blecha, M.L. (1994) Management considerations for returning CRP lands to crop production. *Journal of Soil and Water Conservation* 49(5), 420–437.

McBeth, D. (1995) CRP extended, funding still in question. *SWCS Conservogram.* March, 1–2.

Myers, P.C. (1988) Conservation at the crossroads. *Journal of Soil and Water Conservation* 43(1), 10–13.

National Audubon Society (1995) Birds and conservation in the 1995 Farm Bill. (Mass-mailing flyer.)

National Wildlife Federation (1995) Farming and the environment. *National Wildlife EnviroAction.* 13(1), 16–18.

Nellis, M.D. (1992) In: Bowler, I., Bryant, C. and Nellis, M.D. (eds) *Contemporary Rural Systems in Transition: Agriculture and Environment.* CAB International, Wallingford, UK.

Osborn, C.T., Schnepf, M. and Keim, R. (1994) *The Future Use of Conservation Reserve Program Acres: A National Survey of Farm Owners and Operators.* Soil and Water Conservation Society, Ankeny, Iowa.

Padgitt, M. (1989) Soil diversity and the effects of field eligibility rules in implementary soil conservation programs target to highly erodible land. *Journal of Soil and Water Conservation* 44(1), 91–95.

Ribaudo, M.O., Piper, S., Schaible, G.D., Langner, L.L. and Colacicco, D. (1989) CRP: what economic benefits? *Journal of Soil and Water Conservation* 44(5), 421–424.

Saltiel, J. (1994) Controversy over CRP in Montana: implications for the future. *Journal of Soil and Water Conservation* 49(3), 284–288.

Soil and Water Conservation Society (SWCS) (1994) Legislative notes. *SWCS Conservogram*, Aug./Sept., 4, 6, 8.

Soil and Water Conservation Society (SWCS) (1995) NRCS, SWCS 'listen to the grassroots' to determine public opinions about the environment. *SWCS Conservogram*, Apr., 3.

Soil and Water Conservation Society (SWCS) (1995) NRCS poll shows public's environmental opinions. *SWCS Conservogram.* May, 1, 3.

Steiner, F.R. (1990) *Soil Conservation in the United States: Policy and Planning.* Johns Hopkins University Press. Baltimore, Maryland, 249 pp.

Young, D., Bechtel, A. and Coupal, R. (1994) Comparing performance of the 1985 and the 1990 Conservation Reserve Programs in the West. *Journal of Soil and Water Conservation* 49(5), 484–487.

Young, C.E. and Osborn, C.T. (1990) Costs and benefits of the Conservation Reserve Program. *Journal of Soil and Water Conservation* 45(3), 370–373.

Zinn, J.A. (1988) Monitoring the conservation title: a congressional perspective. *Journal of Soil and Water Conservation* 43(1), 57–59.

15 Something Old, New, Borrowed and Blue: The Marriage of Agriculture and Conservation in England

Nick J. Evans
Department of Geography, Worcester College of Higher Education, Henwick Grove, Worcester WR2 6AJ, UK

Introduction

The 1990s have witnessed a consolidation of agrienvironmental policy in England as a feature of postproductivist agriculture. The 1992 European Union (EU) Common Agricultural Policy (CAP) reform Regulation 2078/92 provided official confirmation that conservation is an important medium-term issue. Indeed, it represents the main current political mechanism that supports agricultural restructuring while simultaneously directing the sector towards a more sustainable existence in the countryside. Nevertheless, resources devoted to a 'greener' CAP are limited, at about 2% of the agricultural budget, in comparison with that still supporting prices of major foodstuffs. This reflects a general concern among policy-makers to favour a short-term maintenance of farmers' incomes over long-term sustainability, leaving the established agroindustrial system of development basically intact (Potter, 1993). It is little wonder that policies are predominantly reactive, dealing with obvious environmental problems stemming from intensive farming systems rather than proactive towards a lasting sustainability. Individual farm businesses following different 'paths of development' (see Bowler, 1992) are now in a position to utilize strategically money offered through agrienvironmental schemes. Three situations appear to be emerging. First, agrienvironmental policy offers an additional avenue for farms to diversify their income where CAP reform constrains restructuring based on increasing output. Second, extending this notion, farmers can now undertake business adjustment solely using conservation schemes, effectively opting to become environmental managers. Third, because agrienvironmental policy is designed to operate alongside price support, a possibility exists for farmers adjusting by expanding food production to take advantage of conservation provisions.

The range of agrienvironmental measures is complex and, contrary to official claims, there is a lack of coordination between them (Colman *et al.*, 1993; Potter, 1993). This chapter takes a cross-section of three schemes to examine uptake among farmers, indicating the extent to which conservation, and thus sustainable practices, form part of business restructuring. First, it focuses upon Environmentally Sensitive Areas (ESAs) that have just completed a major phase of review and expansion. Second, it deals with Countryside Stewardship (CS), about to complete its experimental stage. Third, it examines the Farm and Conservation Grant Scheme (FCGS) which is set for closure in its present form.[1] A fourth section then briefly turns attention to the controversial issue of public access, a feature of all schemes examined. Monitoring uptake serves as an initial assessment of scheme significance during a period of rapid agricultural change. The aim is to raise a series of research questions warranting closer attention through farm survey work. In so doing, it is recognized that future research must address the nature and extent of any shift of farmers' attitudes, shaped by decades of productivist policy, towards environmental concerns (Morris and Potter, 1995). A combination of unpublished data and discussion with Ministry of Agriculture Fisheries and Food (MAFF) officials is used to generate results. Time periods considered vary, but in all cases commence from the date of scheme launch and end with the most recent data available. Thus, ESA data refer to the total number of agreements operating in 1994/95, CS data cover the period 1991 to 1993, and FCGS figures illustrate grants made in ten English Midlands counties between 1989 and 1994.

Something Old - ESAs

First introduced in 1987, ESAs represent the oldest measure in England persuading farmers to take a positive interest in conservation. It was the first scheme to divert agricultural finance away from supporting food production towards conservation (Potter, 1988). It has become transformed from an experiment to a cornerstone of UK agrienvironmental policy, reinforced by its status as an official 'accompanying measure' of EU CAP reform. Details of the scheme are well known (see Potter, 1988; Smith, 1989; Baldock *et al.*, 1990; Whitby, 1994). A brief summary of key features appears in Table 15.1. In essence, ESAs are geographically targeted measures,

Table 15.1. Key features of ESA, CS and FCGS schemes in England.

Features	Scheme		
	ESAs	CS	FCGS
Voluntary for farmers	yes	yes	yes
State can exercise entry discretion	no	yes	yes
Geographically restricted	yes	yes	no
Experimental	no	yes	no
Part-farm entry possible	some areas	yes	n.r.
Capital grants available	yes	yes	yes
Standard annual payments available	yes	yes	no
Tiered payments available	yes	in some	no
Maintains nature conservation	yes	yes	yes
Enhances nature conservation	some tiers	yes	yes
Restores nature conservation	some tiers	yes	no
Supports landscape conservation	yes	yes	yes
Supports historic interests	yes	yes	no
Supports access	yes	yes	yes
Demands cross-compliance	no	occasionally	no
Fully decoupled from price support	no	no	no
Administered by MAFF	yes	from 1996	yes
ESA payments available	n.r.	no	no*
CS payments available	no	n.r.	no
FCGS payments available	no*	no	n.r.

n.r. = Not relevant.
* Payments for conservation are not usually available; other sections of FCGS are not necessarily affected.

designated in England in four batches or 'waves' (see Fig. 15.1). Within their boundaries, financial incentives (management payments[2]) are made available for farmers to undertake environmentally friendly ('traditional' or 'sustainable') farming methods. The aim is to conserve wildlife, flora, landscape and archaeological features of designated areas as appropriate.

The underlying principles of ESAs have survived unaltered since their formulation, although there is a history of continual adjustment through the 'bolting on' of more detailed tiers, a conservation plan and an access tier. Curry (1994) argues that policy-makers are pursuing a 'fallacy of creeping incrementalism', believing that continual modification will yield the objectives the policy is not achieving. This is because ESAs fail to challenge the overcapitalization of agriculture, a root cause of environmental damage. Instead, they are designed to treat the symptoms of a system based on price support for food production (Baldock *et al.*, 1990). It is

First Wave 1987:	Broads (1); Pennine Dales (2); Somerset Levels and Moors (3); South Downs (4); West Penwith (5).
Second Wave 1988:	Breckland (6); North Peak (7); Clun (8); Suffolk River Valleys (9); Test Valley (10).
Third Wave 1993:	Avon Valley (11); Exmoor (12); South Wessex Downs (13); Lake District (14); North Kent Marshes (15); South West Peak (16).
Fourth Wave 1994:	Blackdown Hills (17); Cotswold Hills (18); Shropshire Hills (19); Dartmoor (20); Essex Coast (21); Upper Thames Tributaries (22).

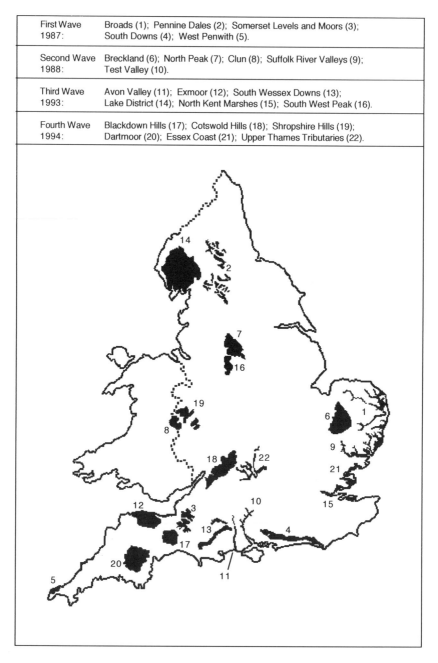

Fig. 15.1. The location and year of designation of ESAs in England.

tempting to account for the proliferation of ESAs in the 1990s as offering a convenient contribution towards government's prime objectives of reducing food surpluses and budgetary expenditure.

In 1995, the area of land eligible for ESA payment stood close to one million ha, whilst over 15,000 farms (approximately 10% of the English total) can apply to enter the scheme. Farmers' general uptake of ESAs has remained good over time, so that their early distinction as a *cause célèbre* of conservation has persisted (Baldock *et al.*, 1990). A correspondence exists between the eligible percentages of area (land) and holdings (farmers) currently enrolled in the scheme, amounting to 42% and 43% respectively. However, the general figures disguise large variations in holding and areal uptake *within* individual schemes. West Penwith ESA attains the highest rate of uptake both in terms of eligible land (89%) and farmers (90%). Breckland, a first wave designation, has the lowest relative area in the scheme at just 10% of that eligible yet this is distributed among 55% of eligible holdings. In fact, the lowest number of farmers signing an ESA agreement occurs in the North Peak ESA (13%), although 88% of the eligible area is covered representing the second highest agreed area. This is due to the landholding structure of large estates within the ESA (see Colman *et al.*, 1993; Froud, 1994). It is the Blackdown Hills ESA (fourth wave), still in relative infancy, which returns the lowest proportion of both land (14%) and farmers (15%) in the scheme.

Following Brotherton (1991), a useful indicator of the variation between percentage land area and percentage holdings in the scheme can be calculated by dividing one into the other to produce a simple average area-holding ratio (Table 15.2). The ratio demonstrates whether farms with larger or smaller than average eligible areas are participating. To explain wide variations in the average area-holding ratio obtained for ESAs in Table 15.2, three case studies are selected and the influence of their specific management prescription regimes on uptake figures established. From Table 15.2, these are West Penwith (0.99), the Cotswold Hills (1.36) and Breckland (0.18), representing a spectrum of average area-holding ratios across lowland ESAs designated in different waves. Table 15.3 picks out the salient management features of each example discussed in the remainder of this section.

West Penwith

The closeness of the average area–holding ratio to 1.00 in West Penwith cannot be explained simply by the high percentage of farmers participating in the scheme. Farms in this locality tend to be small in size, economically marginal and constrained by the imposition of milk quotas which, since 1984, have had a disproportionately adverse impact on the small dairy

Table 15.2. Average area-holding ratios for the 22 English ESAs.

ESA	Wave	Average area–holding ratio	Eligible area under agreement (%)	Eligible holdings with agreement (%)
North Peak	II	6.67	87.8	13.2
The Broads	I	1.75	63.6	36.2
South West Peak	III	1.37	54.4	39.7
Cotswold Hills	IV	1.36	47.8	35.2
Exmoor	III	1.23	56.9	46.5
South Wessex Downs	III	1.17	49.8	42.5
Test Valley	II	1.16	31.8	27.5
Lake District	III	1.08	49.1	45.5
Shropshire Hills	IV	1.01	27.6	27.5
West Penwith	I	0.99	88.9	89.5
Blackdown Hills	IV	0.91	13.9	15.3
Somerset Levels and Moors	I	0.88	52.0	59.4
Clun	II	0.87	71.2	82.1
Pennine Dales	I	0.80	65.2	81.1
Avon Valley	III	0.61	18.4	30.0
Dartmoor	IV	0.55	10.9	19.8
Essex Coast	IV	0.50	11.0	22.0
Upper Thames Tributaries	IV	0.49	19.8	40.7
Suffolk River Valleys	II	0.44	26.8	60.6
North Kent Marshes	III	0.43	35.2	81.4
South Downs	I	0.24	21.5	88.8
Breckland	II	0.18	9.7	54.5

Source: Unpublished MAFF data.

producers that characterize the region. Farmers in this locality are keen to consider any potential source of alternative income, and many have indeed followed a 'diversification' path of business development.[3] Together with the simple management prescriptions associated with entry into a straightforward scheme (Table 15.3), the 'balanced' average area–holding ratio is understandable.

Cotswold Hills

The Cotswold Hills have a greater area entered than the overall percentage of participating holdings would suggest. The key to explanation lies in the 'first tier' offered to farmers. Arable producers have merely to pledge not to increase their cropped area while maintaining field boundaries (especially drystone walls), watercourses and woodland in order to claim £12/ha

Table 15.3. Management tiers in the West Penwith, Cotswold Hills and Breckland ESAs.

ESA	Tier	Purpose of management prescriptions	£ per ha
West Penwith	1	to maintain landscape and field patterns (with additional prescriptions for 'rough' land)	70
Cotswold Hills	1A	not to increase in the existing arable area	12
	1B	to maintain improved permanent grassland (five years old fertilized grassland)	30
	1C	to maintain extensive permanent grassland (five years old unfertilized grassland)	60
	2	to revert arable land to extensive permanent grassland	260
Breckland	1	to maintain heathland	110
	2	to revert arable land or improved grassland to heathland	300
	3	to maintain river valley grassland	125
	4A	to introduce uncropped wildlife strips on arable field edges	350
	4B	to introduce conservation headlands on cereal field edges	110
	plus	to observe a set of 'ALL LAND' conservation guidelines under all tiers	

Source: *MAFF Environmentally Sensitive Areas 'Guidelines for Farmers'* (1994, 1995).

(Table 15.3). Average enrolled areas are high because arable farmers occupy holdings double the average size of dairy producers and five times that of livestock farmers. The area in Tier 1A currently amounts to 62% of that under agreement, whereas just 3% is under a Tier 2 agreement. Competition from the Arable Area Payments Scheme (AAPS), devised to soften the blow of successive cuts in CAP price support (see Robinson and Ilbery, 1993; Wynne, 1994), is still too strong to persuade arable farmers to adopt grassland reversion required under Tier 2. Further, land converted from arable to grassland requires grazing, yet livestock quotas also introduced under CAP reform restrict the ability of farmers to switch profitably to livestock systems. It is yet to be proved that reserve quota will be allocated for this purpose.

Breckland

The low average area–holding ratio for Breckland is undoubtedly influenced by farmers being able to enter parts of their holdings into the scheme due to the fragmented nature of its conservation interest (heathland, wetland and woodland). Nevertheless, Table 15.3 shows that all four tiers have sharply focused management prescriptions. This is because many desirable conservation features require substantial remedial action for maintenance or re-establishment. Compared with the Cotswold Hills ESA, farmers cannot adopt a 'do little' approach in order to claim payment. According to O'Carroll (1994), farmers have tended only to express an interest in the river valley (wet) grassland tier (Tier 3), representing a fraction of the eligible land a farmer has on the holding. Once again, payment levels seem unable to compete with arable and intensive grassland systems still supported through the reformed CAP.

Something New – CS

Countryside Stewardship is set to become MAFF's newest agrienviron-mental scheme in April 1996. It was originally initiated in England from 1991 as a five-year countryside management experiment by the Country-side Commission (CC). The details of the scheme have been reviewed by Bishop and Phillips (1993) as it represents the prime example of a 'market-led' or 'incentives' approach to conservation (see Table 15.1). The basic principle is that the state becomes a buyer of practices which benefit the environment and enhance the provision of recreation services to the general public (CC, 1994). Hence, the state offers incentives for *selected* landowners to adopt specific conservation-oriented farming methods, a basic difference from ESAs. Further, eligible areas are not defined geographically but restricted spatially to 'target landscapes'. Eight target landscapes are identified by the CC and are listed in Table 15.4 together with their main characteristics. CS can, therefore, support conservation interests that are dispersed widely and fragmented.

Uptake of CS has been largely overshadowed by concern for the ESA scheme. CS is frequently reported to be 'popular' and 'successful', reinforced by its impending MAFF adoption, yet there has been minimal examination of scheme data. Bishop and Phillips (1993) presented figures for target landscapes covered by agreements in 1992, but emerging patterns of uptake over time and the geography of uptake have not been addressed. At the end of 1993, some 2963 CS agreements covered 80,174 ha in England, representing about 0.8% of farmland. A direct comparison with ESAs is problematic, but the number of agreements made does highlight the significance of CS as an environmental provision. The amount of land entered differs for each target landscape, so three examples representing relatively large, moderate and small areas enrolled in CS are taken to explain variations in adoption.

Table 15.4. A summary of the main aims of target landscapes under Countryside Stewardship.

Target landscape	Purpose
Chalk and limestone grassland	to conserve, improve or restore species-rich chalk and limestone turf
Lowland heath	to manage and re-create heathland vegetation in lowland England whilst maintaining the tradition of access
Waterside land	to restore water meadows, grazing marshes and wet pastures
The coast	to re-create flower-rich pasture along coastal fringes or cliff top, manage salt-marshes, restore coastal heaths and stabilize sand dunes
Uplands	to restore moorland vegetation (especially heather), conserve and create flower-rich hay meadows
Historic landscapes	to restore historic parklands, conserve landscapes reflecting the development of a locality, preserve archaeological earthworks and historic irrigated water-meadows
Old traditional orchards	to reintroduce management to old orchards
Old meadows and pasture	to protect the species-rich unimproved grassland of Devon and Cornwall, and Hereford and Worcester

Source: Countryside Commission (1994).

Uplands

The highest initial area entered into CS was for upland landscapes and, unsurprisingly with an emphasis on moorland, the three northern CC Regions (Yorkshire and Humberside, Northern and North West) accounted for the majority of land entered. A conspicuous feature over the three years of adoption shown in Fig. 15.2 is the burst of initial uptake followed by decline in interest. This pattern appears to emerge from the interaction of three factors; scheme objectives, landholding structure and the situation of individual farmers. First, the promotion of CS amongst upland farmers could have been more vigorous in the early years. This would quickly attempt to conserve moorland not covered by ESA designations. Although the Pennine Dales ESA has functioned since 1987, the conservation support it offers is restricted to hay meadows in valley bottoms (Gaskell and Tanner, 1991). More cynically, promotional effort could have been inputted to ensure a successful start for CS. Second, some large institutional landowners (such as the National Trust) have made a strategic decision to enter all land into CS where feasible, thereby influencing uptake. Third,

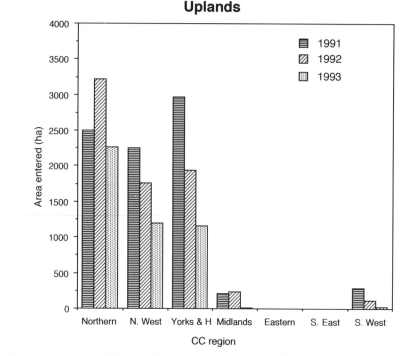

Fig. 15.2. Uptake of 'Uplands' target landscape under Countryside Stewardship, 1991-1993.

hill farmers experienced severe downward pressure on incomes through-out the 1980s, generating just £5900 net income on average by 1991. Faced with increasing business marginalization, the financial rewards of a CS agreement may have proved attractive given the limited opportunities to follow agroindustrial or farm diversification business development paths (Bowler, 1992). The launch of a Lake District ESA in 1993 does seem to limit the future scope of CS in the uplands.

Coasts

Coastal landscapes show a modest absolute hectareage (8545 ha) of land in CS and, like upland landscapes, Fig. 15.3 demonstrates a third year 'tailing off' in area entered for the dominant Northern and South West CC Regions. The activities of the National Trust again seem important as it has acquired extensive stretches of coastline through its Enterprise Neptune programme, many of which will have been entered into CS. The ability of CS to strengthen the non-statutory commitment to conservation within Heritage Coast (HC) planning initiatives is worthy of investigation. HCs are

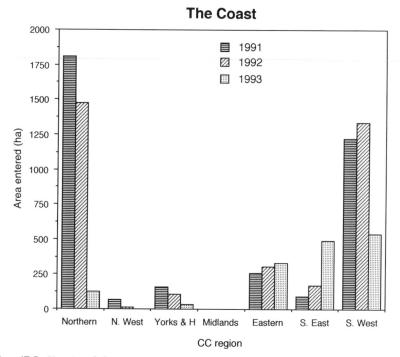

Fig. 15.3. Uptake of 'Coasts' target landscape under Countryside Stewardship, 1991–1993.

most extensive in the South West and Eastern Regions, yet markedly different adoption patterns are apparent. The launch of two ESAs designed specifically to conserve coastal salt and grazing marshes (North Kent Marshes, 1993; Essex Coast, 1994) again seems to limit the relevance of CS in the very region (South East) which experienced most growth in agreed area over the period considered.

Old Orchards

Uptake under the old orchards landscape is measured in numbers entered due to the small areas involved. Over two years of operation, 126 orchards were enrolled in CS. This is a very conservative achievement as there are approximately 3000 commercial orchards in England according to agricultural census data. From Fig. 15.4, entry is surprisingly low from traditional 'top fruit' producing areas in the South East (Kent and Essex) and Midlands (Hereford and the Vale of Evesham). The interplay between scheme conditions and forces of agricultural modernization appears influential. CS supports only 'standard' sized trees of increasingly scarce old

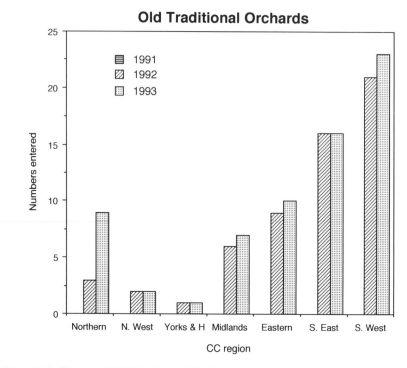

Fig. 15.4. Uptake of 'Old Traditional Orchards' target landscape under Countryside Stewardship, 1991–1993.

varieties rather than modern dwarf varieties. In the highly productive South East and Midlands, a greater proportion of older trees have been replaced to increase output, reducing the eligible orchards. However, longer monitoring of the geography of CS orchard uptake is required.

Something Borrowed – FCGS

Introduced in 1989 throughout England, the FCGS has proved to be MAFF's main capital grant scheme for agriculture in the 1990s. The scheme borrows principles from the former Agricultural Improvement Scheme (AIS)[4] insofar as FCGS was designed to help farmers maintain efficient agricultural systems. The difference was that FCGS recognized the cost of combating pollution (in particular) and conserving the countryside and its wildlife (MAFF, 1989). It represented a policy shift (Ilbery, 1992; Bishop and Phillips, 1993), but there was little scrutiny of the appropriateness of capital payments (a successful mechanism for raising food output) to achieve conservation outcomes. A sign of stress became evident after just two years when a major modification introduced longer-term 'Improvement Plans' to accompany the hitherto 'one-off' grants. Also in 1991, farm diversification grants were curiously incorporated before disappearing entirely in 1993. Finally, FCGS experienced a sudden scaling down in 1994 pending its closure. The range of capital grants-aided projects supported under FCGS (January 1994) are summarized in Box 15.1. Despite its status, FCGS has attracted minimal research attention in terms of rates of uptake and farmers' views on the scheme. A deterrent to investigation, as Crabtree and Chalmers (1994, p. 104) note, has been that 'a detailed appraisal of the FCGS scheme would require data on the additionality element (the extra conservation that took place over and above that which would have taken place in the absence of capital grants) in the grant aid'.

In the English Midlands from April 1989 to March 1994, MAFF made FCGS awards totalling £23.2 million to 4009 holdings, an average of £5800 per application. Figure 15.5 indicates the pattern of uptake of the main FCGS components in this period. A small number of farmers obtained grant aid in the first year, followed by a substantial increase in the second year, providing an interesting contrast with the adoption pattern for CS. Due to its evolution from the AIS, farmers may have more sympathy with a productivist-based ethic of improving efficiency through grant aid for

Box 15.1. Grant type, description and rate under FCGS as at January 1994.

Environmental and countryside　　　　　　　　　　　　　　　　　grant rate 15–50%
Support for capital works on hedges, stone walls and shelter belts, repairs to traditional buildings, heather burning, bracken control and fencing livestock out of broadleaved woodlands or heather moors and heaths.

Waste handling facilities　　　　　　　　　　　　　　　　　　　　　grant rate 50%
Support for facilities for the handling, storage and treatment of agricultural effluents and waste including safety fences and fixed disposal equipment.

Land improvements and energy saving　　　　　　　　　　　　　grant rate 15–25%
Support of fencing, reseeding and regeneration, lime and fertilizer, flood protection and energy saving facilities and the renewal and replacement of field drainage. Available only as part of a farm 'Improvement Plan'.

Horticultural buildings and orchard replanting　　　　　　　　　grant rate 25–40%
Support for the replacement of heated glasshouses, the installation of glasshouse heating systems and the replanting of apple and pear orchards.

Other grants　　　　　　　　　　　　　　　　　　　　grant rate as appropriate
Support for permanent thermal insulation paper ceiling for temperature or atmosphere control in agricultural buildings, plan preparation by ADAS and others, and supplies and insulation of wind and water-powered pumps and generators, solar and other forms of permanent durable energy-saving agricultural investment.

Source: MAFF Farm and Conservation Grant Scheme Handbook.

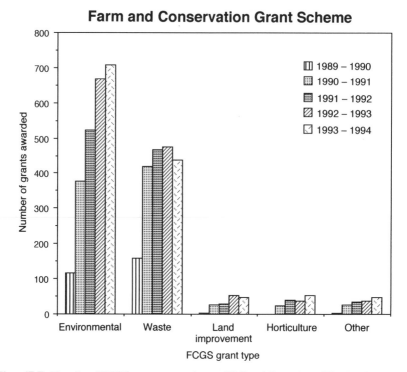

Fig. 15.5. Uptake of FCGS grant types in ten Midland Counties of England, 1989–1994.

works than for payments to follow prescriptions. Research into farmers' views and understanding of capital grant support for conservation seems appropriate. Grants for conservation (environmental and countryside) were most numerous, but financial contributions towards waste handling were greatest at £10,400 per application, reflecting government concern to meet European water pollution standards. With over 3600 grants made for conservation and waste handling in Midland counties, uptake compares favourably with that of the more spatially restricted ESA and CS schemes. These two main aspects of FCGS are considered further.

Environmental and countryside ('conservation') grants

The quantity of conservation grants awarded shows continual growth over the study period, although the rate of growth slows towards 1994 (Fig. 15.5). This is significant given the introduction of CS into the region, the launch of a countrywide Hedgerow Incentive Scheme[5] (now part of CS) and establishment of two ESAs (Clun, 1988; South West Peak, 1993) during the lifetime of the figures. The apparent success of FCGS in the Midlands when 'higher profile' schemes are available is interesting, especially as exploratory farm survey information indicates a low farmer awareness of FCGS. Evidently, more detailed analysis is needed.

Grants for waste handling

Figure 15.5 shows that waste grants reach a plateau in 1992, perhaps because farmers undertake anti-pollution work only when necessary to conform with new legislation. The 1989 Water Act and 1990 Environmental Protection Act both introduced more stringent regulations for the storage and application of livestock manure (see Hanley, 1991, Garner and Jones, 1993). The waste section of FCGS was therefore designed to enable farmers to comply with them. Consequently, there is a clear bias within the region towards grants being awarded in those counties dominated by livestock production. Although withdrawn in 1994, waste handling grants are to reappear in the forthcoming Nitrate Vulnerable Zones, making them spatially restricted.

Something 'Blue' – Increasing Public Access to Farmland

The general lack of public access to farmland held under private property rights in England has been lamented by certain non-governmental organizations, the Ramblers Association (RA) being most vociferous. A criticism increasingly aimed at the three agrienvironmental schemes discussed is their limited contribution towards a more accessible countryside in the face of growing recreational demand (Glyptis, 1991). ESAs originally contained no access provision, CS has access as a major objective, whereas the contribution of FCGS has been minor through grant aid for gates, stiles and footbridges where associated with field boundary renovation[6].

Not until the introduction of CS in 1991 did the deficiencies of ESAs come to the fore. In February 1994, MAFF subsequently 'bolted on' a Public Access Tier to all ESAs as part of Regulation 2078/92. At the discretion of MAFF, an additional flat rate payment of £170 per km (£274 per mile) is available for 10-metre wide 'access routes' on enclosed non-arable land, plus up to 80% capital grant aid on infrastructure (more generous than FCGS offered). The purpose is to provide new opportunities for 'quiet recreation' at locations of historical interest, viewpoints and areas with poor existing access. Its effectiveness is unclear at this early stage, but anecdotal evidence from the Cotswold Hills ESA is less than encouraging. For 1994/95, only two out of the 334 agreements signed involve the access tier. Stringent criteria and the ineligibility of arable land to qualify for access payments, a particular problem in areas dominated by mixed farming or arable systems (such as the Cotswolds), have constrained uptake. Whether the new association of access with ESAs will increase the resistance of non-adopters already concerned about the issue remains to be seen (Morris and Potter, 1995). Work is currently in progress to explore farmers' attitudes to increasing public access to their farmland (for example, Deaville, 1994)[7].

Access has been an integral part of CS since its inception, so a more substantial contribution might be anticipated. Unfortunately, it is the access provision that has received most criticism. Payments should be for new access only and unavailable for rights of way, land under access agreement, or land with a history of *de facto* access (CC, 1994). By the end of 1993, open access had been agreed over 12,925 ha and linear access totalled 303.6 km. In all cases, the amount of new access created each year has declined from the first year of eligibility. By 1993, only 139 new agreements had been made. Apart from the small quantity of access created, work by the RA has revealed four main deficiencies with CS access arrangements. First, contrary to scheme rhetoric, payment is being made for access that already exists and in extreme cases formerly open access land is now restricted to linear usage! Of special concern has been the payments given to public bodies or charities, such as the National Trust, which has a long tradition of permitting open access to its land. This duplicity has been estimated to occur on half of all sites (Smith, 1993). Second, the scheme element of cross-compliance, which demands that all existing access routes are in good order, is not being implemented strongly enough. Third, a persistent complaint is that access sites have been poorly publicised. The CC has pledged to give more locational details, but the RA still point to their inadequacy (Kempe, 1995). Fourth, the RA are unhappy that the scheme has established a precedent of paying landowners for private property rights.

Conclusions

With agrienvironmental policy assuming greater significance in the political arena and among farmers themselves, this chapter has demonstrated that much remains unexplained about patterns of uptake. Research is needed to ascertain the role agrienvironmental schemes now play in paths of farm business development. Only then will an understanding of the geography of agrienvironmental policy and its subsequent contribution to sustainability in farming systems emerge. Farm survey work relating the macro-level evidence presented to the attitudes, opinions and motives of farmers towards schemes should be conducted to achieve this objective. For example, the 'unconventional' pattern of CS adoption should be examined to determine if there is a business advantage in 'being first' (see Freeman, 1985). The impression that conservation is a peripheral outcome of political and economic adjustment from productivism remains. This is demonstrated by the strength of competition exerted by the AAPS of a reformed CAP on ESAs when attempting to recruit farmers to more exacting conservation tiers. The environmental outcomes of current policy are questionable, particularly in respect of long-term sustainability. Indeed, expansion of the ESA programme has been welcomed as increasing conservation potential, but evidence suggests that this serves to constrain the potential future uptake of CS. With MAFF tending to measure the success of conservation schemes in terms of number of participants (Morris and Potter, 1995), dangers for the long-term existence of the CS scheme are already apparent before its relaunch!

The extent to which agrienvironmental schemes are changing farmers' outlook on conservation as a worthwhile practice or business opportunity, extending the work of Morris and Potter (1995), deserves attention. Even if greater numbers of farmers become committed to conservation by entering schemes, there is no guarantee that their view of what they should be doing will move closer to the expectations of conservationists, as Carr and Tait (1991) have demonstrated. Greater integration between schemes is progressing slowly and should theoretically produce better environmental results in terms of sustainability. Disappointingly, savings for the exchequer appear to be the driving force behind such moves. For example, results imply that it is the administrative cost and inflexibility of FCGS rather than poor uptake that has precipitated the recent move to link capital grants with schemes offering standard payments (Colman *et al.*, 1992; Crabtree and Chalmers, 1994). Similarly, government keenness to promote access from within conservation schemes could represent a shift towards obtaining more tangible public benefits from money spent (Crabtree, 1994). For many conservationists, no matter how popular the agri-environmental policies examined prove to be, they will remain unsatisfactory devices for achieving sustainable agriculture compared with direct environmental management payments decoupled from a farm support system based on food production (Potter, 1992).

Notes

1. Much of FCGS now replicates provisions in ESAs and CS, so it will be merged imperceptibly with the latter under its capital works section for target landscapes when adopted by MAFF in 1996.
2. Management or 'standard' payments are defined as 'a voluntary arrangement in which payments are offered as an inducement to accept constraints on farm operations or to engage in additional activities that are beneficial for

conservation' (Crabtree and Chalmers, 1994, p. 94).

3. For an indication of the extent of farm diversification in the county of Cornwall, see Davies (1971) and Evans and Ilbery (1992).
4. It is interesting to note that in the 'new' age of agrienvironmental schemes and conservation, MAFF has continued to finance farm improvements (such as draining and ploughing) in the productivist mould under the AIS. Although closed in 1988, some farmers had ten-year improvement plans approved under the scheme that do not finally expire until 1998.
5. Crabtree and Chalmers (1994) demonstrate that it is grants for hedgerow investment that typically form the greatest part of the 'conservation' grants category of FCGS.
6. A new scheme aimed at increasing public access to set-aside land has been introduced, known as the Countryside Access Scheme. For details, see MAFF (1994).
7. This work is being conducted at Worcester by Jennifer Deaville through a College Research Studentship entitled 'Farm businesses and public access to agricultural land for recreation in England and Wales'.

References

Baldock, D., Cox, G., Lowe, P. and Winter, M. (1990) Environmentally sensitive areas: incrementalism or reform? *Journal of Rural Studies* 6, 143–162.

Bishop, K. and Phillips, A. (1993) Seven steps to the market – the development of the market-led approach to countryside conservation and recreation. *Journal of Rural Studies* 9, 315–338.

Bowler, I. (1992) 'Sustainable agriculture' as an alternative path of farm business development. In: Bowler (ed.) *Contemporary Rural Systems in Transition, Volume 1: Agriculture and Environment.* CAB International, Wallingford, UK, pp. 237–253.

Brotherton, I. (1991) What limits participation in ESAs? *Journal of Environmental Management* 32, 241–249.

Carr, S. and Tait, J. (1991) Farmers' attitudes to conservation. *Built Environment* 16, 218–231.

Colman, D., Crabtree, J., Froud, J. and O'Carroll, L. (1992) *Comparative Effectiveness of Conservation Mechanisms.* Department of Agricultural Economics, University of Manchester.

Colman, D., Froud, J. and O'Carroll, L. (1993) The tiering of conservation policies. *Land Use Policy* 10, 281–292.

Countryside Commission (1994) *Countryside Stewardship: Handbook and Application Form.* CCP 453. Cheltenham.

Crabtree, J. (1994) Developing market approaches to the provision of access. Paper presented to the Rural Geography Study Group, Nottingham, Sept.

Crabtree, J. and Chalmers, N. (1994) Economic evaluation of policy instruments for conservation. *Land Use Policy* 11, 94–106.

Curry, N. (1994) Current priorities for countryside planning. Paper presented to the Geography Department, Worcester College of Higher Education, Worcester UK.

Davies, E. (1971) *Farm Tourism in Cornwall and Devon – Some Economic and Physical*

Considerations. Report No. 184. Agricultural Economics Unit, University of Exeter, Exeter.

Deaville, J. (1994) Farm businesses and public access to agricultural land: a review. Geography Department working paper, Worcester College of Higher Education, Worcester, UK.

Evans, N. and Ilbery, B. (1992) The distribution of farm-based accommodation in England and Wales. *Journal of the Royal Agricultural Society of England* 159, 67–80.

Freeman, D. (1985) The importance of being first: preemption by early adopters of farming innovations in Kenya. *Annals of the Association of American Geographers* 75, 17–28.

Froud, J. (1994) Upland moorland with complex property rights: the case of the North Peak. In Whitby, M. (ed.) *Incentives for Countryside Management: the Case of Environmentally Sensitive Areas.* CAB International, Wallingford, pp. 81–104.

Garner, J. and Jones, B. (1993) *Countryside Law* 2nd edn (with supplement). Shaw & Sons, London.

Gaskell, P. and Tanner, M. (1991) Agricultural change and Environmentally Sensitive Areas. *Geoforum* 22, 81–90.

Glyptis, S. (1991) *Countryside Recreation.* Longman, Harlow, Essex.

Hanley, N. (1991) The economics of nitrate pollution control in the UK. In: Hanley, N. (ed.) *Farming and the Countryside: An Economic Analysis of External Costs and Benefits.* CAB International, Wallingford, UK, pp. 91–116.

Ilbery, B. (1992) From Scott to ALURE – and back again? *Land Use Policy* 9, 131–142.

Kempe, P. (1995) What a swiz! *Rambling Today,* spring.

Ministry of Agriculture, Fisheries and Food (1989) *The Farm and Conservation Grant Scheme.* MAFF Publications, London.

Ministry of Agriculture, Fisheries and Food (1994) *The Countryside Access Scheme.* MAFF Publications, London.

Morris, C. and Potter, C. (1995) Recruiting the new conservationists: adoption of agri-environmental schemes in the UK. *Journal of Rural Studies* 11, 51–63.

O'Carroll, L. (1994) Competition with other environmental designations on a lowland heath: the case of Breckland. In: Whitby, M. (ed.) *Incentives for Countryside Management: The Case of Environmentally Sensitive Areas.* CAB International, Wallingford, UK, pp. 61–79.

Potter, C. (1988) Environmentally Sensitive Areas in England and Wales. *Land Use Policy* 5, 301–313.

Potter, C. (1992) Recranking the engine. *Ecos* 13, 1–2.

Potter, C. (1993) Pieces in a jigsaw: a critique of new agri-environment measures. *Ecos* 14, 52–54.

Robinson, G. and Ilbery, B. (1993) Reforming the CAP: beyond MacSharry. In: Gilg, A. (ed.) *Progress in Rural Policy and Planning,* Volume 3. Belhaven Press, London, pp. 197–207.

Smith, E. (1989) Environmentally Sensitive Areas: a successful UK initiative. *Journal of the Royal Agricultural Society of England* 156, 30–43.

Smith, R. (1993) Better access plan. *Country Walking.* June 1993.

Whitby, M. (ed.) (1994) *Incentives for Countryside Management: The Case of Environmentally Sensitive Areas.* CAB International, Wallingford.

Wynne, P. (1994) Agri-environment schemes – recent events and forthcoming attractions. *Ecos* 15, 48–52.

16

Farmer Reaction to Agrienvironment Schemes: A Study of Participants in South-west England and the Implications for Research and Policy Development

ANDREW W. GILG AND MARTIN R.J. BATTERSHILL
Department of Geography, University of Exeter, Rennes Drive, Exeter EX4 4RJ, UK

Introduction and Chapter Outline

Modern agriculture in Europe has been subjected to a variety of forces that have led to major restructuring and have rendered it increasingly unsustainable. The unifying factor behind all these forces, according to Le Heron (1993), is the world food system which contains two key trends: first, the globalization of production and distribution; and second, the integration of agriculture into the wider economy. Bowler (1992a) has termed this second trend the 'industrialization' of agriculture and a 'third agricultural revolution', which has resulted in a 'food supply system' and which has the food chain as its spine.

The food chain begins with agricultural inputs, which lead into farm production, followed by product processing, food distribution and consumption. External factors that influence the food chain are: state farm policies; international trade in food; the physical environment; and the

© 1997 CAB INTERNATIONAL. *Agricultural Restructuring and Sustainability*
(eds B. Ilbery, Q. Chiotti and T. Rickard)

financial markets. Collectively, the food chain and these external factors make the food supply system. The development of a food supply system does not mean that agriculture itself has yet been fully industrialized, but that it has been integrated into wider industrial processes so that 'production on the farm' now forms only a part in the process.

Thus, agriculture has been subject to both a trading and technological revolution in the postwar period, and this has elicited according to Bowler (1992a) a set of primary responses and a set of secondary consequences, in particular, industrialization and the three related processes of intensification, concentration and specialization.

In more detail, industrialization has involved the creation of scale economies at the farm level (larger farms), increased reliance on purchased inputs (machinery, fertilizers, feed, agrichemicals), resource substitution (capital for land and labour), the implementation of organizational features associated with the concept of the 'firm', specialization of the labour function within the farm business and mechanization of the production process. Intensification has witnessed the use of rising levels of purchased non-farm inputs in agriculture (capitalization) with resulting increases in output per hectare of farmland. Concentration has led to a process whereby productive resources, and the output of particular products, have become confined to fewer but larger farm businesses, and to fewer regions and countries. Finally, specialization has seen the proportion of the total output of a farm, region or country accounted for by a particular product rise rapidly. The secondary consequences of industrialization according to Bowler include several features that have made contemporary agriculture potentially unsustainable, in particular: over dependence on fossil fuels; destruction of the environment and agrosystems; too much food consumed outside the region where it was produced; and an increased risk of system failure.

The industrialization of agriculture has been aided and abetted by support programmes at the national and international level, which have provided high support prices and guaranteed markets, and thus the safety net for the massive investment in new technology that the industrialization of agriculture has entailed. For example, the Common Agricultural Policy of the European Community and its system of Intervention purchases encouraged farmers to produce as much as they could, in the knowledge that they could sell it into intervention at a profit. The net result was overproduction and the creation of surpluses, and a growing cost to the taxpayer.

Thus the world food system has arisen from a variety of forces, and the resulting patterns and modes of production have major implications for the environment. For example, Goodman and Redclift (1991) claim that the system implies a major change in the natural environment because key decisions are now effectively made by capitalists in places that may be continents away from where farmers activate these decisions as pawns in a global game. However, Pierce (1990) reminds us that agriculture still depends very heavily on the natural environment, and is still characterized by very large numbers of relatively small producers. None the less, these small producers have been forced into production systems that many commentators have claimed are unsustainable, because they are so dependent on non-renewable fossil fuels, toxic pesticides and chemicals, a few species of plant and animals and monocultures rather than mixed farming systems. Therefore, many analysts (Bowler, 1992b; Gilg, 1992; Pierce, 1992) have begun to call for a return to more traditonal farming and for national and international farm policies to contain an environmental element so that as agriculture further restructured itself it could become more sustainable. Policy-makers also responded by developing a number of schemes aimed at modifying farm production systems. For example, the UK Environmentally Sensitive Areas scheme, and the French Prime L'Herbe scheme, which compensate farmers for farming in less intensive ways. Between 1985 and 1995 over 20 such schemes were developed (Swales, 1994) and collectively they became known as 'agri-environment schemes'.

Tarrant (1992) in an analysis of the genesis and impact of the schemes has argued that during:

> the 1980s a convergence took place between three strands of the policy-making process in many developed countries. Firstly, the costs of existing farm support programmes became financially unsustainable requiring a revision of agricultural policy. Secondly, the persistence of food surpluses drew a recognition that less farmland was needed in production – that is, that a 'surplus' of farmland existed. Thirdly, concern with the conservation of the environment turned to the damaging impacts of modern agriculture.
>
> Policy-makers began to perceive a linkage between these three strands: 'surplus' farmland could be turned to conservation uses, with farm subsidies redirected from price supports to encouraging 'environmentally friendly' farming practices (Gilg, 1991). Attempts to merge agricultural and environmental policies during the 1980s have, nevertheless, proved disappointing. The agricultural interest has proved resistant to the switching of funds from price supports to the production of 'environmental goods', and the adoption of environmental measures has remained voluntary for the individual farmer rather than mandatory. Most environmental measures for agriculture, therefore, appear to be 'bolted on to' existing farm policies which remain essentially unreformed (Robinson, 1991, p. 266).

This chapter, which is based on ongoing research that began in 1991 into the uptake of agrienvironment schemes, seeks to ascertain how well Tarrant's assertions stand up in the 1990s, notably:

1. the reaction of individual farmers given the voluntary nature of the measures;

2. the degree to which not only government policies, but also farm management practices have remained rooted in the concepts of conventional or industrialized farming; and

3. the extent to which market demand in the 'food chain' for 'regional' and 'environmentally friendly' food could in the longer term replace government policies for sustainable agriculture.

In order to do this, this chapter is divided into three sections. The first section outlines the original research methodology and objectives employed, the second section discusses the first set of research findings in the context of contemporary research, and the third section discusses the implications of the original study for the evolution of agricultural policy and how this research has been extended into a second research project in France.

Research Methodology and Objectives

In order to study the reaction of farmers to environmental measures 14 agrienvironment schemes and initiatives in Devon, Cornwall and west Somerset were identified across a range of mechanisms.These included: incentive schemes; management agreements; conservation advice; food marketing standards; and landlord/tenant arrangements. A full list of the schemes and a brief outline of their main attributes is provided in Box 16.1.

Between May 1992 and August 1993 122 scheme participants were visited. The research sought to illuminate how geographical, attitudinal, socioeconomic and farm system circumstances influenced the nature of environment-friendly farming or activity, and in particular which circumstances were most influential on that activity. The sample was dominated by farms where the whole farming system could be defined as environment-friendly, including some farms which also contained specific areas of uniquely high conservation value. A smaller number of farms were involved with grant-aided creative conservation or public access work, in an otherwise less environment-friendly system. Others were engaged in the single-minded pursuit of premium food prices.

Box 16.1. Outline of the schemes studied.

---

1. Countryside Stewardship scheme: a Countryside Commision scheme offering standard rate acreage and capital works payments for conservation and public access work in certain landscapes.

2. ELMS: a Devon County Council scheme offering ten-year agreements for conservation management (especially of Culm grassland) and public access work.

3. and **4.** ESAs: the Somerset Levels & Moors and West Penwith ESAs were visited, the former protecting a distinct landscape of wetland and drainage dykes, the latter a landscape of Celtic fields and coastal moorland.

5. Exmoor Farm Conservation Scheme: an Exmoor National Park scheme offering shared costs between the Park, the Farm and Conservation Grant Scheme and the farmer for landscape, hedging and conservation work.

6. Dartmoor National Park Management Agreements: agreement holders were visited with agreements for moorland and wetland conservation and for public access work.

7. Wildlife Enhancement Scheme: a pilot scheme run by English Nature to fund conservation improvement work on SSSI status areas of the rare and fragmented Culm grassland habitat.

8. FWAG: offering conservation advice to farmers who request it.

9. Royal Association of British Dairy Farmers Conservation Prize: a national prize scheme with the award based on the successful incorporation of conservation work into commercial farming.

10. Conservation Grade: a food marketing standard encouraging lower-input practices and the use of benign agrochemicals.

11. Soil Association: the largest verification scheme for organic produce.

12. West Country Lamb: a food marketing scheme encouraging extensive and 'traditional' husbandry in sheep farming.

13. and **14.** Duchy of Cornwall and the National Trust: promotion or support of conservation work in a landlord/tenant context.

Abbreviations: ESA, Environmentally Sensitive Area; FWAG, Farming and Wildlife Advisory Group; SSSI, Site of Special Scientific Interest; ELMS, Environmental Land Management Services.

The Research Findings in the Context of Contemporary Research

The detailed research findings are available elsewhere (Battershill, 1995; Battershill and Gilg, 1996a, b, c), but are summarized in Box 16.2. With reference to contemporary research the findings can be divided into those that are broadly consistent, those that differ slightly and those that are markedly different.

Certain findings were broadly consistent with observations contained in previous work. Examples included the fact that most farmers in the sample shared 'intrinsic' farming values (Gasson, 1973); that most of the environment-friendly farming activity and many of the farms of highest

Box 16.2. Two key findings and ten detailed findings.

First key finding: Environmentally friendly farmers are not distinguished by their socioeconomic status and business characteristics but by behavioural or attitudinal traits rooted in their personal histories and circumstances: most are either practically committed to 'traditional' farming or have strong proconservation attitudes. This finding undermines a tradition of structural analysis in this field.

Second key finding: Most conservation schemes have only a marginal impact on farming practices and environmental outcomes compared with what the participants would have been doing in any case.

Ten detailed findings

1. There was no relationship between farmers' attitudes to, or participation in, environment-friendly activities and the degree of economic constraint on the farm.

2. There were very few examples of 'wealthy', 'accumulator' or 'expansionist' farmers engaged specifically in grant-based or creative conservation activity. Such farms appeared more likely to be associated with the single minded pursuit of food marketing premiums.

3. There was no apparent relationship between elderly farmers without successors and high farm landscape conservation value.

4. There was no apparent relationship between the prescence of an agricultural successor and the type of environment-friendly activity.

5. 'Occupancy events', in particular the purchase of a new farm, appeared more likely to lead directly to new conservation work rather than adverse landscape change.

6. Many 'survivors' and part-time farms were organic farms, who retained high levels of environment-friendly activity despite their marginal profitability.

7. Most small farms were less intensive in terms of agrochemical use than were larger farms.

8. In high value landscapes there was some evidence that larger farms had higher conservation value farm landscapes because their larger size embraced a larger number and greater range of landscape features.

9. Mixed tenure farms were often extensive, with the extensive system creating a greater need for more farmland and thus opportune letting of parcels of land.

10. Nearly 80 per cent of farmers throught that environment-friendly farming was also good husbandry.

conservation value were associated with poorer quality farm land or areas (Copper *et al.*, 1989; Tarrant and Cobb, 1991); and that selectivity was more commonly associated with 'traditional' rather than expansionist farmers (Lobley, 1989; Whitby, 1994).

However, other observations differed from earlier research, thus raising varying degrees of doubt about their applicability or continued validity. Examples included the observations that occupancy events, especially new farm purchase, seemed more likely to stimulate conservation than landscape damage (contrasting with Marsden and Munton, 1991); that whereas very few scheme participants had high economic constraint, few were highly enabled either (contrasting with Potter and Gasson, 1988); and that whereas farmers' definitions of 'conservation' were often different from the more sophisticated definitions employed by conservationists, there was no real conflict inherent in the differences (contrasting with Carr and Tait, 1991).

Other findings showed more marked and significant differences. These included the observations that high economic constraint, or 'survivor' circumstances, did not appear to inhibit environment-friendly farming or scheme participation (contrasting with Marsden *et al.*, 1986; Munton *et al.*, 1989); that the presence or otherwise of a farming successor was not a major influence on farming activity or conservation value (contrasting with Potter and Lobley, 1992; Ward and Lowe, 1994); and that there were few examples of 'programmer' or 'expansionist' farmers engaged in creative conservation work having already damaged the conservation interest of their farm (contrasting with Westmacott and Worthington, 1984; Potter, 1986; Munton and Marsden, 1991).

It is possible that aspects of these differences stemmed from the fact that the sample was different in its construction and its character from many of those on which previous research observations had been based. First, the sample was not drawn as a random and stratified sample of ordinary farms, but was constructed purposively to examine farms in environmental schemes. Second, the sample was drawn entirely from the south west of England, whereas most of the literature is based on research conducted in central, southern and eastern England. However the final sample did contain a wide range of farming characteristics in terms of both socioeconomic and geographical structures, a range largely typical of farms in the south west as a whole, and thus the results may indicate a tendency for the human and physical environments of south west England to be more conducive to environment-friendly farming schemes than other regions of lowland England.

Another possible reason for the significant differences between this and previous surveys could be that a significant amount of time has passed since many of the research projects referred to were carried out, during which time the direction of agricultural policy and the nature of farming culture and values has shifted away from production and towards a greater awareness and consideration of environmental needs and objectives. Comparisons with more recent research (for example Whitby, 1994) suggests that such considerations might be important.

However, beyond such considerations, the many differences, when compounded, also point to and describe a situation of altogether more profound significance for attempts and approaches at understanding farm and farmer behaviour. The research findings often suggested that among farms in the sample the nature and style of environment-friendly farming systems had more complex origins than those anticipated by the review of

⌐∪ᴀᴛᴇ UNIVERSITY
∪NIVERSITY PARK
IL 60466

previous research work; in particular, numerous examples existed of farmers practising environment-friendly farming in circumstances where previous research would have predicted they would not. The whole issue of environment-friendly farming was seen to persuasively and consistently manifest one crucial claim: that socioeconomic and even geographical indicators could not alone determine or explain farm and farmer behaviour in a consistent and predictable way. Throughout the investigation, similar structural constraints or stimuli were associated with very different behavioural responses, responses that resulted time and again from individual farmer's attitudes and intentions. The final conclusion drawn was that no simple rules or statistical tests could be said to exist that could accurately anticipate how farmers respond to socioeconomic pressures or policy mechanisms, as it was only at the level of individual attitudes and values that behaviour could be fully comprehended. If our knowledge of the farming world is to be further developed and applied, this is one vital field to which future research approaches and methods should be directed.

Some Research Implications of the Findings

The implications for the evolution of research are twofold. First, there is a need to find out if the circumstantial influences discussed in this chapter are also dominant elsewhere, and so it would be valuable to apply some of the questions to a study of a different set of farmers. This set could include a sample of scheme participants from a contrasting area of Britain, or indeed another country. It would also be interesting to examine a comparable sample of non-scheme farmers in the south west of England (or a similar region), to test whether the circumstances of the farms were regionally based or unique to the fact that the farms were participants in agrienvironment schemes.

Second, since the importance of 'traditional' farming was emphasized by the survey more information is needed about 'traditional' farmers, both as to why they have not responded to structural influences such as economic or family constraints, and what are the attitudinal, sociological and psychological influences which have influenced their decision-making, their behaviour and their continued status as 'traditional' farmers? In addition more information is needed about the characteristics and features by which 'traditional' farmers can be identified, and how they can be most effectively recruited into agrienvironment schemes.

In an attempt to address these implications the authors have extended the research into France, where a two-year project (November 1995–October 1997) is examining two groups of farmers and their relationships with environment-friendly farming. The first group of farmers are those who are involved in the production of 'regional' and 'traditional' foods that are implicitly marketed as being favourable to the environment. The second group of farmers are 'normal' farmers in the mainstream of 'industrial' agriculture.

The research is also addressing the extent to which market demand for the products of sustainable food systems could eventually replace the need for government provided agrienvironment schemes. This theme has been developed from an analysis by Bishop and Phillips (1993), which picked out six policy instruments that have so far been used to induce environment-friendly farming, namely: regulation; advice and information; grants; compensation; support for environment-friendly farming; and cross compliance. They concluded that all of these have some shortcomings and have thus opened the door to a seventh mechanism, the market-led approach. Bishop and Phillips argue that this market-led approach is distinctively different from the other policy tools for reconciling conservation and agricultural interests. First, there is no obligation on either side to participate. Second, payment is for the product, not the process. Third, the thrust is moving away from protection of existing sites, to re-creation of certain landscape or habitat types; and fourth, they are being used to promote multipurpose land management. However, Bishop and Phillips' market-led approach is still dependent on government policy, and so the research in France is attempting to assess the extent to which a truly free, albeit highly regulated, market can provide an alternative to the transformation of food into a sanitized and uniformly bland product across the world or the cheap food policy at any price, which led to the BSE débâcle in the UK. In a nutshell the French research is contrasting quality with quantity, regional differences with global uniformity, and food as a cultural asset compared with food as being merely a fuel.

In attempting to make this assessment one needs to return to the core concept of the world food system that was outlined at the beginning of the chapter. All the evidence we have at the moment suggests that the world food system will continue to grow in influence but will become bipolar. At one end of the spectrum accounting for maybe 80% of production and 70% of value, day to day foods will continue to be produced and distributed by multinational corporations via supermarkets. At the other end of the spectrum, niche food products for weekend meals, quality restaurants and special event meals will continue to be produced by traditional regional systems based on the concept of traceability and the French 'appellation' certificate. Being of higher value these products could account for only 20% of production, but 30% of sales. It is forecast that these niche products will expand for a number of reasons. Primarily, consumers are more aware than ever before of animal welfare, environmental issues, and food quality rather than quantity and price. Second, and somewhat speculatively, developments in modern communications could enable consumers to bypass the supermarket and wholesaler and contact the producer direct. For instance, a specialist home food storage system could monitor the rate at which foods were consumed and feed the information back to a cooordinating warehouse that would replenish stocks via home delivery on a regular basis, A computer prompt could then ask the consumer to order less mundane products for say Christmas up to two years ahead. So a consumer wanting a particular breed of turkey, fed in a certain way, and killed and dressed and stuffed to order, could do so as part of their regular computer housekeeping. In time such computer systems could obviate the need for both public and private planning of agricultural production by providing farmers with long term order books.

Some Concluding Thoughts on Policy Evolution

The authors began this chapter by observing that current policies have been 'bolted on to' existing policies. Their research and an analysis by Gilg (1996) has confirmed the intrinsically incremental nature of policy evolution, and a strong tendency for farmers to only adopt those policies that opportunistically fit in with their own perception of farming and current management practices. Therefore they propose three incremental policy reforms:

Give more support for 'traditional' farming

Britain is fortunate that some farmers in certain areas have resisted the pressures to intensify their farming, and still manage farms and farming systems of high environmental value. The importance of supporting such 'traditional' farming is recognized by the ESA scheme and concept.

However, many 'traditional' farms outside existing ESAs remain unsupported and their systems are still under threat. More ESAs should be designated where a high proportion of 'traditional' farmers remain in an area. Otherwise efforts must be concentrated into identifying and supporting the 'traditional' farming system of farmers in the wider countryside. Financially, it will be necessary to ensure that the only farmers receiving windfall payments are those 'traditional' farmers who have never damaged the conservation value of their whole farms (nor in effect contributed to surpluses). Other farmers must be obliged to change their wider farming practices in order to receive incentive monies to protect specific areas or features. Other than on the most 'traditional' farms, financial assistance for creative conservation work or improvements to farm infrastructure should be on a shared cost basis.

Concentrate more efforts into educating farmers

There is evidence that many farmers have a basic sympathy for conservation, and some a keen interest in creative conservation. Thus in the short term more conservation and general environment-friendly farm management could be achieved simply with strong policy statements and commands, for example using farming colleges, the agricultural media and agricultural advisors.

After a reasonably short period of education and exhortation more planning or regulatory controls could be placed on agricultural practices and landscape features. Such a move will come as no surprise to the farmers who have observed the new policy messages.

Push European agriculture towards organic farming

Ultimately agriculture will not be environmentally sustainable until it returns to the essentially organic practices from which all 'traditional' practices have been derived. Guiding such a change can be the only logical long-term objective for agricultural policy. This will inevitably mean a return to non-chemical based husbandry practices. This process can be assisted by high taxes on sprays and independently funded and disseminated research into organic methods. Whereas this return to organic farming may seem revolutionary and impracticable, many of the farms visited in this research have shown such a revolution to be entirely practical if it is tackled in an incremental gradual way in both agricultural and personal terms.

References

Battershill, M. (1995) *Environmentally Friendly Farming in South West England*, Unpublished PhD thesis, University of Exeter.

Battershill, M. and Gilg, A. (1996a) Environmentally friendly farming in South West England: an exploration and analysis. In: Curry, N. (ed.) *New Directions in Rural Land Use*. Countryside and Community Foundation, Cheltenham, pp. 200–224.

Battershill, M. and Gilg, A. (1996b) Traditional farming and agro-environment policy in south west England: back to the future, *Geoforum* 27, 133–147.

Battershill, M. and Gilg, A. (1996c) New approaches to creative conservation on farms in south west England. *Journal of Environmental Planning and Management* 48, 321–340.

Bishop, K. and Phillips, A. (1993) Seven steps to the market – the development of the market-led approach to countryside conservation and recreation. *Journal of Rural Studies* 9, 315–38.

Bowler, I. (1992a) The Industrialisation of agriculture. In: Bowler, I. (ed.) *The*

Geography of Agriculture in Developed Market Economies. Longman, Harlow, pp. 7–31.

Bowler, I. (1992b) Sustainable agriculture as an alternative path of farm business development. In: Bowler, I., Bryant, C. and Nellis, D. (eds) *Contemporary Rural Systems in Transition: Volume 1: Agriculture and Environment,* CAB International, Wallingford, pp. 237–253.

Carr, S. and Tait, J. (1991) Farmers' attitudes to conservation. *Built Environment* 16, 18–23.

Copper, A., Murray, R. and Warnock, S. (1989) Agriculture and the environment in the Mourne AONB. *Applied Geography* 9, 35–56.

Gasson, R., (1973) Goals and values of farmers. *Journal of Agricultural Economics* 24, 521–542.

Gilg, A.W. (1991) *Countryside Planning Policies for the 1990s,* CAB International, Wallingford.

Gilg, A. (1992) Policy options for the British countryside. In: Bowler, I., Bryant, C. and Nellis, D. (eds) *Contemporary Rural Systems in Transition: Volume 1: Agriculture and Environment,* CAB International, Wallingford, pp. 206–218.

Gilg, A. (1996) *Countryside Planning: The First Half Century.* Routledge, London.

Goodman, D. and Redclift, M. (1991) *Refashioning Nature: Food, ecology, culture.* Routledge, London.

Le Heron, R. (1993) *Globalized Agriculture: Political Choice.* Pergamon, Oxford.

Lobley, M. (1989) A role for ESAs? *Ecos* 10(2), 27–29.

Marsden, T., Whatmore, S., Munton, R. and Little, J. (1986) The restructuring process and economic centrality in capitalist agriculture. *Journal of Rural Studies* 2, 271–280.

Marsden, T. and Munton, R. (1991) The farmed landscape and the occupancy change process. *Environment and Planning A* 23, 663–676.

Munton, R., Whatmore, S. and Marsden, T. (1989) Part-time farming and its implications for the rural landscape: a preliminary analysis. *Environment and Planning A* 21, 523–536.

Munton, R. and Marsden, T. (1991) Occupancy change and the farmed landscape. *Environment and Planning A* 23, 499–510.

Pierce, J. (1990) *The Food Resource.* Longman, Harlow.

Pierce, J. (1992) The policy agenda for sustainable agriculture. In: Bowler, I., Bryant, C. and Nellis, D. (eds) *Contemporary Rural Systems in Transition: Volume 1: Agriculture and the Environment.* CAB International, Wallingford, pp. 221–236.

Potter, C. (1986) Processes of countryside change in lowland England. *Journal of Rural Studies* 2, 187–195.

Potter, C. and Gasson, R. (1988) Farm participation in voluntary land diversion schemes. *Journal of Rural Studies* 4, 365–375.

Potter, C. and Lobley, M. (1992) The conservation status and potential of elderly farmers: results from a survey of England and Wales. *Journal of Rural Studies* 8, 133–143.

Robinson, G. (1991) EC agricultural policy and the environment: land use implications in the UK. *Land Use Policy* 8, 95–107.

Swales, V. (1994) Incentives for countryside management. *Ecos,* 15 (3/4), 52–57.

Tarrant, J. (1992) Agriculture and the state. In: Bowler, I. (ed.) *The Geography of*

Agriculture in Developed Market Economies. Longman, Harlow, pp. 239–274.

Tarrant, J. and Cob, R. (1991) The convergence of agricultural and environmental policies: the case of extensification in eastern England. In: Bowler, I., Bryant, C. and Nellis, D. (eds) *Contemporary Rural Systems in Transition: Volume 1, Agriculture and the Environment.* CAB International, Wallingford, pp. 153–165.

Ward, N. and Lowe, P. (1994) Shifting values in agriculture: the farm family and pollution regulation. *Journal of Rural Studies* 10, 173–184.

Westmacott, R. and Worthington, T. (1984) *Agricultural Landscapes: a Second Look.* Countryside Commission CCP 168, Cheltenham.

Whitby, M. (1994) What future for ESAs? In: Whitby, M. (ed.) *Incentives for Countryside Management: the Case of Environmentally Sensitive Areas.* CAB International, Wallingford, pp. 253–272.

17 Achieving Sustainability in Rural Land Management Through Landowner Involvement in Stewardship Programmes

STEWART G. HILTS
Department of Land Resource Science, University of Guelph, Guelph, Ontario, Canada N1G 2W1

Introduction

In 1979, a large woodlot on the edge of London, Ontario, was annexed, re-zoned and developed for residential use in spite of being initially zoned as conservation land, and being documented as a nationally significant natural area. Known as Warbler Woods, it was the largest woodland left in this region of southern Ontario, and is now a subdivision. This is merely one of thousands of land-use controversies that have pitted development against conservation over the past two decades in Canada – in this case, a controversy that the author was personally involved in.

Efforts to protect such natural habitats as woodlands, wetlands and other features have typically been approached through the land-use planning process in Canada, especially in urban or near-urban regions. This process usually operates through regional or local municipal governments. It is usually governed by provincial legislation such as a provincial Planning Act, and unfortunately often ends in direct conflict between competing views over land-use. In turn this leads to lengthy, expensive land-use hearings, and long delays in development approvals. Land-use issues are often among the most heated local political debates, and a focus of protest for many environmental groups.

However, in rural areas, where development is not an immediate concern, decisions to destroy, keep, or rehabilitate natural habitats are most often actually management decisions by private landowners. As environmental concerns become more widely accepted in Canada, attention is shifting to this landowner level of decision-making as an additional important influence on conservation achievements. Efforts to assist or support rural landowners in making environmentally sustainable land-use decisions on their own properties are leading to new forms of government organization, new roles for government extension staff and for non-government groups, and to new demands for relevant, understandable information.

Such programmes are becoming widely known as 'stewardship programmes', and are emerging out of both the conservation community and the farming community. In many ways they reflect a 'privatizing' of the entire conservation movement. In this chapter the traditional approach to achieving sustainable conservation practices through the land-use planning approach is first reviewed. Then several major new 'stewardship' initiatives that focus on the private landowner are described. The conclusion provides a brief evaluation of this evolving approach to achieving sustainability in rural land management.

Background

Traditionally, conservation-oriented land-use decisions have been approached through public land-use planning. Land-use planning in Canada is primarily directed to the rural–urban fringe, where land-use change is relatively rapid. Planning regulations attempt to place some order on this urban growth, for financial as well as social and environmental reasons. Major land-use pressures exist around all major urban centres in the country.

In recent years important new attempts have been made to increase the environmental component of such land-use planning. Citizens' groups have for years protested the destruction of various natural areas through urban development, and worked to persuade provincial governments to develop policies to protect areas such as wetlands, stream valleys and other natural areas. Municipalities have used a variety of approaches to protect natural areas, zoning them as 'Environmentally Sensitive Areas' or taking them as public parkland during subdivision development (Hilts *et al.*, 1985; Richardson, 1989).

Recently a major Royal Commission in Ontario, the 'Crombie Commission' on the Toronto Waterfront has had a major influence on public opinion in this province. It emphasized the concept of 'ecosystem planning' – which in Crombie's words essentially means recognizing that 'everything is connected to everything else'. In this case, the water quality that determines land-use along the waterfront is entirely a function of activities upstream in the watershed, or 'bioregion' (Royal Commission on the Future of the Toronto Waterfront, 1992). This is but one example of greater public concern over environmental impacts of land-use decisions.

A third major influence is concern over aquatic habitats. Perhaps the fisheries issues in British Columbia illustrate this concern best. With the salmon population depending on the ability to swim upstream to spawn, water quality in rivers and coastal estuaries is critical. Both the collapse of the east and west coast fisheries, and the new federal fisheries Act have contributed to increased public concern over aquatic resources. In B.C. this has led to the Fraser River Estuary Management Programme among other innovative attempts to build water resource protection planning into the land use planning process (Tomalty *et al.*, 1994).

In Ontario, very detailed sub-watershed studies are being undertaken before development. These are remarkable for the extent of inter-disciplinary environmental components included. On watersheds only a few miles in length, they have included extensive hydrogeological model-ling, ecological inventories of natural habitats and examination of aquatic features. The basic principles of landscape ecology have been applied, with suggestions to replant connecting corridors and buffers to protect existing natural landscape features (Ministry of Environment and Energy, 1993). In at least one municipality, entire new forests are proposed, to re-establish a minimum level of forest habitat, particularly larger forested areas.

In 1995, these influences culminated in a newly revised Planning Act in Ontario, extending the protection of environmental features to the entire landscape, and including conservation of virtually all environmental features – river and stream valleys, significant woodlands, shorelines, significant wildlife habitat, rare and endangered species habitat, and wetlands (Ministry of Municipal Affairs, 1994).

Unfortunately, a change in government in mid-1995 has brought a weak-ening of these measures – illustrating the politically controversial nature of such issues. At the time of writing, it remains to be seen how many environ-mental policies will remain as requirements. In other provinces this evolution of planning has led to the incorporation of environmental concerns in a vari-ety of specific projects and some legislation (Tomalty *et al.*, 1994).

In the urban–rural fringe, as development occurs, it is possible to see these many environmental concerns being a very positive balance to urban growth. Even here though, municipalities are now raising questions about who is going to own and manage all the land that is conserved! Who is going to do the rehabilitation of ecosystems to re-create woodland and landscape corridors? With ever-greater budget cuts looming, municipal parks departments do not want to be responsible for acres of woodland, wetland and stream valley.

In rural areas, the fit between these environmental objectives and landowner decisions is even less comfortable. Zoning in rural areas may tend to prevent any form of urban development, including rural residential estates, but is still fairly general in terms of the rural land-uses allowed. The most common land-use planning designations are simply 'rural' and 'agricultural'. Such zoning leaves plenty of room for landowner decisions to keep, destroy or rehabilitate natural habitats, and more specific zoning soon runs into political opposition, often strong enough to defeat the policies proposed.

As a result, a new level of concern is being recognized through programmes to work directly with private landowners. In some ways this is a revival of old-fashioned rural extension work, but it is now seen in terms of environmentally sustainable land-use at the landowner scale.

Government agencies that work with rural land and resource issues are changing the emphasis they place on supporting the land management decisions of rural landowners. Urban municipal governments are beginning to look at such non-government initiatives as private stewardship programmes and non-government land trusts as vehicles for land-use management.

Landowner-based Land-use Decisions – Stewardship Initiatives

The past 10 years have seen the evolution of a number of programmes designed to encourage direct conservation of natural habitats by rural landowners. Whereas there have been many earlier attempts to involve private landowners in conservation, dating back to the dust bowl of the 1930s, today's programmes are a more widespread attempt to influence landowners' management decisions as a complementary approach to influencing land-use through planning regulations (Hilts, 1990).

The term 'stewardship', or more specifically private stewardship, has come to be used to describe the programmes targeting private landowners. Many of these have arisen out of the conservation community, based on a wish to protect natural habitats. These have included, among others:

1. landowner contact by the Island Nature Trust on Prince Edward Island;
2. corporate conservation agreements in Nova Scotia;
3. local programmes run by a number of Quebec community groups;
4. landowner agreements negotiated by the Muskoka Heritage Foundation in Ontario;
5. the Natural Heritage Stewardship Programme, also in Ontario;
6. landowner agreements to conserve wildlife in all three prairie provinces;

7. stewardship agreements with farmers in the Fraser Delta and with landowners in the Cowichan estuary areas of B.C.

Some of these programmes, particularly those on the prairies, aim to influence farm landowners. Others have emerged directly in the farming community. Though efforts to encourage individual farmers to conserve soil and water resources have been around for years, the breadth and extent of environmental concerns included in farm oriented programmes today is much more extensive.

Most notable of all farm-oriented programmes in Canada today is the Environmental Farm Plan programme in Ontario. It is the most comprehensive environmental farm planning programme in the world, and the only major programme of its type run by farmers themselves. Success of this programme will be a major contribution to sustainability in rural land management in Canada. The Ontario programme is already being copied elsewhere in Canada, as well as in other countries.

Other programmes are targeted at conserving natural areas and involve numerous rural non-farm landowners as well as farmers. This is particularly true in the more intensively settled landscape of the country such as the valleys and islands of southern British Columbia and rural southern Ontario. The following sections of this chapter discuss a number of these 'stewardship' initiatives, and their potential contribution to sustainable rural land management.

The Natural Heritage Stewardship Programme

One of the earliest among recent stewardship programmes, the Natural Heritage Stewardship Programme, was designed by the author, and implemented through pilot studies beginning in 1983. This programme was modelled after landowner contact efforts carried out by the US Nature Conservancy, aiming to encourage conservation of specific natural areas by private owners. Sponsored by a cooperating group of government and non-government agencies known as the Natural Heritage League, this programme has involved contact with over 6000 rural landowners in southern Ontario. Owners are asked to make a voluntary, or 'handshake', stewardship agreement to protect the natural area on their land. In total approximately 2000 landowners have made these verbal agreements to protect approximately 24,000 hectares of woodland and wetlands (van Hemessen *et al.*, 1994). This work was carried out in three different programmes, targeting landowners in the Carolinian region of southern Ontario, and along the Niagara Escarpment, as well as landowners of many wetlands.

The Natural Heritage Stewardship Programme is largely inactive today, although some ongoing follow-up work is being done with landowner groups. As well, specific government agencies and some non-government groups such as local Land Trusts have begun to develop similar programmes (Hilts and Mitchell, 1993). One aspect of this work is the provision of new information for landowners, with an emphasis on integration (or the 'ecosystems' approach), ease of understanding, and practical value. An example is a handbook recently prepared by the author and Peter Mitchell, entitled *Caring for Your Land: A Stewardship Handbook for Niagara Escarpment Landowners* (Hilts and Mitchell, 1994).

Other Canadian Stewardship Programmes

Numerous other stewardship programmes have been developed and tested in the past decade, several using the Natural Heritage Stewardship Programme as a prototype. Others evolved at the same time; these used fairly similar techniques of direct landowner contact, but focused on financial incentives rather than handshake agreements. Today some provinces are increasingly considering stronger legal techniques such as conservation easements.

In Nova Scotia, a private forestry company has been a major player in stewardship programmes, developing a range of specific conservation practices to protect salmon populations in local streams, wetlands, and other significant wildlife habitat. Scott Maritimes Ltd owns and manages over 1,000,000 acres of forest land in Nova Scotia (MacLellan, 1994).

Landowner contact programmes in the prairies have been focused on farm practices, and particularly aimed at restoring duck habitats. The prairie pothole landscape has been described as the nursery for North America's duck population. In most cases farmers are asked to set aside not only the potholes or wetlands but also a nesting area of natural vegetation around these natural habitats. The primary crop grown in this region is wheat. Sponsored in part by the North American Waterfowl Management Plan, these programmes have paid a direct financial incentive to farmers to conserve wildlife habitat. A shortcoming is the need to maintain payments, but it has been found that relatively low payment levels are sufficient incentive (S. Lord and J.D. Henry, Calgary, Alberta, 1994, personal communication).

A separate prairie landowner contact programme aims specifically to protect Burrowing Owl habitat by negotiating handshake agreements with landowners whose farms support nesting pairs. This programme, sponsored by a number of partner organizations including the World Wildlife Fund, gained considerable publicity through the visit of Prince Phillip to a participating farm in 1988. Over 500 landowners have protected over 16,000ha of land in Saskatchewan alone through agreements under this programme (Schroeder, 1994).

Numerous other specific projects could be described. These include the work of the Muskoka Heritage Foundation in contacting landowners in the Muskoka Lakes region of Ontario, the work of the Delta Farmland and Wildlife Trust with farmers on the southern margin of the city of Vancouver, and many others. Although much of this effort is so new that it remains undocumented in published literature, several of these are described briefly in the Proceedings of the 'Stewardship '94' conference held in Vancouver (Layard and Delbrouck, 1994). As one author states, in a theme re-echoed in many projects, the response of landowners to such targeted landowner contact programmes has been 'gratifyingly positive' (Doane, 1994, p. 75).

Many of these programmes are now moving beyond individual landowners to deal with community groups in various ways. One of the most innovative of these has been the decision by Ontario's Ministry of Natural Resources to establish Community Stewardship Councils. From a past emphasis on timber management in woodlots, this Ministry has moved to recognize a more holistic view of forestry that approaches integrated rural land stewardship. The emphasis is on small private forest owners, and on sustainable management. Now they are moving to work more directly with local communities through their proposed Stewardship Councils. The Ministry will provide a staff coordinator, and hopes to help integrate the work of various government agencies with local municipalities and the needs of landowners.

The Stewardship Council programme is just being initiated, but three pilot 'one-window' stewardship offices have been running for two to three years. In these offices, representatives of three key government agencies are brought together in one place to respond to landowner inquiries. The purpose is to provide more integrated information, but directing landowners quickly to the correct agency for assistance. Needless to say, there is still considerable conflict among these agencies, as government departments fight for survival in a time of severe budget cuts. Nevertheless, the three pilot studies have revealed the strong influence of cooperative work leading to improved integration of services.

The involvement of local community representatives as advisors is a further step toward allowing landowners to define their own needs from government programmes. Thus the Stewardship Council programme emphasizes both integration of programmes and community control – two factors that are equally important in the Environmental Farm Planning Programme.

The Environmental Farm Planning Programme

Perhaps most notable among landowner initiatives in Canada has been the Environmental Farm Planning Programme developed by a coalition of farm organizations in Ontario. Partly motivated out of genuine environmental ethics among farm leaders, and partly motivated out of fear of being regulated into more sustainable farming practices, 'The Farm Coalition' has developed the most comprehensive package for Environmental Farm Plans in the world to date, and is the only such programme primarily controlled by farm organizations themselves.

Both the provincial Ministry of Agriculture, Food and Rural Affairs, and the federal agency Agriculture and Agri-food Canada have been very supportive of the Environmental Farm Plan initiative, but have refrained from interfering in the farm leadership of the programme. This programme calls for farmers to evaluate their operation using 23 worksheets, each of which has numerous specific related questions. Topics range from chemical handling, manure storage and the septic system around the farmstead through cropping practices and tillage, to conservation of woodlands, wetlands, and wildlife. The comprehensive nature of the farm planning programme is revealed in the list of worksheet topics (Table 17.1).

Administered at the county level by the Ontario Soil and Crop Association, this programme is currently running workshops for farmer participants, and has involved 5000 farm families to date. After completing their farm plans using the detailed workbook provided, farmers are beginning to take action on their farms to minimize environmental risks (George *et al.*, 1995). Little formal evaluation has yet been completed, but a survey of participants in one county revealed that 89% of respondents

Table 17.1. Environmental Farm Plan Worksheets.

Soil and site evaluation	Water wells
Soil management	Pesticide storage
Nutrient management	Fertilizer storage
Manure management	Petroleum storage
Field crop management	Disposal of farm wastes
Pest control	Household waste water
Streams and floodplains	Agricultural wastes
Wetlands and ponds	Livestock yards
Woodlands and wildlife	Silage storage
Energy efficiency	Milkhouse washwater
Water efficiency	Noise and odour

who had identified needed actions on their farms had begun to implement these, and 17% of proposed actions had been completed (Bidgood, 1994).

Such practical on-the-ground results are important, but organizers also describe a fundamental change in attitude among participants, and a high level of satisfaction with the process. Participants clearly feel they learn a great deal through the environmental farm planning exercise, and this directly influences their farming practices (H. Rudy, Guelph, Ontario, personal communication).

The farm planning exercise also leads quickly to information needs by participants. With such a comprehensive programme, some obvious gaps in both information and policy have been identified. Government agencies are working to support the programme with both more appropriate policies and new, more comprehensive and integrated information packages. Among these, the series of eight Best Management Practice Manuals being produced by co-operating federal and provincial agricultural ministries is perhaps the most notable. Attractive, easily readable and practical, they cover the range of topics from cropping and livestock waste management through soil, water and nutrient management, to farm forestry and wildlife habitat (George *et al.*, 1995).

Conclusion

The pattern that has been described here is not unique to Ontario. In both the United States and the United Kingdom, many programmes exist to influence and support the land-use decisions of rural landowners, particularly farmers (Wright and Hilts, 1993; Sheail, 1995). What appears significant in this case is the influence this change is having on government agencies, and on provision of information for rural landowners.

Both the demand for practical information and the increasing level of community control is pushing agencies to provide more complete information of a practical nature for rural landowners. It is also pushing agencies to integrate their work and programmes more effectively, although this effort is still faltering. The nature of the environmental issues themselves demands an integrated approach, and government agencies dealing with rural landowners will likely evolve further toward this goal in the near future.

At the same time, allowing greater community or direct landowner control in programmes is a response of government agencies to landowner demands. This in turn will likely lead to new initiatives based more directly on local landowner needs. Changes may include rationalization of programmes and services, as well as revisions to appropriate policies, in the light of the landowner's perspective. Together these trends introduce a major focus on a new scale of land-use decisions, at the property rather than the municipal or regional scale. This focus provides new opportunities to approach the achievement of environmental sustainability in a very practical sense.

References

Bidgood, M. (1994) *A Study of Actions by Simcoe County Environmental Farm Plan Participants.* Ontario Soil and Crop Improvement Association, Guelph, Ontario (mimeo).

Doane, J. (1994) Cowichan/Chemainus Stewardship Pilot Project. In: Layard, N. and Delbrouck, L. (eds) *Proceedings, Stewardship '94.* Ministry of Environment, Parks and Lands, Victoria, British Columbia, pp. 56–59.

George, R., Markus, J., Allison, B. and Magee, J. (1995) *Our Farm Environmental Agenda: What's Been Achieved – 1992 to 1995.* The Farm Environmental Coalition, Toronto, Ontario, 217 pp.

Hilts, S.G. (1990) Private stewardship: its beginnings and use across Canada. In: Nelson, J.G. and Woodley, S. (eds) *Heritage Conservation and Sustainable Development.* University of Waterloo, Waterloo, Ontario, pp. 191–195.

Hilts, S.G. (1994) The Natural Heritage Stewardship Program. In: Beavis, M.A. (ed.) *Environmental Stewardship: History, Theory and Practice.* Institute of Urban Studies, University of Winnipeg, Winnipeg, Manitoba, pp. 108–117.

Hilts, S.G. and Mitchell, P. (1993) Bucking the free market economy: using land trusts for conservation and community building. *Alternatives* 19, 16–23.

Hilts, S.G. and Mitchell, P. (1994) *Caring for Your Land: A Stewardship Handbook for Niagara Escarpment Landowners.* Centre for Land and Water Stewardship, University of Guelph, Guelph, Ontario, 64 pp.

Hilts, S.G., Reid, R. and Kirk, M. (1985) *Islands of Green: Natural Heritage Protection in Ontario.* Ontario Heritage Foundation, Toronto, Ontario, 199 pp.

Layard, N. and Delbrouck, L. (1994) *Proceedings, Stewardship '94.* Ministry of Environment, Parks and Lands, Victoria, British Columbia, 151 pp.

MacLellan, J. (1994) Corporate forestry on private lands in Nova Scotia. In: Layard, N. and Delbrouck, L. (eds) *Proceedings, Stewardship '94.* Ministry of Environment, Parks and Lands, Victoria, British Columbia, pp. 33–36.

Ministry of Municipal Affairs. (1994) *Comprehensive Set of Policy Statements.* Queen's Printer for Ontario, Toronto, Ontario.

Ministry of Environment and Energy. (1993) *Water Management on a Watershed Basis: Implementing an Ecosystem Approach.* Queen's Printer for Ontario, Toronto, Ontario.

Richardson, N. (1989) *Land Use Planning and Sustainable Development in Canada.* Canadian Environmental Advisory Council, Ottawa, Ontario, 104 pp.

Royal Commission on the Future of the Toronto Waterfront (1992) *Regeneration: Toronto's Waterfront and the Sustainable City: Final Report.* Supply and Services Canada, Ottawa, Ontario, 529 pp.

Sheail, J. (1995) Nature protection, ecologists and the farming context: a U.K. historical context. *Journal of Rural Studies* 11, 79–88.

Schroeder, C. (1994) Operation burrowing owl: a communication plan. In: Layard,

N. and Delbrouck, L. (eds) *Proceedings, Stewardship '94.* Ministry of Environ-ment, Parks and Lands, Victoria, British Columbia, pp. 37–42.

Tomalty, R., Gibson, R.B. Alexander, D.H.M. and Fisher, J. (1994) *Ecosystem Planning for Canadian Urban Regions.* Intergovernmental Committee on Urban and Regional Research, Toronto, Ontario, 84 pp.

van Hemessen, D., O'Grady, L. and Martin, R. (1994) Report on landowner contact information for the Carolinian Canada, Niagara Escarpment, and Wetland Habitat Agreement Programs. Ontario Heritage Foundation, Toronto, Ontario (mimeo).

Wright, J.B. and Hilts, S.G. (1993) An overview of voluntary approaches to landscape conservation in the United States and Canada. *Operational Geographer* 11, 10–14.

18 Scale Change, Discontinuity and Polarization in Canadian Farm-based Rural Systems

MICHAEL TROUGHTON
*Department of Geography, University of Western Ontario,
London, Ontario, Canada N6A 5C2*

Introduction: The Farm-based Rural System

This chapter explores the ongoing relationship between agriculture and rural systems in Canada. The central premise is that agriculture was and is central to the definition and sustainability of rural systems and, thus, it remains important to explicitly link the condition of agriculture, especially farming, to that of the rural system. From an agricultural geographic point of view this involves a somewhat traditional emphasis of farm-based description. However, while one can agree with the need for agricultural (and rural) geography to engage in discussion within the wider context of political economy, including explicit recognition of and investigation within the framework of agrifood systems (Bowler and Ilbery, 1987), an understanding of the current status and problems faced by rural systems also requires the more traditional focus on the condition of farming within the system.

Two models that emphasize the ideal of a farm-based rural system are outlined in Box 18.1a and b. For definitional support, what might be termed 'the ideal model of rurality', drawn from Cloke and Park's (1984) definition of the countryside is cited (Box 18.1a). This simple model defines the rural system in terms of the combination of three farming-related characteristics; the extensive farmed landscape, its farms and the farm population, and the actual and perceived linkages between farming and local agriculturally based rural communities. In the past, rural systems developed and were maintained through the integration of farm area, operation and community relationships; the status of the relationship remains pertinent to current farm-based rural systems.

The second simple model (Box 18.1b) was developed by Lowrance (1990) and modified by Troughton (1993). In it rural system sustainability rests on the integration of five sustainability objectives, of which the first three define farm and rural community-based rural-system viability. Together, these models support the need for and utility of continued evaluation of the condition of farming as an integral part of the rural system.

Today, rural systems are in decline in all developed countries. Decline, in both relative and absolute terms, is frequently related to the breakdown of essential farming relationships. Nowhere, it is argued, is this breakdown

Box 18.1. Two models that emphasize the farm-based nature of rural systems.

---

a. Definition of the rural countryside
The Rural Countryside is defined as an area which:

(i) is dominated by extensive land uses, notably agriculture and forestry;
(ii) contains small, lower order settlements which demonstrate a strong relationship between buildings and extensive landscape, and which are thought of as rural by most of their residents;
(iii) engenders a way of life which is characterized by a cohesive identity based on respect for the environmental and behavioural qualities of living as part of an extensive landscape.

(*Source*: Cloke and Park, 1984, p.13.)

b. Elements of agriculture and rural system sustainability
Given a goal of agricultural and rural system sustainability the prevailing concern should be on the identification of and adherence to a set of *Sustainability Objectives*, namely:

1. *Agronomic Sustainability* – the ability of the land to maintain productivity of food and fibre output for the foreseeable future,
2. *Micro-Economic Sustainability* – the ability of farms to remain economically viable and as the basic economic and social production unit,
3. *Social Sustainability* – the ability of rural communities to retain their demographic and socioeconomic functions on a relatively independent basis,
4. *Macro-Economic Sustainability* – the ability of national production systems to supply domestic markets and to compete in foreign markets,
5. *Ecological Sustainability* – the ability of life support systems to maintain the quality of the environment while contributing to other sustainability objectives.

(*Source*: Troughton, 1993; modified after Lowrance, 1990.)

more pronounced than in Canada, where the distinctions between what is still a very extensive farm landscape and the much reduced number of farms and associated rural communities is acute. A description of the development of this discontinuity, over the last 50 years, is a primary focus of the chapter.

The sequence of processes that have operated to alter Canadian agriculture, especially since 1945; notably its industrialization and the shift in control and decision-making beyond the farm gate have been outlined elsewhere (Troughton, 1986, 1992). However, these processes have also altered the integrated nature of the previous farm-based rural system. The second part of the chapter presents some of the more critical characteristics of the current farm-based system that demonstrate not only its reduction but also polarization and consequent lack of integration. One conclusion is that the marked shifts away from the former, integrated or 'ideal model' rural system represent the reality of the current breakdown and a key problem *vis-à-vis* system viability.

The Development and Characteristics of the Two Major Canadian Farm-based Rural Systems

Looking first, briefly, at the development of the Canadian farm-based rural system, this, essentially European transplant, was the primary agency in the creation of the Canadian *ecumene*. Its beginnings were limited, such that in 1800, after nearly 200 years, rural farm settlement was still confined to the central lowlands of the St Lawrence and to pockets of land in Nova Scotia (Wynn, 1987). However, the century between 1780 and 1880 saw the almost complete development of predominantly rural settlement in the Maritime Provinces, southern Quebec and southern Ontario (Harris and Warkentin, 1976). Although consisting of several geographically and administratively distinct sections, and distinguished by regional ethnic and cultural differences, 'the eastern Canadian rural system' was remarkably homogeneous as to general characteristics, both at the peak of rural dominance (i.e. *c.* 1880) and for the next 50–60 years.

If one includes the attempts to enlarge the system through northern colonization, particularly after 1900, the eastern Canadian rural system involved the occupance of over 20,000,000 ha of farmland, incorporating over 500,000 farms, and supporting a rural population of over 4,000,000, including farm and non-farm rural components. Although eastern Canada as a whole shifted from a majority rural to urban population early in this century, large areas remained predominantly rural in economic and cultural terms. The rural farm economy was based on the combination of farm employment plus local manufacture of both the inputs to farming (e.g. most early machinery) and the processing of farm produce. The latter, for example, included several thousand cheese and other dairy plants, mostly in Ontario and Quebec, along with many grist and flour mills, almost all located in rural villages and hamlets. The farm and non-farm components of rural society were closely aligned within an active and self-contained cultural milieu.

The system was not static, nor without its problems. The latter included a physically fragmented and often marginal land-base, rigid survey systems and a general pattern of dispersed farmsteads versus nucleated non-farm villages. This structure, coupled with an almost total emphasis on economic criteria as the basis for success, often led to considerable local failure, both at the time of settlement and as farming became economically more competitive. Thus, between 1880 and 1940 eastern Canada experienced a net loss of over 100,000 farms and over 1,200,000 ha of farmland. Nevertheless, these losses, mainly in peripheral areas, were atypical of the system as a whole.

Up to the 1940s, and notwithstanding the 1930s Depression, the system remained rather homogeneous, stable and integrated. Throughout eastern Canada the predominant farming typology, as described by Baker (1926/1928), was of medium scale, mixed livestock enterprises with a dairy emphasis, based on hay and small grains. Whereas the pattern was modified locally by more intensive horticulture, extensive cash cropping or beef cattle raising, the operational characteristics and resultants were very uniform. Virtually all farms were full-time operations utilizing a mix of family and hired labour. Yields and farm receipts were modest, as were farm capital values and expenses. As noted, the primary linkages were with locally based (and owned) input supply and processing firms, even though by the 1920s, agribusiness consolidation was beginning to be apparent. Thus, in eastern Canada, one had the first example of an 'ideal' farm-based rural system which, after rapid development in the nineteenth century, underwent a period of relative stability for the next 50 years.

In contrast, the second major farm-based Canadian rural system, that of western Canada, located predominantly in the southern Prairies, grew up more rapidly and on a larger scale. Despite major differences, however, at its mid-1930s peak the western system shared many of the characteristics of the eastern system. Between 1880 and 1930 the system expanded from small bridgeheads of less than 300,000 ha and a few thousand farms, into a huge farmland base of more than 50,000,000 ha, based on over 300,000 farms and an associated 'community system' of over 2000 farm villages and hamlets, virtually all established as delivery points for 5700 elevators located on the 30,000 km network of Prairie rail lines (Zimmerman and Moneo, 1971).

The Prairie system was based upon a more rigid pattern than eastern Canada, with a totally dispersed pattern of farms. Furthermore, despite the regionally uniform and generally fertile land base, conditions were not always suitable for successful grain farming. However, despite a hetero-geneous mix of settlers with very different levels of experience, and a considerable failure rate, especially during periods of drought (e.g. 1916–1918) and depressed economic conditions (e.g. 1920–1922), the overwhelmingly rural farm system developed to an initial peak in the early-1930s. At that time, despite growing mechanization and economies of scale, the system was dominated by medium sized farms which, abeit emphasizing wheat or other small grains, had a subsidiary mix of livestock (Baker, 1926/1928). Farm incomes and capitalization were generally modest. Despite the youthfulness of the system (the majority of farms were still first generation), the common thrust to a commercial farm economy and the linkages of the rail system produced a strongly integrated rural economy and society; the farm majority was supported by the range of services, almost all located in the dense network of local rural service communities. In common with rural folk in eastern Canada, those in the Prairies were cohesive, self-reliant and politically active, especially with respect to creation and support of local populist political parties (e.g. Social Credit and the Commonwealth Cooperative Federation) and critical cooperative institutions, notably the Prairie Wheat Pools.

Despite farm and farmland losses in eastern Canada before 1930, the raid growth of the Prairie system, which continued up to the onset of the Depression, meant that the peak of the rural-farm system in Canada was recorded in the 1941 Agricultural Census. Table 18.1 presents a summary of the situation of the two major farm-based rural systems which serves to emphasize their common and widespread characteristics as late as 1940. However, this peak situation represents a watershed, between the culmina-tion of a period of over 150 years of continuous and rapid expansion and largely rural dominance, and the next 50 years, which saw both absolute and relative decline and an increasing discontinuity with respect to the elements that had combined to create the two somewhat similar farm-based rural systems.

In 1940 (Table 18.1) the farmland base was 70,240,097ha, of which about 30% was in eastern Canada and 70% in the west, almost all in the Prairie provinces. The 732,832 farms were divided 56:44% between the two halves, which meant a somewhat higher average farm size in western Canada (49:156ha). This, however, was generally offset by the lower crop yields in the latter and reflected in lower land values. Another common factor was the relatively modest level of improved land per farm, both in the Prairies, where it reflected the recency of much settlement and in the far eastern areas where it reflected a poor soil base and, in some cases the competing labour demands of forestry or fishing. Levels of farm capital-ization were similar and within a narrow range. Higher land and machinery values per farm in the west were offset by somewhat higher investment in buildings in the east, while investment in livestock per farm was exactly

Table 18.1. Canadian farm-based rural systems 1941.

	Canada (total)	Eastern Canada	Western Canada
Rural population	5,239,094	3,366,327	1,872,767
(% total pop.)	(45.6)	(40.8)	(57.8)
Farm population	3,152,449	1,901,763	1,250,686
(% total pop.)	(27.4)	(23.1)	(51.6)
(% rural pop.)	(60.0)	(56.5)	(66.8)
Number of farms	732,832	409,968	322,863
(x pop./farm)	(4.3)	(5.6)	(3.9)
Farmland area (ha)	70,240,097	19,991,970	50,248,122
X farm size (ha)	95.8	48.8	155.6
Improved land (ha)	37,084,607	10,202,859	26,841,278
(% farmland)	(52.8)	(51.0)	(54.6)
X Improved area/farm (ha)	50.6	24.9	83.1
% Farms rep. cattle	81.8	84.5	78.3
% Farms rep. dairy	79.5	82.7	75.5
% Farms rep. pigs	65.2	66.9	62.9
Capital values/farm			
Total ($)	5,787	5,233	6,492
Land ($)	2,665	2,053	3,442
Buildings ($)	1,469	1,698	1,179
Mach./Equip. ($)	813	642	1,031
Livestock ($)	840	840	840
Expenses/farm ($)	631	507	760
Gross income/farm ($)	1,491	1,432	1,565
Gross income/ha ($)	15.60	29.30	10.10
% Income = non-farm	9.4	12.1	6.6
% Farms = part-time	5.5	7.0	3.5

Source: Dominion Bureau of Statistics (1946).

similar. These similarities were reflected in the levels of gross farm receipts both on a per farm and per hectare basis. Overall, however, well below 10% of farm operators were classed as part-time and non-farm sources were a minor contribution to income.

Although the documentation is less clear cut with respect to associated rural communities, they were part of a large rural population in which the farm component was still numerically dominant. The majority of smaller places were rural-farm service centres whose employment included a combination of those engaged either in rural processing or in the range of economic, social and administrative services that supported quite dense rural farm and non-farm populations. Perhaps the greatest contrast between the two systems, apart from the distinctive physical and human landscapes, was that between the longer-settled and more established farms and rural settlements of the east, and the much more recent and, therefore, less well-established western farms and villages.

Changes to the Farm-based Canadian Rural Systems, 1940-1990

In overall terms, the 50 years since 1941 produced changes of as great a magnitude as those of the period of most rapid growth, but have included precipitous declines, especially in numbers of farms and farm population and with respect to rural community linkages. In the process both the eastern and western systems have lost their homogeneity and cohesiveness and, today, exhibit a series of discontinuities between the defining characteristics. The contrasting situations are set out in Table 18.2.

With respect to the farmland base and the farm landscape, although nationally the net decline between 1940 and 1990 has only been 3.5% (2,490,000ha), it hides some major changes and significant differences. The major contrast is between the eastern system, which has experienced huge losses and considerable contraction, and the western system whose areal extent has continued to grow.

The seemingly stable condition in eastern Canada up to 1940 masked a situation whereby large numbers of farms, and especially those on the eastern and northern peripheries, were marginal in terms of a poor land base which could not support a shift to more highly capitalized commercial agriculture, and became part of a large scale post-war situation of rural farm (and non-farm) poverty (Buckley and Tihanyi, 1967). Consequently, when freed from the constraints of the Depression, these areas responded to the huge postwar demand for urban-industrial labour, creating a rapid out migration of farm and non-farm populations, coupled with widespread farm and farmland abandonment. Those trapped in rural poverty were assisted to leave by government programmes which professed rural rehabilitation but contained no maintenance strategies.

Already by 1970 the eastern Canadian rural system had lost 40% of its 1940 farmland extent and some 150,000 farms. In this first stage of retreat, the region most affected was the Maritimes, especially Nova Scotia and New Brunswick, each of which lost over 60% of the farmland and over 70% of farms. Other major losses occurred in the Eastern Townships in southern Quebec, the Shield in southern Ontario, and the most recently settled areas in the north of both Quebec and Ontario. Although the rate of loss of land slackened somewhat after 1970, overall, between 1941 and 1991, the system has been reduced by 50% in farmland area, including nearly 40% of the improved farmland base. The largest proportional losses have been in the Maritimes (over 80% in both Nova Scotia and New Brunswick), but the largest absolute losses have been in the two central provinces; in Quebec

Table 18.2. Canadian farm-based rural systems 1991.

	Canada (total)	Eastern Canada	Western Canada
Rural population	6,389,985	4,515,155	1,826,885
(% Total pop.)	(23.4)	(23.4)	(23.1)
(% Change 41/91)	(22.0)	(34.0)	(−3.0)
Farm population	807,325	360,680	446,655
(% Total pop.)	(3.0)	(1.9)	(5.6)
(% Change 41/91)	(−74.4)	(−81.0)	(−64.3)
(% Rural pop.)	(12.6)	(8.0)	(24.4)
Number of farms	280,043	117,027	163,016
(x Pop./farm)	(2.9)	(3.1)	(2.7)
(% Change 41/91)	(−61.8)	(−71.5)	(−49.5)
Farmland area (ha)	67,753,700	9,959,879	57,314,407
(% Change 41/91)	(−3.5)	(−50.2)	(14.0)
X farm size (ha)	242	85	352
(% Change 41/91)	(152.5)	(74.3)	(118.8)
Improved land (ha)	45,569,951	6,260,014	39,307,933
(% farmland)	(67.3)	(62.9)	(71.0)
Improved/farm (ha)	163	53.5	241
% Farms rep. cattle	52.0	54.7	50.1
% Farms rep. dairy	14.0	23.9	6.8
% Farms rep. pigs	10.6	12.1	9.5
Capital values/farm Total ($)	468,535	476,962	462,487
Land ($) bldgs ($)	345,455	376,067	323,479
Mach./Equip. ($)	83,172	64,372	96,668
Live. & poultry ($)	39,908	36,522	42,340
Gross farm Receipts/farm ($)	88,665	98,731	81,438
Gross receipts/ha ($)	366	1,161	240

Source: Statistics Canada (1993).

3,600,000 ha of farmland (51%), and nearly 100,000 farms (72%), with losses in excess of over 80% land and farms in the eastern and northern peripheries; similarly, in Ontario, an overall loss of 3,800,000 ha (41%) and 120,000 farms (64%), also leaving the northern and Shield areas at levels close to the earliest periods of settlement.

Although some losses have been associated with urbanization, including some of the highest capability lands in southern Ontario, the major losses have been in the predominantly rural areas. In each case losses of land have been outstripped by even higher percentage losses of farms and farm population. Eastern Canada has experienced a loss of 71.5% of its farms and 81% of its farm population since 1940. Again there have been regional variations, but the current totals represent less than a quarter and one twentieth of the peak farm and farm populations, respectively. Although these have gone hand in hand with increases in farm size (+75%) and improved land area per farm (+115%) and the system still contributes over 35% value of gross farm receipts, its role within the total rural system has shrunk markedly (Table 18.2).

The effects on rural communities have reflected both the declining numbers of farms and the changing characteristics of the agricultural system, especially the growth of agribusiness and the concentration and consolidation of farm input supply and output processing in fewer and increasingly urban locations. In many peripheral areas there no longer exist sufficient farm numbers to support local services. While in remaining core areas, most of the local functions and agricultural employment have drastically declined. Virtually no local cheese or other dairy plants now exist and most rural meat, fruit and vegetable packing and preserving operations have closed. Although feed mills and farm machinery dealers still exist, they tend to relocate in the open country or in the larger towns. While many rural settlements have survived, their current functions are less and less farm related and the level of decoupling has been marked in virtually every case. Although, as Bryant and Johnston (1992) have shown, agriculture is still active and diversified in the urban fringe zones, its role and that of the associated communities is part of an 'invaded countryside' of changing economic and social values that reflect the distintegration of a farm-based rurality (Walker, 1987).

In western Canada a somewhat different set of changes has occurred, but the shifts have been just as marked, especially as regards farms, farm populations and rural communities (Table 18.2). The greatest distinction is in the matter of the land base. Whereas eastern Canada's has declined substantially, that in the west has expanded. Since 1941 the farmland area in western Canada has grown by 7,066,000 ha (14%). Although some of this is in frontier areas (Peace River District) it has included growth in southern, already settled areas. More significant, perhaps, has been the huge increase in improved land that has risen from 54% to 71% of total farmland. This growth has meant that the western farmland component has risen from 71% to 82% of the Canadian total. In contrast, while farmland areas have increased, the number of farms and the farm population has declined at rates similar to those in eastern Canada. Since 1941 western Canada has seen a decline of 187,000 farms (50% of the peak total). Whereas the shifting balance of area and numbers led to modest increases in average farm size in the east, in the west and especially in the Prairies, the average size of farm has increased very rapidly (from 156 to 352 ha) and is currently averaging over 400 ha in Saskatchewan. The key to farming on this scale is mechanization which means less labour. Consequently, the western farm population has declined by over 800,000 or 64%, to less than half the rural total.

Although the western rural population has not declined in overall terms, it has experienced a marked shift away from a majority of the smallest rural service centres, especially in the Prairies. Loss of farm population has left a small and very dispersed farm clientele. Even more impacting have been changes to the transportation system, including the shift from rail to roads, and the demise of two-thirds of the country grain elevators. Insofar as the elevator was the key economic *raison d'être* for most small centres, their loss has led to decline and disappearance. The consolidation of a whole range of other commercial and social services into the small set of larger places has hastened the loss of the majority of the former rural communities. A recent study (Stabler, 1993) suggests only 62 out of a previous total of over 700 rural communities in Saskatchewan remain commercially viable. Thus, in the Prairies, a community system that grew rapidly on the basis of grain farming has largely collapsed due to changes at the farm level and is much more dispersed than in eastern Canada.

Whereas each of the changes is significant in its own right, the key lies in the new combinations that have developed. In overall (Canadian) terms one now has a farmland base which is even more extensively operated, especially in the west; a much reduced set of farms which rely on a small set of operators, their families and an equal number of hired operatives. Together, the farms and their operators are increasingly separate from or form an insignificant proportion of the rural population, the majority of whom live in settlements whose functions and existence owe less and less to the surrounding farm area.

Further Change and Polarization within the Farm-based Rural System

Despite the major changes indicated above, it is still possible to characterize the Canadian farming system as extensive both as to its land and operational base. The 1991 Census recorded 280,000 farms occupying 67,753,700 ha, with significant concentrations of land and farms in the Prairies, in southern Ontario and southern Quebec. These numbers, however, mask another important set of changes that have taken place over the last 50 years; a shift from homogeneity to heterogeneity in terms of enterprise and, more important a polarization or high level of concentration upon a minority of farms *vis-à-vis* the economic function and viability of the system.

Polarization within farming is further evidence of the fragmentation and weakening of the system. It takes two related forms; first, concentration by product or commodity within a minority of the producing units; and second, a concentration of the majority of total farm capital and value of production (gross receipts) within a minority of farm units. Both these tendencies have been associated with the processes of industrialization and both operate to exacerbate the level of discontinuity within the farm-based rural system.

With respect to the concentration by product; whereas 50 years ago the majority of all farms combined a range of crop and livestock elements, with even the majority of Prairie grain farmers keeping some diary cattle and other livestock (Table 18.1), today over 95% of all farms are identified in terms of a specialist enterprise and, in the case of livestock in particular, this tends to be just one major type. Furthermore, increasingly, a minority of farms dominate in terms of production. This applies both to some of the key cash crops (spring wheat, grain corn, canola, and field vegetables) where between 55% and 70% total acreage is grown by less than 30% of farms reporting the crop, and especially for the main livestock types. The most widespread livestock, cattle, are now concentrated to the extent that 75% of animals are on 35% of farms reporting; the figure rises to 85% for sheep, while 75% of all pigs are on 18% of farms and 94% of laying hens are on just 6%.

With respect to concentration and viability in financial terms; a level per farm of $500,000 in total value of Farm Capital and of $100,000 of Gross Farm Receipts have come to be seen as thresholds of economic viability. By 1991, only 27% and 25% of all farms met these capital and receipt thresholds, respectively. These minorities of farms, with a very high level of overlap, accounted for a majority of all capital and gross receipts. As Table 18.3 indicates, although the numbers and percentages of farms in the higher capital value (over $1,000,000 and over $1,500,000) and gross receipts (over $250,000 and over $500,000) decline, their proportional share of the respective totals actually increases. In terms of farm numbers

Table 18.3. Percentages of Gross Farm Receipts (GFR) and Total Farm Capital Value (TFCV), Canada 1991.

	% Farms	% TFCV	% GFR
GFR categories			
$0–25,000/farm	42.0	20.0	4.3
$25,000–100,000/farm	35.5	29.8	20.7
$100,000+/farm	24.5	50.2	74.9
TFCV categories			
$0–200,000/farm	33.0	8.2	7.3
$200,000–500,000/farm	39.6	27.4	27.7
$500,000+/farm	27.4	64.3	65.1

Source: Statistics Canada (1993).

a situation has emerged whereby approximately 70,000 farms utilize 50% of all farm capital and account for three-quarters of all farm receipts. In contrast there also exist some 118,000 farms (42%) that gross less than $25,000 (accounting for under 5% total receipts) and a remaining 100,000 whose gross receipts fall below $100,000 (20% total). This polarization between economically viable and less or non-viable farm operations, with the latter numerically dominant, represents a drastic shift from the previous homogeneity and shared economic involvement. It is worth noting that 38% of all farm operators are classed as 'secondary', with an annual average of 188 days off-farm work, and that the majority contribution to total farm family income has long been from non-farm sources (Statistics Canada, 1995).

Conclusions

This chapter has attempted to show a situation whereby, in Canada at least, the relatively extensive, populated and integrated farm-based rural system which evolved and defined rural Canada, has broken down during the last 50 years. The 'ideal' whereby an integrated system saw farm and non-farm livelihood based on utilization of most of the extensive farm landscape, with communities based on a relatively dense population of farmers and farm-related rural employment concentrated within rural settlements, has disintegrated. The present situation is such that a significant part of the farm base in eastern Canada has been lost and much of the rest may be surplus to production needs. Large numbers of farms have been abandoned and farm people have been dispersed. Almost all agriculturally based rural employment has disappeared and many former rural service centres, especially in the Prairies, have ceased to function. The system, as judged by contemporary economic criteria, is based on a minority of the remaining farms, for which the rural community has lost its economic import, which operate in a series of distinct commodity markets, and whose geographic distribution suggests that large areas of the previously rural-farm ecumene have been effectively marginalized. It follows that for most, if not all rural Canada, the model framework for an agriculturally based sustainable rural system (Box 18.1b) is rapidly disappearing.

References

Baker, O.E. (1926/28) Agricultural regions of North America, *Economic Geography* 2(4), 459–493 and 4(1), 44–73.
Bowler, I.R. and Ilbery, B.W. (1987) Redefining agricultural geography. *Area* 19 (4), 327–332.

Bryant, C.R. and Johnston, T.R. (1992) *Farming in the City's Countryside.* University of Toronto Press, Toronto.

Buckley, H. and Tihanyi, E. (1967) *Canadian Policies for Rural Adjustment; A Study of the Economic Impact of ARDA, PFRA, and MMRA,* Special Study 7. Economic Council of Canada, Ottawa.

Cloke, P.J. and Park, C. (1984) *Rural Resource Management: A Geographic Perspective.* Croom Helm, London.

Dominion Bureau of Statistics (1946) *Census of Canada 1941 – Agriculture.* King's Printer, Ottawa.

Harris, R.C. and Warkentin, J. (1976) *Canada Before Confederation.* Oxford University Press, Toronto.

Lowrance, J. (1990) Research approaches for ecological sustainability. *Journal of Soil and Water Conservation* 45 (1), 51–57.

Stabler, J. (1993) *Rural Community Rationalization: Towards a Whole Rural Policy for Canada, ARRG Working Paper Series, No. 7.* ARRG/Rural Development Institute, Brandon, Man. pp. 11–20.

Statistics Canada (1993) *Census of Canada 1991–Agriculture.* Supply and Services, Ottawa.

Statistics Canada (1995) *Profile of Canadian Agriculture.* Supply and Services, Ottawa.

Troughton, M.J. (1986) Farming systems in the modern world. In: Pacione, M. (ed.) *Progress in Agricultural Geography.* Croom Helm, London, pp. 93–123.

Troughton, M.J. (1992) The restructuring of agriculture; the Canadian example. In: Bowler, I.R., Bryant, C.R. and Nellis, D. (eds) *Contemporary Rural Systems in Transition, Volume 1: Agriculture & Environment.* CAB International, Wallingford, pp. 29–42.

Troughton, M.J. (1993) Conflict or sustainability: contrasts and commonalities between global rural systems. *Geography Research Forum* 13, 1–11.

Walker, G. (1987) *Invaded Countryside; Structures of Life in the Toronto Fringe.* York University Geographical Monograph 17, Toronto.

Wynn, G. (1987) Plate 68 'Eastern Canada in 1800'. In: Harris, R.C. and Matthews, G. (eds) *Historical Atlas of Canada: Volume 1. From the beginning to 1800.* University of Toronto Press, Toronto.

Zimmerman, C.C. and Moneo, G.W. (1971) *The Prairie Community System.* Agricultural Economics Research Council of Canada, Ottawa.

19

Sustainability Issues in the Industrialization of Hog Production in the United States

OWEN J. FURUSETH
Department of Geography, University of North Carolina at Charlotte, North Carolina 28223, USA

Introduction

The global reorganization of food production following World War II has quickly moved agriculture in the USA from a decentralized cottage industry (Opie, 1994) towards an industrial organized agrofood system (Goodman and Redclift, 1991; Friedland *et al.*, 1991). But within this environment, farmers have responded to the restructuring processes in individualistic fashion. Whatmore (1995) reminds one that farming is embedded in locally segmented farming cultures, with strong family-based structures. As a result, the degree to which single farmers or farming communities are integrated into an industrial agrofood system varies geographically and structurally. In some instances, such as California's fresh fruit and vegetable production, the shift to industrialization is complete (Friedland, 1994); while in other cases, including the expanding number of alternative or 'green' oriented farms, the trend is away from the industrial model (Duram, 1995 and Chapter 10).

Recently the evolution of American agriculture along the industrial model has encountered a new obstacle, a growing suite of public policies and societal pressures arising from the principles of sustainable development. Although the term sustainable development and the associated concept of sustainable agriculture are new to farming vernacular, concern for and attention to environmental quality have long been a central theme (if not practice) of the US agricultural community. For example, an increasingly used slogan by agricultural groups is: 'America's farmers, the first environmentalists'. As evidence of this commitment, proponents cite widely practised soil and water conservation strategies, which have over the past half century greatly reduced degradation of the natural resource base.

Although the gains in soil and water stewardship have been impressive, other elements of contemporary farming practices have lead to the current challenges of the expanding industrial agricultural system. The reliance on chemical fertilizers and pesticides and the use of growth hormones in dairy and meat production, as well as other biotechnologies, have heightened consumer concerns over the healthiness of food products. On another level, the substitution of technological innovation for land and labour and the accompanying decline in small farms has weakened the popular image of American farming as the Jeffersonian family farm. This imagery has been critical for creating public goodwill which, in turn, has been essential to maintaining consumer confidence in the quality of American foodstuffs as well as marshalling political support to enhance and defend favourable farm policies at the Federal and State levels.

In the afterglow of the 1992 United Nations conference on Environment and Development (the 'Earth Summit'), political pressure and public expectations that American farmers should become more sustainable have mounted. The current thrust toward sustainable agriculture furthers a limited policy direction begun in 1985. The Food Security Act of 1985 was the first US agricultural policy legislation that directed any federal resources, albeit small amounts of monies, toward 'low-input sustainable agriculture'. Despite vigorous opposition, the federal government's commitment to expanded research and education in sustainable farming was reaffirmed in the Food, Agriculture, and Conservation and Trade Act of 1990 (Helmers and Hoag, 1994). These initiatives have encountered enormous political and bureaucratic resistance. Sustainable agriculture represents a challenge to the way that much of American agriculture has been moving over the past 50 years. Indeed, the grafting of sustainable agriculture onto a larger industrially organized farming system results in an artificial merging of incompatible systems.

One agricultural sector that provides evidence of the contradictions between sustainable and industrial structured farming is the hog industry. Until recently, hog and pig production in the USA was primarily composed of small scale independent farms. Over the past half decade, hog farming has experienced massive reorganization marked by the adoption of an industrial production model.

The research reported in this chapter is concerned with the fundamental tensions between an industrially structured hog commodity system and the sustainable paradigm. Whereas not exclusively using North Carolina as a case study, the geographical emphasis of the analysis is on this south-eastern state. The focus on North Carolina reflects this state's status as the host to the largest hog farmers and processors in the Untied States as well as an arena for widespread public debate surrounding the sustainability of industrial hog farming.

Definition of Sustainable Agriculture

The meaning of sustainable development is subject to considerable interpretation and debate. In broad terms, sustainability refers to a process or state that can be maintained over time. Therefore, sustainable development refers to growth that is directed in such a manner that it can be extended into the future. Sustainable development involves improving the quality of human life while operating within the carrying capacity of supporting ecosystems. Inherent in the objective of sustainable development is the link between economic, social and environmental issues. The connection hinges on the idea that development should respond to the needs of present populations while recognizing the requirements of future generations to meet their needs.

Using the Brundtlund Commission (1987) report, *Our Common Future*, as a guide, sustainable agriculture can be simply defined as an agricultural system that can indefinitely meet demands for food and fibre at socially acceptable economic and environmental costs (Crosson, 1992). The ambiguity in the meaning of sustainable agriculture is focused on what are acceptable economic and environmental costs, and how are they defined and measured. An underlying premise of sustainable agriculture is an intergenerational obligation to manage agricultural resources in order to provide future generations with the opportunity to engage in farming activities at reasonable costs.

After considering the trade-offs surrounding sustainable agriculture, Helmers and Hoag (1994) conclude that this alternative approach has a variety of economic, social and environmental benefits, primarily occurring at the farm level. These include improved returns and reduced risk for farms, increased long-term production efficiency, improved health for farm labour, enhanced quality of soil and water resources, stronger family-farm structure, reduced reliance on external production inputs, revitalized rural communities, and improvement in food quality (Helmers and Hoag, 1994, p. 128).

In addressing the differing interpretations of sustainable development, Cocklin (1995) notes that the discourse surrounding the concept is shaped by alternative conceptualizations and perspectives on sustainability. He goes on to argue that geographical scale has a fundamental role in shaping the process of sustainable development. Thus, sustainable agriculture operates: 'in a nested hierarchy of space, wherein the decisions and actions taken at one level have important implications at other levels' (Cocklin, 1995, p. 254). Accordingly, decisions or actions that may advantage farmers of farming systems on one plain may be detrimental at another level and are not sustainable.

Within this framework, the farm production unit is the most powerful level for shaping sustainability. Differences in farm-level decision-making and environmental conditions, in turn, give rise to a related geographical dimension of sustainability: spatial differentiation. The rural landscape can be seen as a spatial mosaic with nested levels of rural actors and systems operating at various points along a sustainability spectrum. Although a cumulative assessment of rural sustainability for a particular rural area is valuable, it is a course-grained picture compiled from many individual differences operating at a number of geographical scales.

Troughton (1993) has posited a farm-based model of rural sustainability. Within this model, the sustainability of rural systems is structured upon five integrated objectives encompassing economic, environmental and social standards. Again, the individual farms are at the nexus of complex interrelationships. Three of the objectives are farm-based factors (agronomic sustainability, microeconomic sustainability, and social sustainability), while a fourth objective (ecological sustainability) is partially affected by farmers' decision-making. Macroeconomic sustainability is beyond the reach of individual action. In using his sustainable rural system framework, Troughton argues that the decline in rural sustainability in the developed world reflects a breakdown in the interrelationships.

Hog Farming in the United States

From the beginning of European settlement, the raising of hogs and pigs has been an integral part of agricultural life in large portions of the United States. For much of this time, hog production was largely a casual, supplemental activity. That is to say, farmers' production focus was on cash crops such as corn, wheat, cotton, tobacco, or dairy production. Hogs were kept in small numbers on the farm, pasture fed, and raised from farrow to finish. Eventually, they were either butchered for personal consumption or sold off-farm for additional income. The numbers of farmer operators specializing in swine production was modest, yet the number of animals and the number of farms raising hogs in the Midwest and Southeast was always significant.

In the post-World War II era, the production of most meat products in the United States has shifted from independent producers to a vertically integrated system with either contract growers or corporate production (Heffernan and Constance, 1994). However, among livestock commodities, hog production has been unique, the least integrated meat product (Table 19.1). For three decades the shift from independent swine producers has been insignificant, and has lagged far behind not only meat but also most other agricultural products.

Table 19.1. Meat commodity production by labour process (%).

Commodity		1960	1970	1980
Livestock	Independent	68.9	66.3	62.2
(all)	Contract	27.5	29.2	33.0
	Corporate	3.6	4.5	4.8
Fed	Independent	83.3	75.3	85.5
cattle	Contract	10.0	18.0	10.0
	Corporate	6.7	6.7	4.5
Broilers	Independent	1.6	3.0	1.0
	Contract	93.0	90.0	89.0
	Corporate	5.4	7.0	10.0
Turkeys	Independent	66.0	46.0	10.0
	Contract	30.0	42.0	62.0
	Corporate	4.0	12.0	28.0
Hogs	Independent	99.2	98.9	98.4
	Contract	0.7	1.0	1.5
	Corporate	0.1	0.1	0.1
Total farm output	Independent	74.6	72.2	69.2
	Contract	20.6	22.3	24.8
	Corporate	4.8	5.5	6.0

Source: Adapted Heffernan and Constance (1994) from Marion, 1986.

During this period, hog production remained a farm-based activity. Most hogs were produced on multiple-enterprise, crop–livestock farms. The raising of hogs provided an important source of income for hundreds of thousands of farmers, but hog production remained a secondary farm activity. For instance, in 1950, 2.1 million farms or nearly 40% of all American agricultural operations had some hog sales; and among farms in the largest swine producing states, sales averaged only 38 head per farm. By the early 1970s, farms selling hogs were still averaging only 150 head per farm. In contrast, cattle feeding, poultry and egg production were already dominated by large industrial-type production units.

At a time when most American animal meat production was undergoing extensive mechanization and capital–labour substitution, hog farming experienced only minor shifts. Not unexpectedly, the small scale and sideline nature of hog production and the significant capital outlays that were necessary to adopt specialized technology were barriers to most small producers. Over time, as the smallest producers dropped out of hog farming, the proportional investment in specialized equipment, housing and breeding increased. But only the largest and specialized hog producers made capital and technological expenditures equivalent to the larger meat production sector.

Another factor contributing to the absence of capital–labour substitution was the strong association between hog production and cropland. While the pasturing of swine declined rapidly after World War II, with less than 10% of hog farmers maintaining complete field production systems by 1978, most hog farmers continued to raise at least a portion of their herds' ration on the farm. In 1975, almost 80% of the grain fed to hogs was raised on the same farm.

In line with the organizational structure of hog farming, most US hog production was completed by a single farmer on one farm. Farrow to finish enterprises constituted 80% of hog production in 1978. While split phase operations, that is to say, pigs are produced on one farm and finished on a different farm, expanded in the postwar era, they still comprised only 20% of production. Envisioning changes in hog farming, the US Department of Agriculture projected a significant expansion in the split phase production, but expected its share of total output to rise only to 30%.

Geographically, hog farming in the USA was concentrated in the Midwest corn belt states. In the first 30 years following World War II, interregional shifts in production were minimal. Within the corn belt there was some westward shift of production relating to changes in cash crop production. None the less, looking to the future, economists with the US Department of Agriculture expected only minimal changes in the geographical distribution of hog farming owing to '... the economic advantages of having the production of feed grains and hogs relatively close to each other' (Van Arsdall and Gilliam, 1979, p. 209).

As recently as 1978, US Department of Agriculture researchers noted the lack of success in establishing large volume production systems. The Department of Agriculture criterion for large volume hog production was a farm marketing over 5000 head per year. Among the factors that were identified as retarding attempts to set up large volume production units were managerial weaknesses, lack of skilled labour, and disease control.

The typical Midwest hog farm in the late 1970s encompassed 320 acres, of which 120 acres were debt free and 200 acres were cash rented. Corn and soybeans were the principal crops, with 75% of the corn serving as hog feed. This prototypic farm operated a farrow to finish operation producing four groups of hogs (650 head) per year scheduled around the cropping schedule. Technological inputs and mechanization were limited. Capital facilities were similarly affected. Production facilities were limited to a central farrowing house. The farrowing building reduced labour expenses and permitted year round production. Breeding animals were maintained in pastures or in an open-lot setting. This was primarily because of problems associated with confinement breeding. Pig nurseries and finishing facilities were restricted to larger producers and would be considered expensive capital investments for the typical farm operator. The farm provided most of the direct labour for the hog operation, with seasonal labour hired to help harvest the corn and soybeans. Animal waste was contained and collected by the farmer. Manure was spread as supplemental fertilizer on the fields.

Industrial Hog Farming

As recently as 1980 most of the hogs produced and processed in the USA came from hundreds of thousands of small independent farms, like the operation described above. In 1982 there were nearly 330,000 farms producing hogs and pigs. Over the next 10 years the number of hog farmers declined by 42% while the number of hogs sold grew by 17%, from 94,780,000 to 111,330,000. The dramatic shift in these data are expressive of the fundamental realignment in American hog farming during this period.

Structurally, hog farming shifted from low density production systems, characterized by a small number of animals raised over a large area, to high density specialized production systems. The latter technology, pioneered 20 years earlier by the broiler industry, encompasses a massing of large numbers of animals in confinement units. By necessity, the confinement units restrict the movement of animals to individual crates (pens) or small rearing areas containing only a few animals. The costs per animal space is not inexpensive, about $105 per hog in 1995. The structures are developed and engineered for a specific phase in the hog production cycle. This rearing process involves the reliance on mechanical systems to bring rations to the swine and efficiently remove animal waste. Veterinary and other health-related services are still provided by farm labour. Farm-based computer monitoring systems, used to measure growth rates, document breeding and gestation data, and generally gauge operational efficiency, are widespread.

The most important change associated with industrial type swine production is scale. Swine operations with over 3000 animals are not uncommon. The result is a single purpose assembly line type of complex, with large numbers of similarly aged swine restricted to a small area. A bigger is better philosophy is widespread in the industry. Corporately-owned hog farms designed to hold 100,000 head of stock are increasingly widespread. The Circle Four hog farm in Milford, Utah, for example, is planned for a herd of 100,000 sows capable of producing two million animals per year for the West Coast market.

Another structural change accompanying industrial hog farming is specialization. Existing hog farms are increasingly treated as obsolete, and new farrow to finish production operations are unusual. As a rule, production units are designed to handle a single step of a three tier hog production system. Sows breed, gestate, and farrow in one phase. Piglets are weaned at 21 days and 10 pounds and moved to a nursery in the second phase. After a month, the pigs are moved to a finishing operation. As hogs complete each phase, they are shipped by truck to another farm.

Accompanying the increased scale of American hog farming there has been a need for new waste disposal practice. Bearing in mind that a mature hog produces four times more body waste than an adult person, an industrial hog farm finishing 5000 head will produce a waste stream comparable to an urban area with 20,000 residents. Under a conventional regime, with a small numbers of hogs, manure is safely spread on field crops serving as a supplemental or replacement fertilizer. The industrial hog farm requires a large scale waste disposal system, this means an extensive manure lagoon. Lagoons covering several hectares and containing millions of deciliters of liquid waste are not uncommon. These structures are analogous to primitive sewerage treatment facilities attached to a factory. Manure slurry is pumped from the rearing buildings into open air lagoons. Most states have restrictions on the capacity of the lagoons and require a lining material in order to prevent leakage into the ground water. The waste slurry evaporates and bacterial action breaks it down. Periodically, the slurry is tested for nutrient levels and applied by sprayer on pastures and cropland. The rates of application are guided by the absorption capacity of the receiving lands.

Assuming the lagoon systems function properly, they do not contaminate water resources; however, neighbours' complaints about smell accompanying the facilities are common. This had led many states to require setbacks between lagoons and the nearest residence. Hog manure produces a hydrogen sulphide gas, which is detectable by a distinctive 'rotten egg' smell. Ammonia, methane and carbon monoxide are also emitted by the waste. All these gases have documented health risks associated with prolonged and/or high concentration exposures.

Accompanying the change in the scale of individual production units has been an equally significant shift in production controls. Vertically integrated corporations dominate the hog production process. Several large transnational corporations, including Cargill, Tyson and Continental Grain, are active in hog farming. However, the largest corporate farmers, Murphy Family Foods and Carroll's Foods, are firms that were pioneers in the industrialization of the hog industry. When compared with other agricultural commodities, the proportion of production controlled directly by corporate agricultural firms remains small, but it has recently increased sharply. Between 1960 and 1980, the percentage of vertically integrated swine production climbed from 0.7 and 1.5%, but from 1980 to 1990 it jumped to 5% (Manchester, 1992).

Although corporate hog producers have eschewed direct involvement in hog rearing, the autonomous contract grower is fundamental to their operation. In the current restructuring environment, the independent hog farmer has been replaced by the contract grower. Following the organization model of the poultry industry, contract farmers are hired by hog production corporations to tend and feed their stock. The company provides the pigs, feed, management and veterinary services to contracting farmers. The farmer provides the land, buildings, labour, utilities and management. Production targets with associated penalties and bonuses are common. The corporation sets requirements regarding facility design and production operation. At the end of the production cycle, the farmer is paid a fixed sum based upon the contracted production schedule. The market risks of hog prices are assumed by the company, while the liability for low production caused by disease or mortality is assumed by the farmer. A related risk for the contract grower is that production performance or fluctuations in the market may lead to the cancellation of agreements on short notice. In these situations, the contract grower needs to find another corporate partner quickly in order to cover the high costs of production facilities.

A prototype industrial hog farm bears only modest resemblance to the conventional Midwest farm described earlier. The most critical difference is the reduced land area required for the mechanized process. The core of the farm is the complex of buildings housing the animals and supporting equipment and the waste lagoon. Because hog rations are established and supplied by the corporate partner from grain mills, cropland has utility only as a component of the waste disposal system. Conventional concerns for soil quality and the nutrient management of cropland have given way to questions of carrying capacity and nutrient loading. Stated simply, the farmer is concerned that there is enough land to handle the waste stream or how much nitrogen and phosphorus can be absorbed by alternative crops and soil resources. Nutrient saturation represents a critical limitation on the operation of an existing farm or the expansion of the herd. The presence of a nearby waterway or groundwater uptake zone is viewed as a production liability owing to the potential environmental contamination.

Another basic difference between the two different types of farms is labour. While the conventional Midwest hog farm is primarily operated by on-farm labour, the industrial model is run by full time hired workers. Most of the staffing is semi-skilled labour, usually drawn from surplus agricultural workers in the local community. Given the regimented production process and tight operational guidelines associated with production contracts, neither prior farming experience nor day-to-day involvement in the running of the farm are necessary for the farm operators. Indeed, many new contract growers are entrepreneurs attracted to hog farming by the expected high rate of return on capital and the absence of requirements to actively participate in operating the farm.

The principal advantage to the industrial hog production system is efficiency. While quality and consistency of the meat produced in the confinement setting are also cited by proponents as other important considerations, the bottom line is that this type of husbandry system is the most efficient process for converting feed to body weight. Research by Lidvall *et al.* (1980) found that a total confinement system required 3.87 pounds of feed to produce one pound of pork, a conversion ratio of 3.87:1. In comparison, the conversion ratios for pasture production with a hutch is 4.16:1 while the ratio for a partial confinement system is 4.21:1.

One important reason for the greater feed efficiency in industrial units is heavy use of antibiotics. The confinement system has greater risk of disease. Moreover, the consequences of infection in the herd are potentially more devastating. Respiratory diseases, easily spread in a confinement setting, are a major problem in industrial swine production. Therefore, antibiotic regimes are more extensive and fundamental to maintaining the health of the herd. But beyond veterinary concerns, subtherapeutic feeding of antibiotics is commonplace for improving growth rates (National Research Council, 1989). Antibiotics cause hogs to use less feed per unit weight gained. Consequently, the costs of antibiotic services necessary for potential health and environmental hazards are partially mitigated by the enhanced feed conversion effects. Not considered, however, are any potential antibiotic residues in food products associated with the heavy feeding regimes.

North Carolina Experience

For most of this century, the agricultural economy of North Carolina has been built around tobacco. In the East, on the sandy soils of the flue-cured tobacco belt, the cultivation and marketing of tobacco had dominated social and economic relations. Although the tending, harvesting and curing of tobacco was labour intensive, no agricultural commodity could match the income stream that tobacco earned. For many small family operated farms, tobacco was the only dependable cash crop. But over the past decade tobacco's role in the state and regional economy of North Carolina has diminished, and uncertainties surrounding the health risks of cigarettes and the future government tobacco programmes have clouded the future of the crop.

Into this unsettled rural economic environment, large scale, industrially structured hog production has swept across the North Carolina Coastal Plain (Fig. 19.1). According to the *Census of Agriculture* there were slightly more than 2 million hogs and pigs in North Carolina in 1982 and the state ranked eighth among all the states in swine production. By 1992, there were 5.1 million hogs, a 149% increase, and a national ranking of

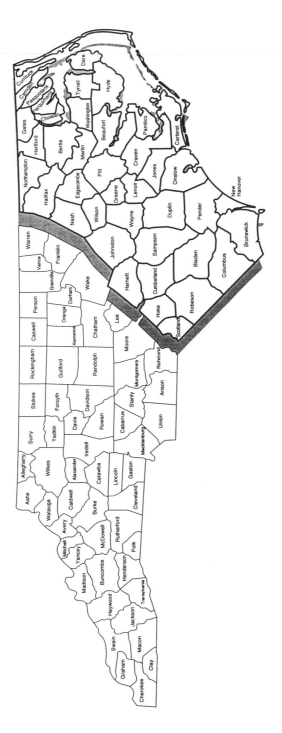

Fig. 19.1. Coastal plain North Carolina.

third. Three years later, there were an estimated 7 million hogs and North Carolina ranks second in the nation in swine production (Warrick and Stith, 1995a).

The growth of industrial hog production is the result of a complex set of factors. At the farm level, many small farm operators fearful of tobacco's future have chosen to become contract hog growers. Non-farm entrepreneurs entering the business have further increased the numbers of new contract growers. Individual farmer acceptance of contract grower arrangements with agrobusinesses and accommodation of the assembly line production systems reflect widespread experience with the poultry industries. This industrial agricultural model was already widely practised from the early 1950s by the chicken, turkey and egg producers (Cook, 1994). The economic success and farmer participation is reflected in recent data showing poultry production as the state's second largest source of farm income.

The adoption of this system has been fostered by the state of North Carolina's political and agricultural institutions as well as indigenous hog production firms. In particular, the state government has taken policy actions and established financial incentives to assist individual contract growers and their corporate partners. In turn, the hog producing firms have worked closely with state government toward a common goal of building an integrated hog production system in Eastern North Carolina.

An overview of the recent geographical pattern of hog production presents contextual insights into the diffusion of this system and the effects of this change on rural communities. A comparison of the most recent *Census of Agriculture* data (1987 and 1992) illustrates the scope of industrial hog farming. The Census classifies the largest hog farms as those with over 1000 head of swine. These are synonymous with the industrial type production system. In 1987 industrial hog farming was heavily focused in the Southcentral Coastal Plain region (Fig. 19.2). The core of this production district was composed of Sampson, Duplin, Greene, Pitt and Wayne Counties. These counties were home to slightly more than one-quarter (26%) of the swine in the state. Five years later, large scale hog farms were found in virtually all non-urban counties on the Coastal Plain. The magnitude of the expansion is reflected by one piece of information; in 1992 there were 2.2 million hogs in Sampson and Duplin Counties. Just five years earlier, the entire hog population in North Carolina had been only slightly larger, 2.5 million head.

During this period, the industrial hog farm has come to characterize swine production in North Carolina as well as dominate the state's hog business. In 1978, the largest farms contained fewer than half of the state's hog and pig population; by 1992 over 90% of North Carolina's swine population was concentrated on 802 farms. The greatest concentration of these intensive production units was focused in Sampson and Duplin

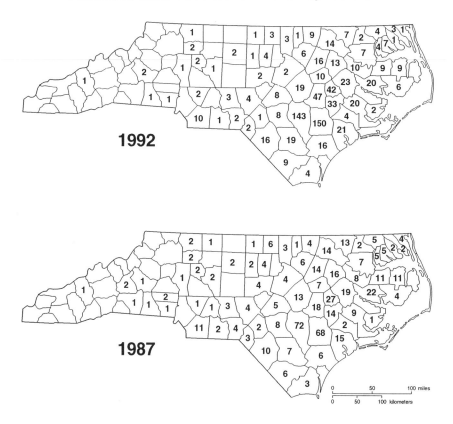

Fig. 19.2. Number of industrial hog farms (+1000 swine) per county.
(Source: Census of Agriculture, 1994.)

Counties. These two jurisdictions contained 37% of the industrial hog farms with herds of over 2 million animals. Viewed areally, there are approximately 480 hogs per square kilometre of land area in these two counties.

The growth in large specialized hog production in North Carolina is the physical evidence of the replacement of independent producers by highly integrated corporate farming entities. Eight of the United States largest hog producing corporations are active in North Carolina. Half of these firms are 'home grown' North Carolina based operations. The latter include Murphy Family Farms and Carroll's Foods, which have expanded in tandem with the growth in corporate hog farming in North Carolina. Currently, Murphy and Carroll's are the two largest US hog producers. The growth of North Carolina hog corporations and their success in the US Food system is reflected by data showing a 10% share of the total American hog market for North Carolina-based firms. Among the larger, diversified transnational corporations operating in the state are Tyson Foods and Cargill, both of which also have broiler production in the state. The high level of concentration in hog production favours cooperation between firms. Integration and conglomeration in feed purchases, hog butchering, and hog production are widespread.

Corporate hog producers in North Carolina are reliant on contract farmers to carry out the production processes. For smaller farmers looking for income producing options beyond tobacco, contract hog production is perceived as an attractive alternative. Farmer acceptance of contract production has been critical to the restructuring of hog farming. During the past four years, hog production in North Carolina has doubled, with nearly all of the increase occurring on contract farms. Currently 82% of North Carolina's hogs and pigs are grown by contract farmers. Contrastly, in Iowa, the largest hog producing state, the proportion of contract growers in only 15%. According to Murphy, Prestage, and Cargill representatives, the popularity of contract farming arrangements is so widespread that it has resulted in a one year wait for farmers wanting to participate in the programme.

The attractiveness of contract hog farming comes in spite of some economic data that challenges the profitability of production arrangements and negative publicity surrounding the recent widespread contract termination for broiler producers in Northwest North Carolina as a result of corporate mergers. At least one North Carolina economist has estimated that over the 15-year life of a confinement unit, a contract grower can expect to earn about $7.00 per hour for his/her labour and approximately 50 cents per pig for return on investment (Centre for Rural Affairs, 1995). The potential risks of contract hog farming are allayed by corporate agents promoting the profitability of hog farming and the support of the state agricultural institutions, including the North Carolina Department of Agriculture; North Carolina State University; and the State Extension Service and the Division of Soil and Water Conservation.

The match between industrial hog production and North Carolina farmers would not have been so complete if the State of North Carolina had not actively recruited and accommodated the industry. Not unexpectedly, the strongest and most direct support within state government is provided by the North Carolina Department of Agriculture and its affiliate organizations. Since the middle 1970s the political leadership and professional staff of the Department have viewed tobacco as a declining crop and actively sought substitute commodities, especially for small land holding farmers. Beyond promoting and sustaining agricultural activities in North Carolina, the protection and encouragement of family owned farms is a pervasive theme in the culture and public policies of the Department of Agriculture. In the eyes of the Department, contract hog farming represents a long term, economically resilient agricultural system that supports family farmers at a time when tobacco's economic impact is waning.

As a result, the Department endorsed the contract hog farming model and has provided substantial financial support to foster the expansion of the industry. The linkage, between the Department, the hog corporations, and the NC Pork Producers Association, the contractors organization, is strong and at times questionable. On a growing number of occasions the Department has seemed to advocate policy positions at odds with other units of state and local government in deference to hog producers and corporate interests. In particular, the Department has shown little willingness to evaluate the environmental costs associated with hog production and processing. It has lobbied to postpone and weaken enforcing water and air quality regulations for hog producers.

In a similar fashion, the State Legislature and Governor also have been strong proponents for the industry. Over the past 15 years, the hog producers and agrobusinesses have been given a plethora of financial and regulatory advantages by the state legislative (Warrick and Stith, 1995b). Financial benefits for hog producers are far greater than the benefits extended to other farm producers. These include sales tax exemptions for equipment and buildings, exclusion from gasoline taxes for feed delivery trucks, exemption from a state feed inspection fee, and the elimination of local property tax on feed used in livestock production. According to the State Fiscal Research Division, the tax benefits awarded to hog producers amount to several millions of dollars annually. Because of the character of these awards, they have tended to favour new contract farmers or those farmers actively expanding their operations, as well as corporate hog interests.

Under North Carolina law there is no restriction on corporate ownership of farms. Moreover, 'family farms', no matter who owns them or how large they are, are given protection from restrictions and regulations placed on other businesses. For example, the Right-to-Farm legislation, passed in the 1970s before the development of large confinement animal rearing operations, protects hog farms from nuisance law suits. More recently, in 1991, the State Legislature exempted 'bona fide' hog and other livestock and poultry farms from local zoning regulations. Thus, local governments cannot control the location of high density hog farms, except for a 750-foot setback from the nearest residential structure. In the area of labour regulations, the legislature exempted hog farms from the requirements of the minimum wage law as well as allowing workers to legally organize unions.

While the state legislature's gratuities to the hog industry have been plentiful, the most far reaching largesse has been in the area of environmental protection. North Carolina's air and water quality standards and enforcement programmes are not considered to be especially rigorous, but they have been weakened further or, in some cases, ignored in order to accommodate hog producers and their hog processing partners. North Carolina leads the US in high density hog farming, yet until the rupture of three large animal waste lagoons, during the summer of 1995, the state essentially had no standards or monitoring programme for hog waste lagoons. Even in cases where a lagoon is suspected to be leaking, state rules limited the inspection powers, penalties and remediation options that could be assessed to polluters.

Industrial Hog Farming: Unsustainable Agriculture?

The sustainable agriculture paradigm requires that capital be passed from one generation to the next, an intergenerational equity (Crosson, 1992). The integrated industrial farm that is the template for American hog production does not meet this test. This efficiency driven system of animal production has sharply redefined interrelationships between hog growers and their neighbours and created widespread conflict at the local, regional and state levels. Human-made capital has replaced value assigned to natural capital and superseded cultural capital. A refashioned capital framework operates for short-term quantitative gain rather than long-term capacity. Advocates note that industrial hog farming and other technology-focused agricultural systems are more profitable. Gokany and Sprague (1992), for example, calculated that without sophisticated technology four times more land would be required to produce the current food output. Within this framework, the critical role that geographical scale plays in differentiating the capitalization of sustainability is integral. At the farm level, the economic success of industrial hog production is evidenced by the growth in farm income and expansion in the number of production units as well as herd size. Beyond the farm gate, however, the effects on the surrounding community are on balance far less favourable. The impacts on air and water resources are perhaps most serious. In locales with large numbers of hogs, waste disposal leads to deterioration in air quality and water pollution. Leakage from waste lagoons to surface and ground water resources represent on-going concerns. Given the herd size of industrial production units, nutrient saturation of soils is likely to occur over time resulting in ground water contamination with potential adverse effects on well water. For example, research in North Carolina suggests that three counties have already exceeded the capacity of cropland in absorbing nitrogen loading from animal waste, and 18 counties have exceeded the phosphorus loading (Barker and Zublena, 1995).

Beyond environmental externalities, contract hog farming contributes fewer economic benefits to the local or regional economy than conventional farming operations. Research carried out in several areas in the

Midwest and Southeast has found that smaller scale, independent hog farmers used more local resources and had a greater positive impact on other businesses in the local area (Centre for Rural Affairs, 1995). Virginia researchers, for instance, evaluated the effects of comparably sized hog farms, containing 5000 sows. When the economic impacts of independent farms were matched against vertically integrated industrial farms, the research showed that conventional hog farming produced 10% more expansion in permanent jobs, a 20% greater increase in local retail sales, and 37% greater use in local per capita income (Thornesbury *et al.*, n.d.). One Missouri economic study found that large scale hog production costs a local community more jobs, both on-farm and off-farm, by displacement than they create (Ikerd, 1994).

For conventional hog farmers, the expansion of contract growers and vertical integration of hog farming means shrinking market opportunities. As the proportion of pork production capacity controlled by agrobusinesses increases, the pork processors buy fewer hogs from independent farmers and pay lower prices to these farmers than contract growers (Azza and Wellman, 1992). Recently, in North Carolina the price per hundredweight for hogs from contract growers was $51.00 while the price garnered for independently produced hogs was $39.00 (Centre for Rural Affairs, 1995).

A related impact developing out of the economic disruptions and environmental degradation is social stress and loss of community cohesion. Local conflicts arising from the deleterious effects of industrial hog farming have pitted neighbours against farm operators as well as conventional hog farmers against contract growers. Longstanding class and racial tensions have been exacerbated. In many areas of Eastern North Carolina, African-Americans and Native American residents have alleged that industrial hog operations represent environmental racism. Poorer rural communities near confinement operations suffer from odour problems as well as potential air and water pollution risks. Property values and alternative economic opportunities plummet. Moreover, contract growers are over-whelmingly white even in communities where a majority of the rural residents are black, while the noxious environmental impacts of the operations affect minorities disproportionately.

Taken together, the operational characteristics and effects of industrial styled hog farming have created new agricultural enterprises on the rural landscape. The links between farm operations, the natural resource base and local agricultural infrastructure have been marginalized. Industrial engineering and substitutionism have created pseudo-farms, more factory than farm, that are not reliant on the local environment for resource inputs or locational advantage. Rather, regulatory and political advantage are far more powerful determinants for farm location. Accordingly, as criticism of industrial hog production has developed across Eastern North Carolina, corporate efforts to penetrate independent hog farming environments of the Midwest, most notably Iowa, have been expanded, and efforts to discover other states and rural communities where the institutional environment is friendly to large scale, confinement hog production have similarly increased (Warrick and Stith, 1995b). At this juncture, the continued expansion of industrial hog production units is challenged by the unsustainable effects of the process. The ultimate status of the restructuring of US hog farming is, however, uncertain, whether or not questions of sustainability shape public discourse and political decision-making are still unanswered.

References

Azzam, A.M. and Wellman, A.C. (1992) *Packer Integration into Hog Production: Current Status and Likely Impacts of Increased Vertical Control on Hog Prices and Quantities.* Research Bulletin 315-F, Institute of Agriculture and Natural Resources, University of Nebraska-Lincoln, Lincoln, USA.

Barker, J.C. and Zublena, J.P. (1995) Livestock manure nutrient assessment in North Carolina. Paper presented at Seventh International Symposium on Agricultural and Food Processing Wastes, Chicago, June.

Centre for Rural Affairs (1995) *Spotlight on Pork II, Corporate Farming Update.* Center for Rural Affairs, Walthill NE, 16 pp.

Cocklin, C. (1995) Agriculture, society and environment: discourses on sustainability. *International Journal of Sustainable Development and World Ecology* 2, 240–256.

Crosson, P.R. (1992) Sustainable agriculture. *Resources* 106, 14–17.

Cook, R.E. (1994) Vertical integration boosts N.C. poultry. *Carolina Farmer,* July, 46.

Duram, L.A. (1995) Great plains agroecologies: the continuum from conventional to alternative agriculture in Colorado. Paper presented at 1995 Anglo-Canadian–US Rural Geography Symposium Charlotte, NC 30 July–5 August.

Friedland, W.H. (1994) The global fresh fruit and vegetable system: an industrial organization analysis. In: McMichael, P. (ed.) *The Global Restructuring of Agro-Food Systems.* Cornell University Press, Ithaca, pp. 173–189.

Friedland, W.H., Barton, A.E. and Thomas, R.J. (1981) *Manufacturing Green Gold: Capital, Labor, and Technology in the Lettuce Industry.* Cambridge University Press, New York.

Gokany, I. and Sprague, M. (1992) *Sustaining Development and Biodiversity: Productivity Efficiency and Conservation.* Policy Analysis No. 175, CATO Institute, Washington DC.

Goodman, D. and Redclift, M. (1991) *Refashioning Nature, Food Ecology and Culture.* Routledge, New York, 279 pp.

Heffernan, W.D. and Constance, D.H. (1994) Transnational corporations and the globalization of the food system. In: Bonnanno, A., Busch, L., Friedland, W.H., Gouveia, L., and Minquione, E. (eds) *From Colombus to ConAgra: the Globalization of Agriculture and Food.* University Press of Kansas, Lawrence, pp. 29–51.

Helmers, G.A. and Hoag, D.L. (1994) Sustainable agriculture. In: Hallberg, M.C., Spitze, R.G.F. and Ray, D.E. (eds) *Food, Agriculture, and Rural Policy into the*

Twenty-First Century, Issues and Trade-offs, Westview Press, Boulder, Colorado, pp. 11–134.

Ikerd, J. (1994) *The Economic Impacts of Increased Contract Swine Production in Missouri: Another Viewpoint.* Report, Sustainable Agriculture Systems Program, University of Missouri-Columbia, Columbia, USA.

Lidvall, E.R., Ray, R.M., Dixon, M.C. and Wyatt, R.L. (1980) *A Comparison of Three Farrow-Finish Pork Production Systems.* Tennessee Farm and Home Science No. 116, University of Tennessee-Knoxville, USA.

Manchester, A.C. (1992) Transition in the farm and food system. Paper prepared for the National Planning Association, Committee on Agriculture.

Marion, B.W. (1986) The Organization and Performance of the US Food System. Lexington Books, Toronto.

National Research Council (1989) *Alternative Agriculture.* National Academy Press, Washington DC, 448 pp.

National Research Council (1988) *Designing Foods: Animal Product Options in the Marketplace.* National Academy Press, Washington DC, 394 pp.

Opie, John (1994) *The Law of the Land: Two Hundred Years of American Farmland Policy,* 2nd edn. University of Nebraska Press, Lincoln, 253 pp.

Thornsbury, S., Kambhampaty, S.M. and Kenyon, D. (n.d.) *Economic Impact of a Swine Complex in Southside Virginia.* Report, Department of Agricultural and Applied Economics, Virginia Polytechnic Institute and State University, Blacksburg, USA.

Troughton, M.J. (1993) Conflict or sustainability: contrasts and commonalities between global rural systems. *Geography Research Forum* 13, 1–11.

U.S. Department of Agriculture (1994) *1992 Census of Agriculture, Volume 1: Geographic Area Series.* US Government Printing Office, Washington DC.

Van Arsdall, R.N. and Gilliam, H.C. (1979) Pork. In: Schwartz, L.P. (ed.) *Another Revolution in U.S. Farming?* US Department of Agriculture, Washington DC, pp. 190–254.

Warrick, J. and Stith, J. (1995a) Lacking Staff, DEM finds it hard to challenge status quo. *The Raleigh News and Observer,* 19 February, p. A–18.

Warrick, J. and Stith, J. (1995b) Midwest farms fear NC-style expansion. *The Raleigh News and Observer,* 22 February, p. A–7.

Whatmore, S. (1995) From farming to agribusiness: the global agro-food system. In: Johnston, R.J., Taylor, P.J. and Watts, M.J. (eds) *Geographies of Global Change, Remapping the World in the Late Twentieth Century,* Blackwell, Cambridge, Massachusetts, pp. 36–49.

20 Sustainable Agriculture and Its Social Geographic Context in Ontario

GERALD WALKER
Department of Geography, York University, Toronto, Ontario, Canada M3J 1P3

The Problem and the Setting

Agriculture in Ontario has continued its evolution, from a mixed cultivation and livestock regime to a series of specializations, highly capitalized and occupying very few farmers and labourers. This chapter examines the social context of this evolution. The settlement system has been fundamentally altered through the combined processes of ruralization (population decline, occupational simplification and demographic transformation) and urbanization of the countryside (associated with the larger processes of counter-urbanization). The social and cultural context of agriculture is intimately linked to the processes of capitalist transformation. Several aspects of the transformation have been argued in economic terms. Here, the character of the transformation as it occurs in the social and cultural realm will first be considered. This chapter will then reflect on the relations these processes have to sustainability.

Agriculture, narrowly examined, is composed of a landed surface, under constant transformation through changing land-uses. Agriculture is also composed of a set of capital elements literally planted on the landed surface. Finally, agriculture is composed of a population of active agricultural entrepreneurs, their families, their labourers and managers. This population, using capital on the land, is organized around both the productive process within agriculture, which extends to a wide range of persons involved in the productive process beyond the farm gate, and in the reproductive realm of the life-world. Within that life-world are social networks and interactions among agricultural people and with the rest of the world. The rest of world, however, can very quickly be divided into the locality and its surroundings, intermediate distances to various places in the urban hierarchy, and beyond. Beyond is complexly organized, but the contacts thin as beyond is approached. This is to assert that place, space and geography are important in the composition of agriculture over and above the land surface of cultivation.

The cultural edge of the set of agricultural practices should be considered as significant and problematic. It is conventional to look at only the practices that create the commodity form of farm produce. Though, even here, it has become conventional to deal with the demographic and economic positions of the farming class. Agricultural practices fade into the general cultural ambience of rurality, and, increasingly urbanity. This system of practices as it is simplified through census materials has been examined elsewhere (Walker, 1992, 1993). Geographic differentiation associated with social and cultural differentiation characterizes those practices.

Agricultural regions surely exist, but they exist within the context of other regionalizations of the life-world of experience and events. Within the current conceptualization of rural in Canada is a recognition of the profound impact of the urban system on the settlement system of the countryside (Bryant and Johnston, 1992). Urban regions and regional cities are found throughout Ontario, in constant and intimate contact with agriculture. In Canada, regional cities are defined in two ways. The first follows the practice of Statistics Canada in its regionalization of census metropolitan and census agglomeration areas. This regionalization is based on the commuting fields of the urban core of the regional city. Dual criteria are used in examining and categorizing the complex: size of the urban core, and the commuting field as a percentage, currently 50%, identifying a significant break in spatial association. This is an attractive way of identifying the urban system as a system of interaction.

The second way of defining a regional city is through the hierarchy of municipal types ranging from cities and a borough, towns, townships, villages, unorganized areas and aboriginal reserves. This is the style adopted in presentation of the results of the census of population and the census of agriculture. Additionally, the census records the larger divisions, counties, districts, regions and regional districts for the units of report. The orientation is toward the municipality as a quasi-natural unit. Always a strong force, municipal organization serves as the cellular unit of Ontarian and Canadian regions. Occasionally, politically defined intermediate regions are recognized. In Ontario, the provincial government has recognized the Greater Toronto Area as perhaps its most significant, urban region (see Fig. 20.1). This style of regionalization is very handy and follows the line of least resistance in data collection. There are many municipal statistics available.

Rurality is a central concern in the regionalizing concepts of government agencies, particularly those dealing with the censuses. In the not too

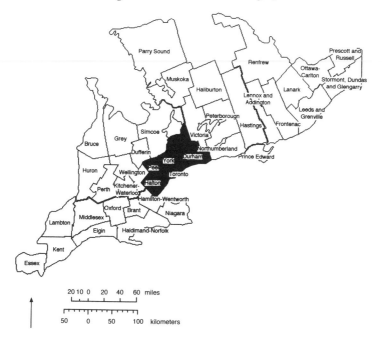

Fig. 20.1. Southern Ontario census divisions and the Greater Toronto Area.

distant past, rurality was taken as a residuum of urban. However urban was defined, and in Canada, this definition has evolved dramatically in the last generation, whatever was left over was rural. On the one hand there was little need for definition. Everyone knew a rural from an urban place. Rural implied low densities, agriculture, and occasionally nature nearby. Sometimes agriculture faded and was supplemented and even replaced by other primary occupations at the base of the social food chain. This older concept of rurality has been reconsidered due to the continued transformation of the urban system.

It is recognized that rurality extends as a cultural category beyond agriculture, to which it has an integral relation. In effect, agriculture is economically and culturally hegemonic in rurality. The enormous expansion and transformation of the urban system, the colonization of the countryside by urbanites and their subsequent transformation into new ruralites, has forced reconsideration of rurality. It is also recognized that rurality is directly related to urbanity and is not a residuum of the definition of urban. Rurality is urbanitys' other. In a recent work *Rural Canada: A Profile* (1994) several classes of geographic areas were presented in relation to the rural. Agglomerations were contiguous groups of municipalities of very high density. Surrounding them were intermediate regions, essentially the rural–urban fringe, a deep zone of interaction between the urban core and the surrounding countryside. Beyond that were three zones of rurality: metro adjacent, non-metro adjacent and rural north. In Ontario the unit of resolution of the analyses that produced this categorization was the county-region. In this system the north, beyond being north, is distinguished by the small impact of agriculture (see Fig. 20.2).

The intermediate zone, beyond the core urban municipalities, but before the first group of rural places, is interesting because it comprises about half of the divisions of southern Ontario. Agricultural places are divided between three zones: intermediate, rural metro adjacent, and rural non-metro adjacent. The conduct of agriculture in the urbanized districts becomes a significant structure of practices in Ontario.

Luckily, this developing agriculture in urbanized zones has been examined in Canada, within a framework that can be extended to other capitalist political economies. The conclusion has been there are real differences in the agricultural practices in urbanized regions and outside of those regions (Bryant and Johnston, 1992). Almost all agricultural activities are carried out on private property, property that is commonly sold through the land market and takes on values that can suggest the overall surface of property values. Regions having high property values are those with the highest degree of urbanization. Agriculture, while having its own surface of value, is subsumed within the larger property market, one which is oriented to urban values. Thus, the property system links rural with urban, and at the same time differentiates between the two. Further, the property system is a mediator between enterprises, such as farms, and residences. This is one of the most important senses in which the contemporary agricultural and rural systems are integrated into the capitalist mode of production.

Linkage between different uses of property in the market is also associated with the processes by which the city has colonized the country-side in this last generation. One of the conventional definitions of rurality in Canada is the lack of a coherent rental market for residences. Rural dwellers, in whatever densities of places, live in owner occupied residences. In one sense, the rural world is particularly integrated, both as property and as agricultural practice, into the capitalist system. Yet, non-capitalist cultural values flourish in rural areas.

An alternative vision of rurality from that based on the capital–labour divide is one that stresses the forms of solidarity and group formation that characterize rural places. Solidarity is intimately related to cultural values. It is those values, and the norms that emerge, which set the membership of collectivities. Solidarity, the life-world, and social reproduction are all linked both by property, the site of life, and communication between interrelated subjects. Communication within the life-world is associated with the formation of groups and the norms of groups.

Farmers, farm labourers, and farm managers are linked in two

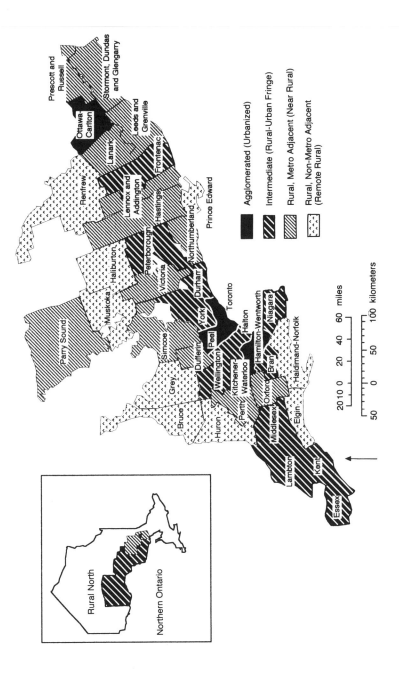

Fig. 20.2. Landscapes in southern Ontario census divisions. (Source: Research subcommittee of the Interdepartmental Committee on Rural and Remote Canada, 1994.)

directions. One is through the productive system of commodity production, the other is through social interactions into the realm of social reproduction. The latter direction includes all kinds of non-farm people, particularly, recently, urban in-migrants. One way to reformulate this is that agricultural interests are losing their hegemonic position. Increasingly, in places close to urban centres such interests must accommodate anew, both economically and socially.

Lurking a bit indistinctly behind agriculture is the question of its sustainability. There are several ways to look at sustainability. It can mean the ability, technically defined, to continue agricultural practices on the same ground indefinitely. This orientation of sustainability is strongly oriented to practices that do not physically degrade the land and render it incapable of cultivation. On this definition, Ontario agriculture is not sustainable. The application of pesticides, fertilizers and the use of a variety of practices that continue erosional processes all point to a relatively short horizon of sustainability. These physical insults to the land, insults that will ultimately lead to the abandonment of agriculture in large parts of Ontario are probably tractable, though the price they will extract is lowering of yields.

One of the more profound contradictions of agriculture, in Ontario and elsewhere, becomes crucial. Agriculture is entirely dominated by capitalist productive and distributive processes through the creation of commodities from the soil. Within capitalist production there is a gradually rising productivity. Part of the rising productivity is located in countries and regions that have relatively recently joined the world capitalist system. They are in the process of early transformation with striking increases in productivity. The net result, ongoing for at least the last two centuries, is to increase productivity on the land faster than consumption of agricultural commodities can be generated.

The core of the contradiction is rising productivity and falling prices generating a continuing and apparently unsolvable crisis in agriculture. The short-term solutions to the crisis have been, and seem likely to continue to be, to accelerate productive increases and hence generate the next iteration of crisis and move in the opposite direction from sustainability.

A second sense in which it is reasonable to consider sustainability is in the social milieu that permits the social reproduction of agriculturalists and agriculture and integrates them into the life-world of civil societies. In this sense, it is an open question whether Ontario agriculture is sustainable. However, as will be seen shortly, there is some reason to hope that social sustainability is a possibility within the current system.

A Preliminary Empirical Investigation in Southern Ontario

The empirical portion of this chapter is going to relate some of the differences and similarities among agricultural practices to census variables of population concerned with demographic and household structures. The units of observation will be municipalities with agriculture practised within them. The data to be used are from the census of population and the census of agriculture for the year 1990, the census being carried out in 1991 (Statistics Canada, 1992, 1994). A set of municipalities, defined by the presence of agriculture, produces 385 places out of a total of 880 municipalities in the province. Of the 385 places 42 are consolidated subdivisions. These are groupings of from two to six municipalities into a single reporting unit. These consolidations include most of the major cities of Ontario, but also include many of the more remote and peripheral agricultural places. The consolidated subdivisions account for a total of 99 municipalities, producing a total of 442 municipalities with agricultural practice, grouped into 385 statistical areas. It is necessary to stress the consolidated subdivision as a particularly heterogeneous collection of reporting places.

From the discussion of the processes operating in the countryside of agriculture and the operational basis of data analysis, several orienting hypotheses can be stated for the following empirical discussion. Agricultural practice will be related to municipal situation. That relation will be mediated through increased values for capital, land and commodities in urbanized agricultural regions, and increasing specialization of agricultural practice in urbanized agricultural places. The social context of agricultural districts within urban fields will be more complex and more heterogeneous than those outside the urban commuting reach.

These hypotheses, general as they are, suggest that urbanized agricultural places will demonstrate practices different from, and partially explicable in terms of, the social surround of urbanity. If these hypotheses can be shown to be substantially correct, the integrity of both 'rural' and 'agricultural' will have been demonstrated. The market demands of nearby urbanity are almost certain to destroy sustainability.

A series of analyses of agricultural and population variables over a set of agriculturally active municipalities will be presented. The aim is to consider urbanized and rural situations for agriculture. Two data sets will be considered, both at the municipal level of resolution. The first data set is taken from three tables of small area data (municipalities and consolidated subdivisions) presented in *Agricultural Profile of Ontario: Part 2* (1992). The data tables identify total area in farms, area owned, land in crops, a series of agricultural innovations to control soil erosion, crop rotation using clover alfalfa, etc., winter cover crops, grassed waterways, strip cropping, contour cultivation and other practices, total farm capital, gross farm receipts, expenses and a series of derived measures. The second data set is derived from the census of population, short form.

Two sets of analyses have been performed on the agricultural data: a series of analyses of variance over municipal types, and a component analysis of the same variables and a few more derived variables. Examination begins with the analyses of variance based on money values for capital, receipts, expenses and derived measures of profit, returns to capital, values by hectare and per farm over municipal type. In these analyses, four municipal types were used: cities (13), towns (26), townships (302) and consolidated subdivisions (42). The highest average capital values were found in towns (averaging over $278,000,000), then in cities (at $164,000,000), then in the townships (at $95,000,000) and finally in consolidated subdivisions (at $43,000,000). Farm receipts and expenses, and the derived profits followed the same pattern of town, city, township and consolidated subdivision. Farm capitalization per farm followed this pattern as well. However, when considering receipts, expenses and profits per farm, the 'city agricultural places' reversed positions with the town and became the leading source of value.

Looking at the derived measures of hectares per farm, owned farms and cropped farms, all followed the order of consolidated subdivisions, township, town, city. Here consolidated subdivisions show the importance of peripheral consolidations. Returning to the profits of agriculture, another pattern emerges. Profits per hectare were highest in cities, then towns, much lower in townships and far and away lowest in consolidated subdivisions. When the hectares per cropped farm is considered, the differences narrow perceptibly and townships emerge as the highest average, followed by consolidated subdivisions, towns and cities.

Returns to capital are not significantly differentiated by municipal type. However, townships have the highest average at 3.3%, followed by cities (2.8%), consolidated subdivisions (2.5%), and towns (2.1%). Even lacking statistical significance, the low returns to capital are very much characteristic of the differences in return to enterprise in agriculture and other businesses. Return to agricultural enterprise, that is independent of details of geography including urban situation, is very low and most certainly accounts for the net decline of farms. Although profits per farm were not differentiated significantly by municipal type, they nevertheless went from $23,000 in cities, to $19,000 in towns, $16,000 in townships, to $15,000 in consolidated subdivisions. Given that, at this time, prevailing per capita average income was $17,000, farmers were doing about the same as other people in the system working for wages and salaries. Of course, it should not be forgotten that most urban households now have two income earners, but this is also true of most agricultural households as well. The differences in incomes were much the same between agriculture and others in their position in the urban hierarchy.

Agricultural innovation presents a more complex picture than land and value relations. The percentage of agriculturalists in the municipality was calculated for each of the innovation variables. Crop rotation, winter cover, and other practices were significantly different over municipal types. Grass waterways, strip cultivation and contour cultivation were not significantly different. In two of the significant innovations the results indicated a gradient of innovation that put township agriculture as the most likely site of innovation, but only small differences between cities and towns. In the third innovation, winter cover, towns had the highest values, but only marginally greater than townships. In all three innovations, consolidated subdivisions were far below the other types in innovativeness.

Two aspects of innovative practice are indicated here. On the one side the heterogeneous consolidated subdivisions are dominated by their peripheral component and are least innovative of agricultural places. On the other side the rural heartland of agricultural townships are centres for innovation. Many of these townships are metro-adjacent, but the differences between these and non-metro adjacent are not significant. Agricultural places are all, to varying degrees, qualitatively innovative. This cultural character of innovation strengthens the importance of rurality in the mix of agricultural practices.

The second set of analyses is related to the component structure of the agricultural and municipal type variables. The municipal types were identified through dummy variables. Otherwise values, innovation percentages, acreages and numbers of farms are used as indicators of agricultural structure. Two component analyses were calculated, one using cities, the other using consolidated subdivisions. In both cases the core component produced an explained variance of about 38%. This first factor was loaded on sizes of agricultural land, and values for expenses and receipts. In the analysis with cities, there was also a link to winter cover percentage, though the relationship with cities was equivocal. After this strong first factor there was a sharp fall in explained variance, dropping to about 15% in each analysis. In the analysis that excluded the dummy variable for cities and included that for consolidated subdivisions, this second factor identified townships with agricultural innovation. In the city-based analysis a dichotomy between towns and townships appeared, but not ordered by innovations. The third factor in each analysis identified the structure of innovativeness, in each case with an explained variation of about 11%. A fourth factor identified the linkage between profits and innovations, particularly winter cover crops and other practices at 6% of variance. In the consolidated subdivision analysis the fourth factor was a dichotomy between townships and consolidated subdivisions at 6%. Finally, a fifth factor in the city analysis was based on the single variable, a dummy at that, of cities, at less than 6% of variance. In the analysis of consolidated subdivisions the profits and innovations factor appeared, at 6% of variance.

The results of these analyses suggest that there is a very considerable degree of coherence in agriculture and the coherence is related, though in complex ways, to the urban hierarchy in the settlement system. One can distinguish the resource base of agriculture, represented by acreages, financial values, output and income to agriculture. This resource base has an especially strong linkage to the urban system and is part of a continuous value gradient. The second aspect of agricultural structure represented in these data is related to agricultural innovativeness. Innovative practices are closely related to rurality and only nominally to the urban hierarchy. The third result from the statistical analyses is that geography, as represented by types of municipality, is significant in the composition and structuration of agriculture.

The second level of analyses deals with the census of population, also distributed over municipalities. In this case the two types of analysis will be repeated, that is, analysis of variance and principal component analysis. The analyses of variance will use both municipal types and a dichotomy of agricultural places into consolidated subdivisions and other agricultural places. In the principal component analysis, dummy variables will be used to identify municipal types and the entire data set will be restricted to those municipalities with agricultural practices.

This discussion will begin with the analyses of variance and to start the analyses by municipal type. Both population change and size of population were, as one would expect, statistically significantly different over municipal type. The highest growth rates were in towns (20%), then cities (12%), then townships (11%), then unorganized municipalities (11%), then the village (10%), and finally the borough, an anomolous survival of Toronto's earlier municipal structure, East York (2%). Population figures pretty self-evidently follow the urban hierarchy as to average numbers with cities as a whole averaging 179,000, towns 33,000, townships 3792, the village 1130 and unorganized places 251 residents.

The majority of the remainder of the variables relate to various characteristics of census and economic households as percentages: marital status, non-family households, female headed single parent households, husband–wife households, household size, children at home, ownership, house type, language used in the household. Finally, percentages of unattached individuals and males were used. These variables seemed to capture the dimensions of household structure and to present some differences over ethnicity, using language as a surrogate.

Household structures were differentiated over municipal types. Average persons per household, average number of non-married children at home, average number of persons per census family and average numbers of persons per economic family were all significantly different over municipal types. The normal gradient was from the largest averages in the rural townships followed by, and occasionally preceded by, towns then cities and finally the borough, village and unorganized places. The percentage of married adults was also differentiated, with unorganized places first, followed closely by townships and towns then a drop to cities, the village and the borough. The percentage of widowed people was also significant, but the direction of the gradient was quite different. The village had over 10% widowed followed by the borough at 7.5%, then the unorganized places at 5.5%, the cities at 5%, the towns at 4.6% and the townships at 4.3%. Husband–wife families as a percentage of all families was also significantly differentiated with the unorganized places the highest followed by the townships, the towns, the cities and finally the borough and village. Female headed single parent families as a percentage of family households was differentiated with the borough and village the highest, followed by the cities, unorganized places, towns and, lowest, the townships. The percentage of unattached individuals followed that gradient with the borough, the village then the cities, and unorganized places, followed by towns and, again at the lowest value, the townships. Non-family households were also differentiated over municipalities. The highest percentage was found in the borough, then the village followed by the cities, the unorganized places, towns and, at the lowest value, the townships. Males were significantly different over municipal types with the highest percentage in the townships followed by towns, the cities, unorganized places, and finally the village and borough.

Housing information presented a similar picture to households. Ownership was the highest in the unorganized places then in townships followed closely by towns, with a break to cities, the village, and much the lowest, the borough. Single detached houses as a percentage of all residential buildings were distributed in exactly the same manner from unorganized places and townships to the lowest value in the borough of East York.

Language cultures were also statistically different over municipal types. English speakers were most concentrated in the unorganized places, townships, towns, cities and the borough. The single agricultural village was largely francophone in character. Two other language groups were also examined. Italian speakers were statistically different over urban municipalities, with very low representation in townships and none in the unorganized places and the village. German speakers did not show statistically significant differences, but were concentrated in the unorganized places then townships followed by towns, cities and the borough of East York.

When the analyses of variance were calculated with the dichotomy of consolidated subdivisions as opposed to other agricultural places, lesser differentiation occurred, but that which did was roughly the same as over municipal types. Average numbers for households were highest in the other agricultural places and lower in consolidated subdivisions. Male percentage and per cent married were not different between consolidated subdivisions and other agricultural places. Widows were more concentrated in the consolidated subdivisions. Husband–wife families, as a percentage, were more concentrated in the other agricultural places than in the consolidated subdivisions. Female headed single parent families, as percentage, unattached individuals, and non-family households were all concentrated in the consolidated subdivisions. In demographic terms, the consolidated subdivisions were predominantly urban in character while the other agricultural places were typically rural.

Language was significantly different in this analysis as well. Italian speakers were concentrated in consolidated subdivisions, English speakers also concentrated there, while German speakers were a higher percentage in other agricultural places.

As a result of these analyses of variance over municipal types and the dichotomy between consolidated subdivisions and other agricultural places, a principal components analysis was calculated. Averages were used for household structures, changes in population between 1986 and 1991, a series of dummy variables for municipal type, and percentages for other household types, language culture and density. The resultant factors gave much the kind of fleshed out picture that the analyses of variance had suggested. The first factor, giving a variance of 36.5%, was loaded on differences between urban and rural places in association with total population, housing ownership and type and household structures. The second factor identified household composition and seems to be similar to what are usually called familistic factors in the social ecology literature. This factor accounted for 19% of variance. The third factor, accounting for 7% of variance was loaded on the dichotomy between towns and townships. A fourth factor, with 6% explanation, identified a gradient of population and changes in population in association with German and Italian speakers on opposite sides. Finally a fifth factor was loaded on English speakers at 5%.

Interpretively, the agriculturally active municipalities displayed very much the expected direction of urban to rural structure on a number of different dimensions. Put another way, there is surely a close association between the urban hierarchy and agriculture. Agriculture certainly operates within and is adapted to that urban hierarchy.

Preliminary Interpretations

The fundamental orienting hypothesis of a link between the urban system and agriculture has been borne out. In terms of agricultural innovation, the hypothesis that innovation would be concentrated in the more urban agricultural municipalities was not sustained. It is precisely in the majority of rural townships that agricultural innovation is most typical. The urbanized agricultural municipalities, including consolidated subdivisions, were not particularly innovative, but the rural townships were. However, the more urbanized agricultural districts were the centres of value, particularly acreage values, receipts and expenses. Nevertheless, when these were converted into profits or returns to capital the differences between urbanized and non-urbanized agricultural contexts waned.

Two preliminary conclusions may be derived from the analysis in this chapter. First, the effects on agriculture, primarily in terms of money values, are concentrated in urbanized areas and represent the rent surfaces of a combined urban and rural competitive market. These effects make it necessary for farmers to operate large properties with high turnover of values in commodities, and bear heavy expenses. The effects of this urban situation for farmers seem to be primarily economic in character. The second conclusion derives from the findings that agricultural innovation is concentrated in the more traditionally defined rural districts. There is a surviving rural culture that can orient farmers in their agricultural practices.

It seems reasonable to expand on this second conclusion. Agricultural innovation is directly related to processes of communication. Geographic research and theorization has stressed the communicative character of the diffusion of agricultural innovations. Communication is anchored on the materiality of social networks. That is to say, communication goes through circuits of prior contact. Farmers talk to one another. They also watch one another, but watching is in a communicative context. Talking to one another is itself based on the commonality of culture between participants in communicative exchanges. Habermas, in his theory of communicative action, makes much of the processes of interaction that lead to the formation of common understandings about situations that can lead, in turn, to the development of strategic action to accomplish goals through the common understandings (Habermas, 1984, 1989).

The impulse to talk about agricultural practices is easy to understand. Farmers are constantly under economic pressures to increase production, to cut costs, to eliminate labour. However, the context within which farmers do the talking is culturally derived, both from the social positions of farmers, the habitus of expectations and experience that has been identified by Bourdieu (1984), and from the prior existence of a culture within which conversations about agricultural practice make sense. This culture is the core of rurality in the non-urbanized agricultural municipalities of southern Ontario.

Rural culture operates much like culture at large. There are groups with the ability to define themselves as dominant and to define their version of the common culture as hegemonic. The farmers as a substantial collectivity within rural areas have held that power for a long time. Within the farmers' population there are also sub-cultures, some of which are sub-cultures of resistance. The sub-culture of resistance was demonstrated very clearly in the early to mid-1980s when masked farmers showed up at forced farm sales, armed and ready to disrupt the sale process. This concentration of cultural elements suggests that rural culture, even with its divisions, dominance–subordination, hegemony and resistance, is still intact in Ontario.

Yet, in substantial parts of Ontario, particularly those where the urbanization process is most developed, farmers have lost their hegemonic position. Urbanites have contested, often successfully, with farmers for political control of presumably rural municipalities that have received many urbanite in-migrants in recent decades. Central elements of the farmers' philosophy have been rejected by the newcomers. The leadership of farmers in the larger rural community has been challenged. In the recent waste-dump siting controversy in the Greater Toronto Area, farmers were used by the upper middle class, service class participants, as an icon, but did not play a strong leadership role. The symbolic centrality of farmers to the rural community was acknowledged, but leadership remained with the newcomers (Walker, 1995).

At this point, the sustainability of the system should be reconsidered. The physical system of farming is not sustainable. As frequently noted (Bowler, 1992) the industrial transformation of agriculture makes it much volatile. It is subject to amplifying feedback that can and probably will blow the industry apart. This is, after all, the fate of most industries. They are not sustainable, and even the notion that they should be is outside the rhetoric of capitalism. Things change and industries come and go. Further, in the GTA context, as in so many other urbanized rural regions, the farmers and agriculture are valued, preserved, but lost their hegemony and, in many ways, their freedom of action. Where there does seem to be some element of sustainability, it is in the rural hinterland. Farmers are embedded in a social system that is, while under threat and stressed, more or less self-sustaining. It is here where one should look for the sustainability of agriculture. However, this observation raises more questions than it solves.

These reflections suggest that the rural world is still alive, but that it is significantly impacted by the expanding influence and population of the urban world. In the deeper countryside, rural culture would appear to be most intact. Whether this means that the older farmers' cultural hegemony is still intact is a reasonable question. Further, what degree of rural culture still exists in the urbanized agricultural municipalities is also open to question. These are some of the questions that will guide the next stage of this research.

References

Bowler, I.R. (ed.) (1992) *The Geography of Agriculture in Developed Market Economies.* Longman Scientific and Technical, Essex.

Bourdieu, Pierre (1984) *Distinction: A Social Critique of the Judgement of Taste.* Routledge and Kegan Paul, London.

Bryant, C.R. and Johnson, T.R.R. (1992) *Agriculture in the City's Countryside.* University of Toronto Press, Toronto, Buffalo.

Habermas, J. (1984) *The Theory of Communicative Action: Volume 1: Reason and the Rationalization of Society,* transl. Thomas McCarthy. Beacon Press, Boston.

Habermas, J. (1989) *The Theory of Communicative Action: Volume 2: Lifeworld and System: A Critique of Functionalist Reason,* transl. Thomas McCarthy. Beacon Press, Boston.

Research Subcommittee of the Interdepartmental Committee on Rural and Remote Canada (1994) *Rural Canada: A Profile.*

Statistics Canada, Agricultural Division (1992) *Agricultural Profile of Ontario, Parts 1 and 2.* Ministry of Industry, Science and Technology, Ottawa.

Ministry of Industry, Science and Technology (1994) *Canadian Agriculture at a Glance.* Ministry of Industry, Science and Technology, Ottawa.

Walker, G. (1992) *Agriculture in an Urban Field: Toronto, Canada.* Commission on Changing Rural Systems, Economic and Social Restructuring of Rural Areas, Kansas State University, Manhattan, Kansas.

Walker, G. (1993) *Adjustment in Ontario Agriculture 1986–1991.* IGU Study Group. The Sustainability of Rural Systems, Conference, 14–20 Aug., Université de Montréal.

Walker, G. (1995) Social mobilization in the city's countryside: 'rural Toronto fights waste dumps'. *Journal of Rural Studies* 11 (3), 243–254.

21 Restructuring for Rural Sustainability: Overcoming Scale Conflicts and Cultural Biases

DARRELL NAPTON
Department of Geography, South Dakota State University, Brookings, South Dakota 57007-0648, USA

Introduction

Farmers, ranchers, and small town leaders in marginal and relatively inaccessible areas are grappling with the problems of achieving economic and environmental sustainability that are related to their locations and to beliefs about the propriety of cooperation and large scale enterprises. These problems are in addition to the sustainability issues that more prosperous regions face. This chapter will examine an area on the western fringe of the US Corn Belt and the eastern margin of the Great plains that is representative of many parts of the subhumid/semi-arid belt that stretches from Texas to the Prairie Provinces of Canada. Rural and agricultural economies located in this region are within a particularly vulnerable part of the nation (Popper and Popper, 1989). This northern plains region is farthest from both domestic and international markets and has a continental climate punctuated by cycles of drought. Farmers, ranchers, and local communities within this transitional belt have struggled to achieve sustainable economies, stable populations and ecological harmony.

Federal efforts to bolster the region's economy began during the New Deal programmes to help states cope with the Dust Bowl drought that coincided with the Great Depression of the 1930s (Arrington and Reading, 1991). North and South Dakota ranked first and second in per capita expenditures from the federal Agricultural Adjustment Administration, the Farm Security Administration, and the Farm Credit Administration. Federal transfer of payments to South Dakota continued after the drought and depression ended. In 1992, South Dakotans received $1.36 for each dollar of federal taxes paid. This was a net gain of about $1,200 per person per year (*Sioux Falls Argus Leader*, 15 Oct. 1993). But 70 years of massive federal payments transfer have failed to stem out-migration or produce healthy farms and towns. Between 1980 and 1990, South Dakota had a net out-migration of 52,000. Seventy-five per cent of those migrants were between the ages of 18 and 40 (Dockendorf, 1994). One result of this emigration was that only 21 per cent (66) of South Dakota's towns and cities grew.

Rural Communities: Issues of Survival and Scale

Increasingly, earnings and material quality of life are related to where people choose to live (Reich, 1991; Amato and Meyer, 1993). This poses a dilemma for those who are relatively immobile; those who do not move out of poor areas are more likely to remain poor. In South Dakota, many have chosen to move. This places those who remain in the precarious position of struggling to cope with fewer taxpayers and higher taxes to provide minimal services, fewer local stores, smaller schools, and reduced leadership capacity. Typically, there has been an accompanying reorganization and redistribution of jobs from scattered small towns to the county seats. Communities, however, are 'rooted in a place and cannot move' (Borchert, 1987).

Today South Dakota farmers, ranchers, and communities are faced with difficult decisions about how to achieve individual and collective prosperity. One way to achieve prosperous farms and adequate public services is to encourage large scale enterprises. Farmers can be encouraged to adopt capital and information-intensive technologies that allow one large scale, and presumably prosperous, farmer to take the place of many. Government service offices can also be encouraged to adopt capital and information-intensive technologies that allow one, well-placed office to provide the services previously offered by several.

This rationalization of the economic landscape allows fewer people to replace many while providing the same or higher level of output. In South Dakota, the transition to larger scale farms and government and community enterprises has not come easily. A significant number of farmers and community residents perceive that their quality of life will decline if larger scale enterprises become the norm. The state's farmers would like to make a better, lower risk living; however, maintaining a small farm and town lifestyle is also important for many of them. Some of these farmers also perceive that they cannot compete with larger, agribusiness-oriented farms.

One result is animosity toward large enterprises. Many of these farmers are also philosophically or morally opposed to some of the technologies that must be used to achieve economies of size. Examples include the effects of bovine growth hormone on dairy cows and perhaps indirectly on people, the effects of biocides on wildlife and water supplies, and the treatment of livestock in confinement facilities.

Many community businesses, churches, schools, and civic organizations are also concerned about regionalizing government, health and other services. One result of regionalization would be a higher quality of service but at a less accessible location and one step farther removed from local community control. Another consideration is the potential impact of such reorganizations upon county and regional population distributions. The redistribution of jobs from many small places that would lose jobs to a few larger places that would gain jobs might be accompanied by a redistribution of population that would be the death blow to many communities that have already been weakened by decades of out-migration and economic scale changes. Finally, community leaders realize that towns that cannot offer medical care, retirement housing, schools, and other services have no long-term future unless they are close enough to a growing city to become a bedroom community.

The struggle to prevent size increases goes against an American world view that bigger is better. This world view may have created a bias in public policies that favours larger units of production and projects (Frederick, 1991). The bigger is better concept has been recently expanded to include the world economy. The argument is that if farmers are to prosper and if the United States is to be globally competitive, then farms should be large enough to reap economies of size (Cochrane, 1993; Hamlin and Shepard, 1993). A normally unchallenged corollary to this is that large size and growth are synonymous with progress (Luten, 1986). Great plains agriculture in particular has evolved under conditions that have promoted growth and intensification (Riebsame, 1994). In South Dakota, to accept or even embrace small-scale economic endeavours also goes against the lingering frontier notion of the absence of limits (Garreau, 1981).

One of the key geographical distinctions in Plains communities is the division between insiders and outsiders (Norris, 1993). Larger scale enterprises usually bring new people into the community. Those who are rigid about the distinction between natives and outsiders may believe that their community is self-sufficient and has little need of the larger world. This distinction often leads to a distrust of those from the larger or outside world. Local businesses sometimes fight to prevent economic development because new businesses lead to competition and perhaps an increase in local wages (Norris, 1993; Luanne Napton, 1995 personal communication).

In South Dakota, scale battles are also about control and distrust of

those from the larger or outside world. Ranchers in south-central South
Dakota started a self-help programme largely because they did not trust any
government agency or anyone from outside the immediate neighbour-
hood. The success of that programme has been difficult to transfer to other
ranching neighbourhoods because these ranchers also distrust any outside
ideas from strangers, even other ranchers. Former US Senator James
Abourezk recalled the distrust of government in rural South Dakota when
a sheep rancher asked him at a political meeting, 'when (are) we going to
get enough gumption to overthrow the federal government' (McLaird,
1989).

Larger-scale enterprises require new ideas and technologies that in
turn require changes in organization and behaviour. South Dakotans are
typically resistant to change. Kathleen Norris thinks that South Dakotans
try to reject change because change has often been unkind (Norris, 1993).
She hints that change and its acceptance may become impossible in some
small towns because 'the luxury of knowing those with whom we do
business has largely atrophied our ability to deal with any issue larger than
personality'.

Scale is also an issue because of differing views about the good life.
Proponents of small towns and farms often talk about protecting and
nurturing a rural way of life as well as a small-scale economic system
(Dakota Rural Action, n.d. A; Satterlee, n.d.). They perceive that this way
of life is threatened by scale changes. In opposition are those who promote
larger enterprises. These folks seldom mention non-economic criteria for
supporting changes and are more likely to discuss things in terms of
efficiency (South Dakota Pork Producers Council, 1994).

Many South Dakotans embrace each of these sometimes conflicting
notions. Ask small town residents of virtually any age or background what
they like about their community and the nearly universal responses include
that it is quiet, peaceful, safe, and that they know everyone. When asked
what they would like to see changed, the typical responses are more jobs,
more excitement, more people, more stores and more entertainment.

Attempts to Achieve Stability and Sustainability

For these reasons scale issues may reach battlefield proportions, especially
when the concerns are family farming or school and government consolidation. Different groups perceive that their members or communities will
gain or lose from scale changes. Typical scale battles in South Dakota have
involved feed lot size, the number and size of local governments, and the
number and size of local schools. Some think that an increase in the size
of farms or schools will result in economic and community stability and
prosperity. An opposing view is that larger farms, fewer schools, and the

consolidation of some government services will exacerbate the death throes of Main Street and will lead to depopulated, socially bankrupt rural areas. These groups struggle in the public arena to either promote scale changes or prevent them from occurring.

This chapter will focus upon several non-governmental or quasi-governmental organizations and some individuals that are attempting to stabilize South Dakota's rural emigration and the struggling farm and ranch economy. These individuals and groups will take on a new importance as federal subsidies and their related services are eliminated, trimmed, or privatized. Some of the organizations attempt to resist outside forces that are encouraging changes in the size of South Dakota farms, ranches, schools, utilities, and local political units, while others embrace these changes. These attempts range from a grassroots ranching organization that promotes its own version of holistic ranching and distrusts organizations larger than the individual ranch to the Farm Bureau that encourages larger farms and corporate farms.

Each of these organizations is grappling with geographic and scale issues in an attempt to improve the success ratio of farmers, ranchers, and communities. Their efforts reflect differing methods of setting priorities for which farms, organizations, and communities should receive assistance, the nature of the help, and how that help should be delivered. The efforts also reflect different values about the proper size of farm, the role of government and the nature of community.

Dakota Rural Action

Dakota Rural Action (DRA) is a grassroots organization affiliated with the Western Organization of Resource Councils and funded through member contributions and grants. A major thrust of DRA is to support legislation and other efforts to help beginning farmers and family farming in general. DRA members think that the trend towards larger farms is a problem. 'Vertical integration of the livestock industry, larger farms and fewer beginning farmers mean the loss of the rural way of life – fewer small towns, consolidated rural schools and fewer businesses to support farming. This means a loss of the "rural community". It also means the United States is losing its foundation' (Dakota Rural Action, n.d. A). To DRA, the change in farm size is a scale issue as well as a community survival and way of life issue. The assumption is that large scale enterprises threaten 'family farms'.

In 1988, National Farms of Kansas City, Missouri, proposed to construct a 300,000 head farrow to finishing hog facility in the state. This operation would have produced 10 per cent of South Dakota's hogs, and DRA estimated that it would have put 300 farmers out of business while greatly reducing the local price of hogs (Dakota Rural Action, Jan. 1988). In response, the legislature passed a bill to prohibit corporations from owning and operating hog facilities, which essentially prohibited National Farms from expanding into the state, but the Governor vetoed the measure (*Sioux Falls Argus Leader*, 8 Mar. 1988). State hog farmers then obtained enough signatures from voters to put the issue to a public vote. Dakota Rural Action was active in resisting the move toward corporate farming (DRA 8 Mar. 1988), the public voted to ban corporate hog farms, and National Farms decided to locate in another state.

In 1993, the governor appointed a Pork Industry Task Force to determine why the number of hogs in South Dakota had not increased in recent years. The group recommended that South Dakota's tough corporate hog law be repealed (South Dakota Pork Producer's Council, 1994). The South Dakota Farm Bureau supported the changes, but Dakota Rural Action opposed the amendment. DRA spokesman Charlie Johnson said that 'We need to place a premium on family pork producers, not just on pork production. Family pork producers are going to spend their money in South Dakota rural communities for clothes and food, while large hog confinement units owned by outside corporations won't' (DRA, n.d. B). The South Dakota Pork Producers' Council rejected the recommendation and is now looking for legal ways to allow large-scale hog farms while continuing to ban corporate ownership (*Sioux Falls Argus Leader*, 20 Nov. 1994; *Brookings Register*, 10 Jan. 1995).

DRA takes both proactive and reactive stances at three political scales to support its agenda. It is active at the local, state, and national levels. The size of feedlots has been of continuing interest in South Dakota. One recent DRA local action was a campaign to prevent Minnesota Corn Processors (MPC), a three-state farmer-owned cooperative, from constructing a 20,000–35,000-head cattle feedlot. The proposed feedlot would have been the largest in South Dakota and was supported by the state Secretary of Agriculture (*Sioux Falls Argus Leader*, 18 Feb. 1994). The presidents of the South Dakota Cattlemen's Association and South Dakota Corn Utilization Council also supported the feedlot. These officers suggested that the feedlot would help local hay farmers and truckers as well as cattle farmers.

During the summer of 1994, there was a series of land use and zoning debates and discussions about the desirability of the proposed feedlot. DRA organized opponents aggressively fought to prevent the feedlot, partially on the grounds that an operation of that magnitude would undermine family farms and harm local groundwater (DRA, Nov. 1994; *Brookings Register*, 10 Oct. 1994). After a protracted battle, MPC withdrew.

DRA is also active at the state level. Recently it has supported state legislation to assist beginning farmers and to allow new farm buildings to receive the same tax benefits that businesses receive. Nationally, DRA works with other state organizations in the Western Organization of Resource Councils to influence federal legislation that affects family farms. Recently

DRA fought to prevent the use of bovine growth hormone to increase milk production. This was partially a scale issue with the drug projected to promote increased herd size, which may put part of the nation's dairy farmers out of business (DRA, May/June 1992).

Bootstraps

In 1988, a group of south-central South Dakota ranchers began meeting to develop a Bootstraps programme to address family and community needs (South Dakota State University, 1995). Most of these ranchers were distrustful of government and the traditional agencies designed to assist them. They encouraged those agencies to try new approaches to providing ranch and farm resources and focused attention on acquiring self-help tools that would make their operations more profitable, more satisfying, and more environmentally benign. Today South Dakota State University has accepted the Bootstraps model and is encouraging other farmers and ranchers to organize similar programmes. The Bootstraps model is holistic by focusing on farm and ranch families and communities in addition to profitability. By helping families and communities to work more smoothly and more profitability, Bootstraps encourages successful intergenerational land transfers, smaller, more profitable farms and ranches, and larger, more successful communities than might otherwise occur.

Community survival and renewal

During the decade of the 1980s, South Dakota had a net out-migration of over 50,000. This had its greatest impact in the rural areas and small cities of the state (Census Data Center, 1993). The population loss threatened the viability of churches, schools, local governments, and farm service centres. Proposals to regionalize or consolidate government services strike a raw nerve in South Dakota. The conflict between maximizing local control and making government more efficient and less redundant has long provided fodder for political debates. As early as the 1930s, governors recommended that the legislature abolish or consolidate townships to minimize inefficient duplication of efforts (Hendrickson, 1994). During the 1970s there was another attempt to consolidate townships. State residents, however, believed that rural interests were underrepresented at the state level and wanted to keep township governments to provide a local democratic forum. Their state lobbying efforts resulted in a law that protects townships from consolidation. Shortly after that, townships formed the South Dakota Township Association to enhance their roles (Hendrickson 1994). By the 1990s, South Dakota had more local governments per capita than any state except North Dakota (*Governing*, 1995). Townships account for more than

950 of South Dakota's 1785 local and regional governments.

Regional cooperation has been proposed as another way to compensate for the state's population loss. James Beddow, Dakota Wesleyan College president, promoted community cooperation and regional approaches during his unsuccessful run for the governor's office (*Sioux Falls Argus Leader,* 16 Aug. 1993). Beddow argued that South Dakota's 'historical, social, and cultural values are based on rugged individualism' and that 'communities are used to competing economically and athletically'. He articulated that these nineteenth-century values may not encourage prosperous communities today. Beddow's 'collaborative regional initiatives' required state-level coordination and empowerment for regional efforts.

Rural sociologist Jim Satterlee has also promoted a rural government and services reorganization model to help communities respond to the erosion of services that accompanies out-migration. Satterlee has stated that regionalization of services is 'all about a new community that reflects some type of vitality. South Dakota must create that vitality in order to keep its youth. We have to commit to educate our children to go back home' (Dart, 1994). In Satterlee's model, called The New Community, several counties would consolidate services cooperatively. Representatives from each service agency would travel to each county to provide local assistance one day per week in Community Government Centres. Each resident would be within a one hour drive of a Community Service Centre. For example, one county seat would provide a regional hospital, with minimal health care in other parts of The New Community provided at Community Service Centres. Other county seats would provide additional services such as the jail and sheriff's department, or a regional school.

The 'collaborative regional initiative' model and the New Community model each face political conflict related to issues of scale and local control in attempting to achieve economic and community vitality. In opposition to these most recent consolidation ideas, Governor Walter D. Miller responded that:

> 'I don't intend to be part of anything that is going to try to force upon the rural people of this state some type of a mandated governmental designed reorganizational, regional process. I'm not going to be part of that. I don't think the government can plan everybody's future out there. They have to determine some of their own destiny'
>
> (Dart, 1994)

One of the tools that communities use to help forge their destiny is the local school. Schools are central to rural communities. Traditionally, schools provided community common space while school sports, plays, and music festivals provided live entertainment. Communities without schools are 'less attractive to potential residents' and they have a difficult time maintaining a retail sector (Rural School and Community Renewal, n.d.). South Dakota State University and the W. K. Kellogg Foundation are trying to revitalize endangered South Dakota rural schools and communities. Historically, schools and their communities were seen as separate. This programme tries to 'help rural residents see the community and the school as unified rather than as separate entities' (Rural School and Community Renewal, n.d.). The Rural School and Community Renewal programme encourages schools to begin conducting research in the community that could lead to a stronger economy that is less dependent upon distant industries. The goal is to prevent school consolidation with the underlying belief that small, local schools have more to offer than larger, distant, consolidated schools.

Rutland is one example of the Rural School and Community Renewal Programme. Rutland is located in east-central South Dakota, 14 miles north-east of Madison, a town of 6000. Rutland has 20 residents and no businesses. The only public areas in town are the post office and the high school, which has only 31 students (*Sioux Falls Argus Leader*, 14 Oct. 1994, 1). High schools that have fewer than 35 students for two consecutive years lose state funding. Generally, small districts are dependent upon state funding, so its loss almost invariably leads to school closure and consolidation. The Rutland community and school are working together to have the high school open a convenience store. The store would be a business laboratory for the students who will staff it. The assumption is that a convenience store will make it easier to attract new residents and the school's enrollment will then increase.

Rural renewal

Recently a bipartisan group of state legislators has met informally under the name Rural Renewal (*Brookings Register*, 6 Apr. 1994). The group believes that South Dakota farms and towns are at a crossroads between large corporate farming that would lead to a smaller rural population and the potential collapse of many farm dependent towns, or a profitable farming system for small, family farmers that would result in a larger rural population and thriving rural schools and towns (Dockendorf, 1994). Local farmers tend to purchase supplies locally while corporations and larger farms normally do business directly with wholesalers. The Rural Renewal legislators held a series of hearings and wrote and passed a package of six bills.

Three of the new laws were designed to assist persons beginning to farm or start a small rural business. The other bills are to encourage more value-added agriculture, to give livestock facilities the same tax breaks that non-farm businesses had been receiving, and to upgrade the state's Adult

Farm Education Programme (*Brookings Register*, 6 Apr. 1994). Rural Renewal
is now investigating additional things that the state can do to help farmers
and rural communities become stable and profitable.

Conclusion

The Northern Plains transition belt may be typical of marginal regions that
are struggling to achieve environmental and economic stability while trying
to compete with other regions. South Dakota communities, for example,
are struggling to take advantage of those culture traits that are most useful
and discard those that are least useful in today's economic setting.
Unfortunately, there is little agreement about which ideas have utility and
which should be discarded. Meanwhile, climatic and economic forces
continue to buffet the region. The critical question is to what extent
farmers, business owners, and communities should embrace or reject large-
scale enterprises. When these issues are discussed, the associated geographical concepts of scale change, regionalization, and central places are
seldom mentioned. The chosen methods of setting priorities are often
related to preconceived notions of whether bigger is truly better and issues
of local control. Commonly these different ideas result in conflict and lack
of coordination.

Concerned citizens try to avoid objectifying the evaluation of regionalization and service consolidation, because to develop a process that
facilitates cooperation and prioritizes assistance with known criteria would
mean admitting the unmentionable. They would have to admit first that not
all communities can be helped. Second, they would have to face the
possibility that 'my' community may not be one that can be helped. Many
South Dakotans are unready to accept that possibility. Nor are local
communities ready to trust outsiders or to cooperate with neighbouring
towns. The Northern Plains transitional belt is still an area of relatively small
family farms and small, rural towns. If the rural values to which the citizens
of this region claim allegiance are to survive, then individuals and
communities must devise plans that engender cooperation and they must
articulate the nature of the forces that encourage fewer and larger farms
and towns. Their future and the survival of these values depends upon it.

References

Amato, J. and Meyer, J.W. (1993) *The Decline of Rural Minnesota*. Crossings Press
    Marshall, Minnesota, 85 pp.
Arrington, L.J. and D.C. Reading (1991) New deal economic programs in the
    northern tier states, 1933–1939. In: Lang, W.L. (ed.) *Centennial West: Essays on
    the Northern Tier States*. University of Washington Press, Seattle, pp. 227–243.

Beddow, J.B. (1993) Cooperation, not competition will keep plains alive. *Sioux Falls Argus Leader*, 16 Aug, A-9.

Borchert, J.R. (1987) *America's Northern Heartland*. University of Minnesota Press, Minneapolis, Minnesota, 250 pp.

Brookings Register (6 Apr. 1994) Passage of six bills sponsored by rural renewal lawmakers said 'Only a start', 1.

Brookings Register (10 Oct. 1994) MPC bids country farewell, 1.

Brookings Register (10 Jan. 1995) Producers reject plan to repeal hog farm law, 14.

Census Data Center (1993) South Dakota population, housing, and farm census facts. Update Series C-229 number 27. Brookings, Department of Rural Sociology, Agricultural Experiment Station, South Dakota State University.

Cochrane, W.W. (1993) *The Development of American Agriculture: A Historical Analysis*, 2nd edn. University of Minnesota Press, Minneapolis, Minnesota, 500 pp.

Dakota Rural Action (n.d. A) Keeping the farm sized for the family: 'What' is being done in South Dakota?'

Dakota Rural Action (n.d. B) The South Dakota Family Farm Act: protection or barrier?

Dakota Rural Action (Jan. 1988) Strengthening the South Dakota Family Farm Act.

Dakota Rural Action (8 Mar. 1988) Letter to state house and senate members.

Dakota Rural Action (May/June 1992) Expanded BGH research needed at SDSU. Action alert, 6(3).

Dakota Rural Action (Nov. 1994) DRA accomplishments and achievements.

Dart, Luann (1994) Which road will South Dakota travel? *High Liner Magazine* 47(1), 12 and 15–19.

Dockendorf, R. (1994). Rural renewal finds legislative success. *Yankton Daily Press and Dakotan*, 5 Mar, 2-C.

Frederick, K.D. (1991) Water resources: increasing demand and scarce supplies. In: Frederick, K.D. and Sedjo, R.A. (eds) *America's Renewable Resources: Historical Trends and Current Challenges*. Resources for the Future, Washington, DC, pp. 23–78.

Garreau, J. (1981) *The Nine Nations of North America*. Houghton Mifflin Company, Boston, 427 pp.

Governing (1995) Ranking the 50 states, 7–8.

Hamlin, C. and Shepard, P.T. (1993). *Deep Disagreement in U.S. Agriculture: Making Sense of Policy Conflict*. Westview Press, Boulder, Colorado, 319pp.

Hendrickson, J.P. (1994) Grass-roots government: South Dakota's enduring townships. *South Dakota History* 24(1), 19–42.

Luten, D. (1986) In: Vale, T.R. (ed.) *Progress Against Growth: Daniel B. Luten on the American Landscape*. Guildford Press, New York, 366 pp.

McLaird, J.D. (1989) From bib overalls to cowboy boots: East River/West River differences in South Dakota. *South Dakota History* 19(4), 455–491.

Norris, K. (1993) *Dakota: A Spiritual Geography*. Ticknor & Fields, New York, 224 pp.

Popper, D.E. and Popper, F.J., Jr (1989). The fate of the Plains. In: Marston, E. (ed.) *Reopening the Western Frontier*. Island Press Washington, DC, pp. 98–113.

Riebsame, W.E. (1994) The historical bias for growth and intensification in Great Plains agricultural development. In: Gilg, A.W. *Progress in Rural Policy and Planning*, Volume 4. John Wiley, pp. 45–53.

Reich, R.B. (1991) *The Work of Nations*. Vintage Books, New York, 339 pp.

Rural School and Community Renewal brochure (n.d.) College of Education and Counseling, South Dakota State University, Brookings, SD.

Satterlee, J. (n.d.) The New Community: A Model for Community Development. South Dakota State University Rural Sociology Department.

Sioux Falls Argus Leader (8 Mar. 1988) Hog producers seek statewide vote on corporate farm ban.

Sioux Falls Argus Leader (16 Aug. 1993).

Sioux Falls Argus Leader (15 Oct. 1993).

Sioux Falls Argus Leader (18 Feb. 1994) 1-B.

Sioux Falls Argus Leader (14 Oct. 1994) Town banks on convenience store to save school.

Sioux Falls Argus Leader (20 Nov. 1994) Groups back repeal of corporate farm ban.

South Dakota Pork Producers Council (1994) South Dakota pork industry task force report.

South Dakota State University (1995) A program called bootstraps.

Index